CONTROL TECHNOLOGIES FOR AIR POLLUTION

CONTROL TECHNOLOGIES FOR AIR POLLUTION

Articles from Volumes 7–11 of
ENVIRONMENTAL SCIENCE & TECHNOLOGY

Collected by **Stanton S. Miller,**
Managing Editor

An ACS Reprint Collection

American Chemical Society
Washington, D.C.   1979

52898234

CHEM

Library of Congress CIP Data

Control technologies for air pollution.
  (An ACS reprint collection)

  Includes index.

  1. Air—Pollution—Addresses, essays, lectures.
  I. Miller, Stanton S. II. Environmental science &
technology.

TD883.14.C66   628.5'3   78-11630
ISBN 0-8412-0450-0
ISBN 0-8412-0451-9 pbk.   1–177 (1979)

Contents

Clean Coal and Energy Sources

Books

Preface

"Control Technologies for Air Pollution" is the seventh in a continuing series of volumes of reprinted articles from ES&T and complements the earlier reprint books on environmental monitoring, air pollution, water pollution, and solid waste. This collection of 50 articles that appeared in ES&T from 1973 to 1977 and the listing of 47 books plus an additional eight ACS titles on pertinent subjects summarize and highlight the developments in this area of control technology. Together with the earlier collections, this volume will bring the reader up to date on the important aspects of what has been happening during the 70s, our environmental decade.

Another source for the newest information on environmental technologies is the Currents department in the monthly issues of ES&T. About a half dozen technology items of news value appear in this section each month. Since over 350 such items appeared during this five-year period, reprinting them would cause extreme problems in organization and in providing access for specific items. In the interest of time and cost, they are not included here, but a serious practitioner should not overlook this source of information on the latest technological developments. The editor is indebted to Julian Josephson, ES&T's associate editor, who prepared both the technology currents and the book listing and to Lois Ember, ES&T's associate editor, who edited and coordinated most of the features included in this volume.

STANTON S. MILLER
Managing Editor, *ES&T*

CONTROL TECHNOLOGIES FOR AIR POLLUTION

Catalytic converters: an answer from technology

ES&T's Lena Gibney finds that these units are a choice of '75 automakers, tame harmful exhaust emissions, and vie for future use

Reprinted from ENVIRON. SCI. TECHNOL., **8**, 793 (September 1974)

To earthlings, cars have been a nearly indispensable mode of transportation, thing of beauty, and sometimes symbol of status. Yet these same wonderful machines are one source of polluted air, exhausting such pollutants as carbon monoxide (CO), hydrocarbons (HC), and nitrogen oxides (NO_x). In 1970, the federal government took action, with the Clean Air Act, against the cars and set forth a set of exhaust emission standards in the hopes of righting the wrongs.

The standards underwent changes and what became the interim 1975 standards have elicited a spectrum of responses and tactics from the automobile manufacturers. The new '75 passenger cars will come equipped with diverse emission control devices. Particularly so are the cars making the scene in California, the forerunner of the clean air cause, the state with even more stringent standards than the rest of the nation. These new California cars will make up approximately 10% of new car sales this fall.

With Congress approving the extension of the interim 1975 standards through 1977, prospects look good for the car industry to improve further their control devices and design even better ones. 1975 will, however, give the public the first inkling of the results of massive research and efforts on the part of the industry to protect our air.

Basic concepts

To comprehend the varied subject of automotive emissions control and how the industry is coping with the emission standards, some fundamentals must be defined.

Emissions within the engine are formed when the hydrocarbon fuel is burned incompletely to HC and CO in the engine's combustion chamber. Ideally, the fuel and oxygen in the air entering the chamber yield harmless exhaust products—carbon dioxide, water vapor, and inert nitrogen. But the generation of the pollutants CO, HC, and NO_x (mainly NO) is a function of the proportional amounts of air and fuel. At lean air–fuel (A/F) ratio, CO and HC emissions are decreased because of the greater quantity of oxygen available for combustion. When the A/F becomes too lean, however, HC emissions will increase again.

NO_x, on the other hand, is an exponential function of flame temperature. At low temperature, nitrogen and oxygen from the air will not unite to form significant concentrations of NO. Low temperature is achieved at A/F mixtures richer and leaner than the optimum, because of the diluent effect exerted by unburned fuel in the former case and the excess air in the latter.

What then is optimum burning within the engine? It is close to the stoichiometric point, where the amount of fuel is matched exactly with the amount of oxygen to burn it completely. For most gasolines, the stoichiometric value falls between 14.5–15 lb of air per lb of fuel, with maximum fuel economy at 16–16.5. It is interesting to note that NO_x formation peaks at about the same A/F ratio as fuel efficiency.

Emissions outside of the engine are another matter. High exhaust gas temperatures are needed for the control of HC and CO. And yet lower exhaust gas temperatures would increase fuel economy and decrease NO_x production. To resolve this incongruity, devices such as the catalytic converters came into being.

Catalytic HC and CO control

For almost all new '75 California cars and a majority of other new cars nationwide, Detroit and many foreign car manufacturers have chosen the catalytic route, using an oxidation catalyst to convert HC and CO to carbon dioxide and water. EPA has stipulated that the catalytically equipped cars must meet interim standards at 50,000 miles, allowing for one catalyst change. Tests by all the car and catalyst makers have shown compliance. For 1975 cars, there will be two types of catalysts—a blend of noble metals in pellet or monolithic form.

The noble metals were chosen after testing many of the chemical elements (outside of the rare gases)

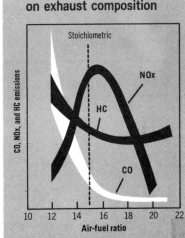

Effect of air-fuel ratio on exhaust composition

Stoichiometric

CO, NO_x, and HC emissions

NO_x

HC

CO

Air-fuel ratio

10   12   14   16   18   20   22

$$(CH_2)_n + O_2 \rightarrow (CH_2)_n + CO + H_2O + CO_2$$

smaller

Emission control systems

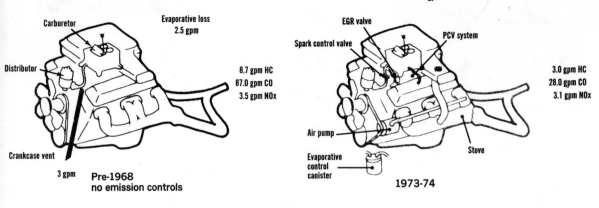

Pre-1968
no emission controls

- Carburetor
- Evaporative loss 2.5 gpm
- Distributor
- Crankcase vent — 3 gpm
- 8.7 gpm HC
- 87.0 gpm CO
- 3.5 gpm NOx

1973-74

- Evaporative loss 1.9 gpm
- EGR valve
- Spark control valve
- PCV system
- Air pump
- Evaporative control canister
- Stove
- 3.0 gpm HC
- 28.0 gpm CO
- 3.1 gpm NOx

in the periodic table, singly or in combination. They placed first among three possible categories— base metals, noble metals, and metal alloys.

Specifically, platinum (Pt), palladium (Pd), or a Pt/Pd blend will be used. The unison of Pt and Pd is particularly desirable because they are found together in the earth and are thus mined at the same time. Even though these noble metals are more expensive than nonnoble types, they are not as subject to poisoning by sulfur compounds in the gasoline. Besides, lesser amounts are needed to achieve the same control as other types. Only about 0.05–0.1 troy oz (1.5–3 grams) of active metal per car will be needed.

The catalyst(s) in their beds are mounted before the muffler, as close as possible to the engine manifold on the exhaust pipe. The selection of one of two geometric configurations —pellet (beaded) or monolith—depends on the operating conditions under which the catalyst functions; in other words, on the design of the individual car itself.

The pellets ($\frac{1}{6}$–$\frac{3}{8}$ in. in diameter) are largely alumina and generally exhibit a larger surface area per unit mass. They do, however, have lower crush resistance than the monolith. This could result in a loss of catalyst and substrate during car operation, according to National Academy of Sciences (NAS) committee on motor vehicle emissions.

The monolith, on the other hand, offers low attrition loss, low pressure drop, and quick warm-up due to its low total mass. In the form of a cylinder (3–6 in. in diameter) or an elliptic cylinder of equivalent volume, containing an internal honeycomb structure extending in a longitudinal direction along the entire length of the cylinder, and surrounded by an integral outer skin (alumina or cordierite), it has good mechanical strength with high resistance to

shock and vibration. The monolith does have breakage problems, suffers from lack of movement, and is more expensive than the pellets as it takes more noble metal loading.

Availability

The aspect of metal loading brings forth the costs and the question of whether the supply of Pt and Pd will meet the demand. At press time, Pt is selling at $190/troy oz, and Pd at $84/troy oz. World production of Pt plus Pd has been 3677 troy oz/yr during the period 1969–72, according to the "Mineral Yearbook." Projected world consumption of these same metals could be as high as 6219 troy

oz in the year 1980, with the U.S. demanding at least 17% of that for automotive and 23% of that for other needs. To thwart the possible unavailability of the metals and unsurmountable prices, there will have to be a more conscious effort to recycle these metals.

During the past few years, about 25% of the total domestic consumption of Pt and Pd has been recycled material. But this has essentially involved the chemical and petroleum industries. To retrieve some 1 million troy oz of metals from the automobiles will be a more complex problem and a more urgent need.

Poisoning

Much is said about the poisoning of catalysts by the additives in the

gasoline and lube oils.

Lead in gasoline is harmful to noble and also to some base metal catalysts. Sulfur in gasoline and phosphorus and other impurities in lube oils are likewise poisonous, sulfur to a lesser extent than phosphorus. What really happens is that the lead coats and insulates the metal, making it inaccessible to the exhaust gases. But all catalyst makers agree that an accidental fillup or two with leaded fuel will not kill the noble metal catalyst, only render it inactive for a period of time. Upon return to unleaded fuel, the catalyst will recover.

This phenomenon leads back to

Exhaust emission reduction

	HYDROCARBONS		CARBON MONOXIDE		NO$_x$
	Level, g/mi	Reduction, %	Level, g/mi	Reduction, %	Level, g/mi
Pre-1968 vehicles (uncontrolled)	8.6	—	87.5	—	3.5
1975 national interim standards	1.5	83	15	83	3.1
1975 California interim standards	0.9	90	9.0	90	3.1
1975 Clean Air Standards	0.41	95	3.4	96	3.1

the catalyst configuration, in that lead tends to migrate to the center of the pellet system and therefore leaves the surface area open for conversion. When unleaded fuel is re-introduced, the poisoned pellets move into a high heat zone, and as they do so, the lead is volatilized and the catalyst cleansed.

The same holds true for sulfur. Much discussion has evolved around the conversion by the catalyst of sulfur to undesirable sulfates. According to GM, at a Senate subcommittee hearing in late 1973, sulfur also migrates to the center of the pellet and stores itself there until the car reaches a high speed. At that point, it burns off. Sulfur content in unleaded fuel is not a problem. It averages about 0.024% (by weight). If it ever

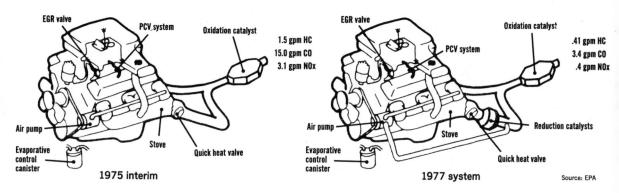

1975 interim

Evaporative loss
1.9 gpm

EGR valve
PCV system
Oxidation catalyst
Air pump
Stove
Evaporative control canister
Quick heat valve

1.5 gpm HC
15.0 gpm CO
3.1 gpm NOx

1977 system

Evaporative loss
1.9 gpm

EGR valve
PCV system
Oxidation catalyst
Air pump
Stove
Evaporative control canister
Quick heat valve
Reduction catalysts

.41 gpm HC
3.4 gpm CO
.4 gpm NOx

becomes too much of a problem, desulfurization of gasoline has been suggested as a solution, which will open up heated debates between the oil industry and the EPA.

The phosphorus limit has been set at 0.005 g/gal by the EPA, as opposed to the 0.05 g/gal for lead. Last year, Matthey Bishop's president, V. W. Makin, testified that phosphorus, in the form of a heavy metal or as zinc dialkyldithiophosphate in lube oil, exerts a negligible effect on the noble metal catalyst. However, present as organic thiophosphate (antiwear additive), it deactivates this catalyst.

This business of unleaded fuel

Since lead fluid (tetraalkylead plus chloride and bromide scavengers) has been found to have health effects and also has proved to be the most harmful to the noble metal, EPA is asking that lead be gradually phased out of all gasoline until the permissible lead concentration is 0.05 g/gal by 1979. Meanwhile, EPA has ruled that some 111,000 gasoline stations across the nation sell at least one grade of unleaded gasoline by July 1, of this year. It also proposed that an additional estimated 10,000 stations, mostly in rural areas, be required to sell unleaded fuel after January 1 of next year.

This could mean that by next year, more than one out of every two stations will be required to pump unleaded gasoline. These stations are expected to serve from 60–85% (estimated 6 million) of new '75 models which are equipped with catalytic converters.

Unleaded gasoline, or as others coin it—lead free or clear, with octane number 91, will be sold at pumps with a hose nozzle somewhat smaller than other pumps. Only these nozzles will fit the catalytically equipped cars and thus are the sure means of preventing an inadvertent fillup with leaded gasoline.

Making unleaded gasoline available has drawn forth certain controversy. Exxon researchers confirm their earlier findings that trace lead in amounts up to 0.10 g/gal will have little effect on activity maintenance of noble metals during long-term aging studies. In the case of the inadvertent contamination of unleaded gasoline by lead, Exxon does have a process which will reduce the lead content from as high as 0.20 g/gal down to the federally specified 0.05-g/gal limit. Their technique uses activated carbon.

In fact, the culprit may not be lead after all. It could be ethylene dibromide, another gasoline additive included in lead fluid, Chrysler Corp. reported. Based on its laboratory findings, Chrysler is going ahead with road test data before drawing a full conclusion.

As far as energy requirements go, Texaco's vice-president of environmental protection, W. J. Coppoc, testified before a Senate subcommittee this Spring that the production of lead-free gasoline and the gradual phase reduction of lead in existing leaded gasolines will cut U.S. gasoline supply by a minimum of almost 800 million barrels between now and 1980. This figure, he said, represents an annual loss of 5% in consumption at zero growth in domestic demand; furthermore, the figure does not include the fuel economy loss with the lower compression engines which must accompany unleaded fuel. He felt that lead traps or filters are feasible alternatives.

The American Petroleum Institute supported Texaco's views in that it felt there would be substantial and expensive modifications of refinery operations to manufacture the higher octane blending stocks needed to replace lead. The cost could run as high as $15 billion, and construction modification could take four years or more.

Arguing that such high costs could

be offset by savings in maintenance are engineers from American Oil. They reported their findings at the Automotive Engineering Congress held in Detroit last year. The use of lead-free gasoline will reduce the need for frequent replacements of spark plugs, mufflers, and other hardware, they said, so much so that a clear-cut cost advantage can amount to $0.05/gal over the lifetime of the average car.

Besides, calculations show that the energy requirement for the production of lead antiknock compounds is equivalent to a major portion of the internal refinery energy requirement during the production of the higher octane lead-free gasoline, Universal Oil Products' V. Haensel, vice-president of science and technology, told ES&T.

Perhaps lead could be replaced by rare-earth metals and their compounds. Researchers have found cerium compounds encouraging as antiknock additivies at Wright-Patterson Air Force Base, but translation of these compounds into everyday use is far off.

Meanwhile, there is the nagging thought that maybe there will not be enough of the unleaded gasoline to go around this fall. American Automobile Association seems to think so.

Health hazards

Apprehensions have been expressed by many, particularly by John Moran of EPA's Environmental Research Center (North Carolina), that Pt and Pd and sulfur (converted by catalyst to sulfur trioxide which reacts with water in the air to form sulfuric acid mists) emissions may be dangerous to man. He confirmed that sulfuric acid is emitted from the catalytic-equipped cars at a rate of 0.05 g/mi, between 9–50 times higher than emissions from noncatalytic cars. This rate could affect people with lung or heart trouble.

3

Moran was also concerned that Pt and Pd, once they escaped to the atmosphere, might undergo methylation in drinking water and pose a threat. There could be degradation of the biosphere when cars are eventually junked, he asserted further.

GM's Engelhard and others have testified that Pt and Pd are not toxic in the metallic form. The president of the National Academy of Sciences has also testified that Pt is rather inert.

A comprehensive survey of open literature would reveal that Pt is only toxic in its water-soluble salt forms. As such, they can cause respiratory or skin irrations. The Pt from the converter, according to some scientists, is changed to the oxide vapor which subsequently undergoes decomposition and condensation to a finely dispersed Pt metal aerosol as the exhaust gas cools.

Not much has been published about the toxicity of Pd or its salts. Studies made at catalytic cracking operations of refineries show that there has been no evidence that exposure to metallic platinum and palladium has caused employees to suffer from pulmonary or cancerous diseases.

Much has been done on toxicity of

sulfur oxides, but not much on sulfuric acid. Hazleton Laboratories (Vienna, Va.) have found that monkeys, exposed continuously to a mixture of SO_2, fly ash, and 100 $\mu g/m^3$ of sulfuric acid, did not demonstrate any significant changes after 18 months.

The divergence of reports from the various authorities has led to the call for information on the part of the Panel on Environmental Science and Technology of the Senate Committee on Public Works. Hearings started this May on the present state of knowledge about the health effects of lead particulates emitted from the cars and the technology available for their control.

Requests for data on sulfur emissions (sulfur trioxide, sulfuric acid, and sulfates) were made by the EPA. They placed emphasis on the magnitude of sulfate emissions, their impact on ambient air quality and health, and the feasibility of control methods.

Benefits

Rubbing out HC and CO with a catalyst, in face of the clamor about the ill consequences, can lead to increased fuel economy and improved driveability. The catalyst itself does

not command that. It simply replaces former controls, such as retarded spark timing, which consume more fuel. According to Eric Stork, EPA's director of mobile source pollution control, gasoline mileage improvement should be better than the 7% predicted by EPA last year and the 11–13% predicted by GM this year. He bases his optimism on EPA's test data on cars submitted by manufacturers for the 4000-mile certification run. As much as 26% better mileage was obtained with one car in the 5500-lb class.

Another plus for the catalyst, said EPA administrator Russell Train, is the 95–98% control of "unregulated emissions," such as benzene, toluene, xylene, and the carcinogenic polynuclear aromatics in engine exhaust.

NO$_x$ control

For 1975 cars and at least for the new cars in the next couple of years, nitrogen oxides (mostly nitric oxide) emissions will be handled by the EGR (exhaust gas recirculation) system. The concept here is to use an inert gas to reduce peak temperature which in turn reduces NO$_x$ formation. The exhaust gases are just the source for a continuous supply of

Many '75 automakers are relying on catalysts ...

U.S. Manufacturer	Type	Supplier	Catalyst plant
General Motors	Pellet[a] Pt:Pd 70:30	25% from Engelhard Minerals & Chemicals Corp.	Newark, N.J.
		25% from Air Products & Chemicals Inc.	Calvert City, Ken.
		25% from W.R. Grace	Curtis Bay, Md.
		25% from The Catalyst Company	Azusa, Calif.
		(joint venture American Cyanamid with Japan Catalytic Chemicals Inc.)	
Ford	Monolith Pt/Pd	60% from Engelhard Minerals & Chemicals Corp.	Newark, N.J.
	Monolith Pt/Pd	40% from Matthey Bishop	Devon, Pa.
Chrysler	Monolith Pt/Pd	Universal Oil Products	Tulsa Port of Catoosa, Okla.
American Motors	Pellet Pt/Pd	Buy from General Motors	
Foreign Imports[b]			
English Manufacturers			
Rolls Royce	Monolith Pt/Pd	Matthey Bishop	Royston, England
British Leyland	Monolith Pt/Pd	Matthey Bishop	Royston, England
French Manufacturers			
Peugeot	Monolith Pt/Pd	Engelhard Minerals & Chemicals	Huntsville, Ala.
Renault	Monolith Pt/Pd	Engelhard Minerals & Chemicals	Huntsville, Ala.
German Manufacturers			
Mercedes	Monolith Pt/Pd	Engelhard Minerals & Chemicals	Huntsville, Ala.
Volkswagen	Monolith Pt/Pd	50% Matthey Bishop	Royston, England
		50% Degussa	Frankfurt, Germany
Swedish Manufacturer			
Volvo	Monolith Pt/Pd	Engelhard Minerals & Chemicals	Huntsville, Ala.
Italian Manufacturer			
Fiat	Pellet Pt/Pd	Universal Oil Products	Shreveport, La.
Japanese Manufacturer			
Daihatsu Kogyo	Pellet Pt/Pd	Universal Oil Products	Shreveport, La.
Nissan	Monolith Pt/Pd	Engelhard Minerals & Chemicals	Huntsville, Ala.
(Datsun)	Pellet Pt/Pd	Universal Oil Products	Shreveport, La.
Toyota	Monolith Pt/Pd	Engelhard Minerals & Chemicals	Huntsville, Ala.
	Pellet Pt/Pd	Universal Oil Products	Shreveport, La.

[a] Pt—platinum, Pd—palladium; substrate suppliers—American Lava, division of 3M (Chatanooga, Tenn.); Corning Glass (Erwin, N.Y.).
[b] Catalysts for most foreign customers are canned overseas; Arvin (Columbus, Ind.) is canning for most U.S. customers. General Motors is canning all its catalysts at its AC Sparks Div. plant in Oakcreek (Wis.), and Flint (Mich.).

inert gases. These gases serve to dilute the combustion chamber charge, slow the combustion process, and lower the peak combustion temperature. The trick is to use the correct amounts of these gases.

The EGR valve does that job of proper metering. Mounted on the intake manifold and operated by a timed vacuum signal from a port on the carburetor throttle body, it ensures calibrated recirculation.

EGR systems are not new. In the past they have been associated with driveability and fuel economy problems. The cause has not been the exhaust gases themselves, but how they were recirculated. 1975 cars will have a much more sophisticated system so that mileage and driveability will no longer be penalized.

Catalysts for NO_x control are not an impossibility to some manufacturers. Reduction could be via a dual-bed or a three-way system. The former consists of an oxidation catalyst placed downstream of the reduction catalyst, with the engine operating at rich. NO_x passing through the reduction catalyst will be reduced to inert nitrogen and water. Air will have to be added before the HC and CO, traveling on through to the oxidation catalyst, can be converted to harm-

less carbon dioxide and water. This extra air causes problems because any ammonia which may form from the reduction will be reoxidized to NO_x by the oxidation catalyst.

Engelhard, in a statement before the Senate subcommittee, reported progress with their ruthenium catalyst which converts very little NO_x to ammonia. Previous difficulties with the loss of ruthenium during engine operation on the oxidizing side for extensive lengths at high temperatures have been minimized.

Matthey Bishop also reported improved ruthenium stability which they obtained by the reaction of alkaline or rare earths with ruthenium in the catalyst bed. This reaction gives ruthenates which have less tendency to volatilize. Synthesizing the ruthenates in position on the monoliths or pellets has resulted in a decreased weight loss of the noble metal.

Gould has been successful with their reduction catalyst, nickel/copper on a metallic monolith, so much so that they were reported to be able to meet the statutory 1976 NO_x standards, during an evaluation test at EPA labs.

Unlike the dual-bed system, the three-way catalyst system performs the oxidation and reduction functions

in one catalyst bed. Controlling all three components at the same time would involve the operation of the engine near stoichiometric, by the use of an oxygen sensor. This sensor would detect how much oxygen is in the exhaust and then feed back a signal to a reliable fuel-metering system. The correct content of oxygen will have to be maintained to be just sufficient to oxidize the HC and CO and yet not enough to impede the NO_x reduction. Another consideration is catalyst durability at such closely controlled A/F ratio. However, the catalyst manufacturers, mum as they are about the composition of the three-way catalyst, are confident that the system is a very viable one.

Retrofit

Pre-'75 cars in California and in other states where transportation control plans are in effect, will, unlike their new catalytically controlled family, need retrofit devices to meet the ambient air quality standards proposed by the Clean Air Act. California has been involved in retrofit programs since the mid 1960's. At the present time, there are two retrofit programs in California. One program, for HC control, affects 2.8 million

...but some prefer other exhaust emission controls

Foreign Imports	Control technique
Honda	CVCC (compound vortex controlled combustion) or 3-valve or stratified charge engine system. There are two combustion chambers—main and auxiliary—instead of one per cylinder. One carburetor meters a rich mixture into the auxiliary chamber which is connected to the main chamber by a little passageway. Another carburetor meters a lean mixture to the main chamber. The spark plug ignites the rich mixture, and the resulting flame shoots through the passageway and ignites the lean mixture. The overall effect is a leaner air fuel ratio. This reduces HC and CO. The low availability of oxygen in the auxiliary chamber, on the other hand, keeps the nitrogen oxides at low levels. Allows use of conventional leaded gasoline
Saab	Improved fuel-metering system, a more advanced fuel injection system and installation of a thermal reactor, alteration of ignition timing
Mercedes	Diesel-powered
Nissan (Datsun)	
Opel	
Peugeot	
Toyo Kogyo (Mazda)	Rotary engine; air-cooled thermal reactor with "modulated" air injection. No controls needed for NO_x. Allows use of low lead or leaded gasoline.

Choice. *It's noble metals, not types, that count*

5

1955–65 cars. The second program covers 4.5 million 1966–70 cars and is intended to control NO_x (42% minimum) by exhaust gas recirculation or vacuum spark advance disconnect.

Dana Corp's Retronox unit is an example of EGR mechanism. It is mounted on the engine with a stainless steel tube that runs down to the exhaust pipe. The spark advance is delayed for a short time at low speeds until a vacuum delay valve allows the EGR valve to open, permitting a small portion of the exhaust gas to recycle back to the engine. The device is bolted-on, requires no change in tuning specifications and does not disconnect the vacuum spark advance. Hence fuel economy and performance are unaffected.

For areas that need HC and CO control (Arizona and Colorado, for example), Retronox also comes with an oversized PCV (positive crankcase ventilation) valve which acts as an air bleed. This air bleed leans the combustion mixture to give about a 35% reduction in HC and 40–55% in CO, depending on altitude conditions. This unit costs $40.

Catalytic retrofit is also available. UOP's Purzaust system fits in the engine compartment just after the exhaust manifold and reduces HC by 68%, CO by 69% and NO_x by 13%. It is the noble metal type and utilizes an air pump for the adequate supply of oxygen in the exhaust. The total cost can be as low as $105 and as high as $260, depending on the requirement for an air pump.

All pollution control retrofit devices have to be accredited by the state of California and the particular state involved. The rule of thumb in these accreditation procedures is that the device, designed to control one or two pollutants, will not inadvertently increase the emissions of the third. Such was the earlier experience in California. Engine modifications made to reduce HC and CO in the 1966–70 vehicles increased NO_x by 30%. The NO_x in turn combined with the HC to form what is prevalent as the Los Angeles smog.

California, the first state to commence accreditation, has sanctioned eight devices at press time, including Air Quality's Pure Power, GM's device, Carter's Kit, Contignitron's Equalizer, Dana's Retronox, Echlin's device, STP's device, and UOP's Purzaust system.

Looking to MECA

Whenever there is a wide range of choices, there emerge some foresighted people who will band together to offer guidance and uniformity in the selection process. And the Manu-facturers of Emission Control Association was formed last September. It was formed primarily out of a need to have a more unified voice in dealing with the automobile emission control programs in California.

MECA then expanded to a multi-purpose organization concerned with:

• achieving a coordinated voice for the automotive emission control industry in promoting their systems as an effective means of reducing air pollution from new and used motor vehicles at a reasonable cost

• providing coordinated lobbying activities at the federal and state level to speak for or against legislation, rules, or regulations which affect the interests of MECA members, the industry, and the public

• monitoring EPA programs related to the setting, promulgation, and enforcement of standards and to the research and development for emission controls

• providing expert testimony at administrative and legislative hearings

• providing general or technical information to consumer, environmental and public interest organizations seeking expertise in automotive emission control

• providing publications and up-to-date information to members.

Basically, the association aids states (17 at present) affected by the clean air program. It sets up evaluation programs with the state to determine the nature of the problem and the suitable type of control technology. Then the floor is opened to suitable device manufacturers who will go about getting their devices accredited in that particular state.

MECA's involvement is timely. This March, EPA proposed to developers and marketers of auto emission control retrofit systems a voluntary evaluation program to be operated by EPA. This program would eliminate unnecessary duplication of data gathering and testing by individual states in that each manufacturer releases the test data of his device to the EPA which in turn evaluates the data and provides the information to all interested states.

In fact, one of the prerequisites for joining MECA is that the manufacturer sells a device that "does what it says it does." Weeding out the "phonies," said Mac McCullough (Dana Corp.), vice-president of MECA, is the objective. There is sufficient expertise in the association and available test data for just that, he pointed out.

Currently, Robert L. Joseph of UOP heads the MECA group of five, all manufacturers of retrofit devices accredited by California. While MECA is primarily made up of retrofit device manufacturers, Joseph is hopeful that other companies in the automotive emissions control industry will join the association. There are monthly meetings and a newsletter. MECA's first concern is modification of EPA's proposed voluntary evaluation test procedures.

Futuristic investigations

A substantial amount of work has been performed on the thermal reactor, a chamber for HC and CO combustion external to the engine. It is usually bolted to the cylinder head instead of the normal exhaust manifold.

There are two basic types of the thermal reactor—fuel rich and fuel lean. The former suffers from poor fuel economy but gives better NO_x control than the latter.

Thermal reactors plus catalyst systems are also being developed. GM's combination system has a bypass mode which is sealed with a valve at low speed. When the speed goes above 55 mph, the catalyst (dual in this case) system is bypassed through a thermal reactor. Questor's approach, at an approximate cost of $125, consists of a partial oxidation of HC and CO in a thermal reactor bolted onto the cylinder head, followed by NO_x reduction over a catalyst, and final oxidation in another thermal reactor. Both met 1975–76 statutory standards.

Improvements are being made to the not-so-new stratified charge and Wankel engines, practiced by Honda and Toyo Kogyo (Mazda), respectively. Ford is targeting its version of the stratified charge engine for 1978. Called Ford Proco, meaning programed combustion, the car will not include a prechamber. Stratification will be created by a fuel injector which shoots a cloud directly into the cylinder, a cloud rich in the center and lean on the outside.

Steam-propeled cars are still in the running. EPA confirmed that a car equipped with steam engine from Jay Carter Enterprises (Burkburnett, Tex.) and a test bed steam engine from Scientific Energy Systems Inc. (Waltham, Mass.) achieved 1975 statutory standards. Electric cars are already on the road (see ES&T. May 1974, p 410).

A vast amount of work remains to solve the problems associated with each of the futuristic approaches: driveability, improved mileage, adaptability to larger cars—to mention a few.

The same holds true for the use of hydrogen as a fuel. The low energy-density of hydrogen gas requires a tank, much larger than a gasoline tank, to keep the gas at 100° below zero. In addition, this very cold material does not store well, leaking off to maintain a constant low temperature. Storage systems are being studied—such as the metal hydrides.

Tests have been made, generating hydrogen on board a vehicle by reforming gasoline. This method seems to defeat the purpose of substituting hydrogen for gasoline.

Another fuel, patented and moving from the tube to the tank stage, is Goodyear's alcohol–water combination. "As much as 65% reduction in specific pollutants have been achieved compared with today's fuels," said researchers K. J. Frech and J. J. Tazuma. They believe that the blend of 25% *t*-butyl alcohol, 3% water, and 72% gasoline is particularly encouraging. This blend, they felt, could be preblended at refineries and could boost the octane number of basic gasoline to a rating at which

virtually any high compression engine could run knock-free, without lead, that is.

In perspective

At a cost of $50–100 and weighing about 10–30 lb, the catalytic converters must gain general acceptance. Industry has made its choice; the public awaits answers to their concerns:

● increased cost-per-gallon of unleaded gasoline arising from the modifications in refining technology, and distribution systems at the stations
● supply of unleaded gasoline
● possible health hazard contributed by sulfuric acid
● odor from traces of hydrogen sulfide which may be formed
● cost of replacing the catalyst after the end of the warranty which

covers one replacement within the first 50,000 mi.

One remembers that catalysts are not new in refining needs, according to UOP's Herman Bloch, recipient of the 1974 Murphree Award from the American Chemical Society. But they are new in the adaptation to the requirements of automotive use. There is further research and probing into this requirement: The National Academy of Sciences plans to release its final recommendations to Congress this fall; car and catalyst makers are aiming to improve catalyst performance and durability; and all are watching the health effects.

Meanwhile—at least in 1975—the catalysts will be coming down on "the side of angels in improving ecology," as Engelhard president M. Rosenthal aptly put it.

Regenerative SO₂ removal process

Reprinted from ENVIRON. SCI. TECHNOL., **11**, 22 (January 1977)

The French process of IFP (Institut Français du Pétrole) for
reducing exit stack gas concentrations of sulfur dioxide
checked out on a power plant near Paris last year

Most of the flue gas desulfurization systems that are in operation, under construction, or under a contract award, are first-generation or throwaway processes. These processes generate sludges and, in most cases, unsalable waste products.

On the other hand, second-generation processes, the so-called regenerative processes, have not come along that fast. Removing SO₂ from industrial stack gases by a process that produces an industrial raw material has come to light. The process is an extension of the IFP Claus tail-gas cleanup process; it can be used without creating a sludge handling problem.

In this process, basically, hydrogen sulfide is made to react with the SO₂ in the gaseous stack environment. The products of this reaction are sulfur, as a handlable solid, and water vapor, which is simply released to the atmosphere.

IFP director general J. C. Balacéanu says, "the antipollution industry is a new industry, and a chemical industry at that." He elaborates, "the IFP stack gas cleanup process is one that does not convert an air pollution problem into a problem of another sort. Rather, it produces a material (elemental sulfur) which has economic value."

Today, there are 30 industrial plants licensed by IFP to use its Claus tail-gas cleanup process; 10 are in the U.S. The extension of it, to cleaning up flue gas, is useful on refineries, electric power plants, non-ferrous smelters, and coke ovens.

The complete process has been demonstrated at the Champagne-sur-Oise power plant near Paris. Owned by the French national electric utility company (Electricité de France), this plant has two oil-fired generating units, each rated at 250 MW. The IFP process was evaluated on a 30 MW slipstream (about one-eight stream) during March–July last year (1976). After all the bugs were removed and the end of a 3-month continuous operation, the demonstration unit was voluntarily shut down.

Those SO₂ removal processes that yield a commercial product should prevail.

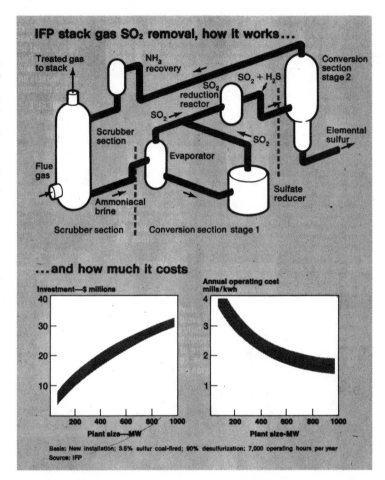

IFP stack gas SO₂ removal, how it works...

Treated gas to stack — NH₃ recovery — Conversion section stage 2 — SO₂ + H₂S — SO₂ reduction reactor — SO₂ — Scrubber section — Evaporator — SO₂ — Elemental sulfur — Flue gas — Ammoniacal brine — Sulfate reducer

Scrubber section — Conversion section stage 1

...and how much it costs

Investment—$ millions / Plant size—MW

Annual operating cost mills/kwh / Plant size-MW

Basis: New installation; 3.5% sulfur coal-fired; 90% desulfurization; 7,000 operating hours per year
Source: IFP

The solid inert sulfur is much easier to handle than either the sludge from throwaway furnaces or liquid sulfuric acid, which is corrosive and difficult to handle, from another regenerative process.

Of course, the moot points are that sulfur may become a glut on the market; its economic value varies widely, (see *ES&T*, February 1976, p 124), and operating costs for the regenerative process must be minimized.

How it works

An industrial plant using the IFP stack gas cleanup process consists of two separate sections. These sections do not have to be close to one another. The first section, a scrubbing section, captures stack gas pollutants in an aqueous ammonia liquor. In the second section, the reaction or conversion section, all sulfur values in the liquor are converted to pure

Owner	Location	Tail gas flow rate (NM^3/H)	Conversion (%)	Start up
Delta Projects	Alberta	730	87	1971
Nippon Petroleum	Japan	28 400	85	1971
Idemitsu Oil	Japan	26 800	75	1972
Showa Oil	Japan	5700	85	1972
Kyokuto Petroleum	Japan	18 600	90	1972
Chevron Standard	Alberta	25 300	60	1972
Mitsubishi Oil[b]	Japan	22 000	85	1972
Mitsubishi Oil[b]	Japan	29 300	85	1972
Phillips	Texas	4800	90	1973
Stauffer	Delaware	c	93	1973
Commonwealth Oil	Puerto Rico	6000	80	1973
Koa Oil	Japan	c	c	1973
Phillips	Texas	4800	90	1973
Kawatetsu	Japan	4400	c	1973
Ministry of Gas	USSR	76 800	75	1974
Ministry of Gas	USSR	76 800	75	1974
Ministry of Gas	USSR	76 800	75	1974
Kawatetsu	Japan	4400	c	1974
Koa Oil	Japan	c	c	1974
Esso	Belgium	39 600	86	1974
Amoco	Texas	40 900	93	1975
Charter	Texas	26 900	89	1976
Phillips	Texas	13 300	93	1976
Phillips/Monsanto	California	19 700	94	1976
FMC	West Virginia	19 500	93	1976
Kawatetsu	Japan	4400	85	1976
BP	Germany	4000	80	1978
Statsforetag	Sweden	16 700	90	1979
Polimex Cekop	Poland	5000	90	1979

[a] The IFP Claus tail-gas cleanup process takes SO$_2$ down to about 1000–1500 ppm.
[b] Associated with IFP flue gas clean-up process, which take SO$_2$ down to about 100 ppm.
[c] Confidential.

molten sulfur, and the ammonia is recycled to the first section, the scrubbing section.

Ammonia scrubbing is an old and proven technology but as people in the sulfite pulp industry know, it can produce "blue plume," a troublesome stack problem developed when submicron particles of ammonium salts escape from the scrubber. IFP has formed an association with several other companies, among them Air Products, whose patented ammonia scrubbing technology produces no opacity in the plume.

When stack gas is scrubbed by ammonia with this technique, the effluent liquor from the scrubber contains ammonium sulfite and bisulfite and a small amount of sulfate. Widely varying concentrations of SO$_2$ can be handled without difficulty. The SO$_2$ content of the flue gas, perhaps 1000–1500 ppm, can be reduced to as low as 100 ppm.

In the conversion section of the IFP plant, the liquor is decomposed into ammonia, sulfur dioxide, and water in a 2-stage operation that uses IFP submerged combustion technology to reduce sulfates.

IFP director general Balacéanu
"antipollution . . . a chemical industry"

Two-thirds of the SO$_2$ is then transformed into H$_2$S by using any sort of reducing gas. In practice, a gasifier is used to produce this gas. But the IFP process can use any gas with hydrogen and carbon monoxide in any proportions.

In the tower, the hydrogen sulfide and sulfur dioxide are injected in the stoichiometric ratio of 2:1 and react according to the Claus equation to form elemental sulfur. The reaction takes place in the presence of the solvent. Commercial quality sulfur, 99.9% pure, is withdrawn in a molten state from the bottom of the tower. The tower is typically about 60-ft tall and 15-ft in diameter.

The heat of reaction from the Claus reaction (63 000 Btu for every pound mole of converted SO$_2$) is recovered and used to produce steam that is used in various parts of the IFP process.

How much does it cost?

Based on the operation of the demonstration plant, IFP estimates that the investment for a 250-MW oil-fired power-plant to be $14.6 million. This figure includes catalyst and solvent and is equivalent to $58.50 per installed kilowatt. Operating costs are 3.3 mils per kilowatt-h, neglecting any income received from sulfur sales.

In perspective, these figures for an IFP unit correspond to an increase of about 20% in the plant investment and 15% in operating costs in comparison with a power plant operating without any SO$_2$ removal system.

Last word

Second generation SO$_2$ stack gas cleanup systems are having a difficult time. The dry scrubbing of utility emissions (*ES&T*, August 1975, pp 712–713) has suspended operations, according to the recent PEDCO report. Earlier it was believed that recovery processes would develop rapidly and supersede throwaway processes. But in the Swedish delegate report, P. O. Alfredsson of the ECE (*ES&T*, February 1976, p 125), said that this has not been the case. Whether the life-time operating costs for recovery systems are higher than throwaway systems, even with credit for recovered products, must await further trials.

Removing SO₂ in heating plant boilers

Research-Cottrell, the U.S. licensee of the Swedish process of A. B. Bahco, cleans up emissions from coal-fired boilers; the process is useful on small installations of the 100 MW size—small boilers, industrial boilers, and small power plant boilers

Reprinted from ENVIRON. SCI. TECHNOL., **11**, 645 (July 1977)

Historians may remember 1977 as the year of the coal lump instead of the snake. The energy crunch hit hard for millions who still face the prospects of dwindling natural gas supplies and soaring electrical costs.

Experts say that the U.S. has an unlimited coal resource within its continental borders that could generate our electrical and heating requirements for the next 500 years. But what about the pollution problem?

Near Columbus, Ohio, is Strategic Air Command, Rickenbacker Air Force Base. The field was built in 1942 and was designed to be heated with coal. The plant houses eight boilers and an impressive number of smoke stacks. It operates much to the dismay of local residents, around the clock, 365 days a year. Only 12 months ago the plant was the source of a 12-mile-long smoke screen.

Civil engineer Jim Rasor, said, "Rickenbacker had plans to convert to oil in 1973. At the time bids were being sent out for the job, the Pentagon was looking at Research-Cottrell and the Swedish Bahco scrubber. R-C obtained a license to use the technology in the U.S. and Rickenbacker AFB is where it went."

The Air Force experimental pollution abatement system with the Bahco scrubber is one of a kind in the U.S. and has been in operation for almost a year. Because of the importance of coal as an alternate fuel, Rickenbacker has become a common name in environmental circles.

SO₂ removal: How the Bahco process works

How it works

In back of the heating plant is a silo filled with lime. The game plan calls for the smoke to be blown into the scrubber, which contains two venturis. The bottom of the scrubber is filled with a lime solution; another pump with lime solution is located halfway up the chimney-shaped tower. Large pumps churn the liquid.

The smoke is forced through the lower venturi into the churning lime, which sponges up harmful sulphur oxides. The process is again repeated through the upper venturi, with the gases passing through a fan that spins them upward like a tornado. The remaining suspended particles are spun out of the updraft, strike the inner walls of the chamber, and fall harmlessly to the bottom of the tank. The sanitized gases then leave the scrubber stack in the form of harmless steam.

"The system is not without its drawbacks," Rasor continues. "There is a sludge by-product, calcium sulfate. This thick black liquid is pumped into a nearby rubber-lined lagoon. The pool will take 5 years worth of sludge. The Japanese have a Bahco system similar to ours and have been adding chemicals to the by-product to make gypsum wall board."

Since the Rickenbacker scrubber is the only one in the U.S., the Environmental Protection Agency keeps close tabs on the operation. The scrubber has far surpassed present EPA emission standards for this type of system, which means that the Air Force experiment is doing the job.

The project at the AF base has been saving money by burning a cheaper, locally available, high-sulphur coal, and even with the added cost of the lime they are still operating below the price of fuel oil. The Rickenbacker project may become a blueprint for future American coal-burning plants and could lead to greater use of our most abundant energy resource.

Civil engineer Rasor
"AF experiment is doing the job"

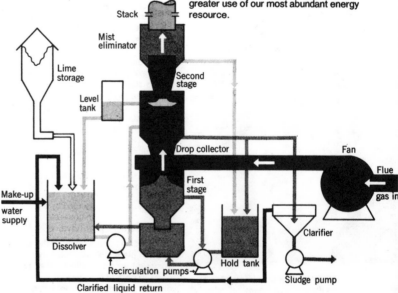

Reprinted from ENVIRON. SCI. TECHNOL., **11**, 226 (March 1977)

A way to lower NO$_x$ in utility boilers

The Exxon thermal deNO$_x$ process reduced emissions
50 % and worked at full load on a refinery in Japan

When fossil fuel is burned within most combustion devices, nitric oxide (NO) is formed both by thermal fixation of atmospheric nitrogen and by oxidation of organic nitrogen compounds present in the fuel. The NO formed in these different modes is often referred to as thermal NO$_x$ and fuel NO$_x$, respectively. While the NO formed within the combustion device is discharged to the atmosphere largely as NO, it is thereafter slowly converted into NO$_2$, a key component of photochemical smog and by itself a highly poisonous substance. Accordingly, the emission of NO and NO$_2$ (NO$_x$) from mobile and stationary sources has been regulated in the U.S., Japan, and to a lesser extent in other countries.

The problem

The prospect in the U.S. is for increasingly severe regulation of stationary sources in those regions where the ambient air quality is a problem. The reason for this is twofold: First, regions with air quality problems tend to be regions with high industrial growth rates. Thus, if the total contribution of all stationary sources

to the air quality burden is to be held constant, the amount each source may emit must be reduced in proportion to the growth.

Second, there is a growing body of opinion to the effect that it is more cost effective to control stationary sources than to control mobile sources. To regulate mobile sources one must regulate all mobile sources nationwide, whereas one can selectively regulate only those stationary sources within air quality critical regions. Further, since one large stationary source has NO$_x$ emissions equivalent to those of many automobiles, there is an obvious economy of scale to regulating the former. In direct conflict with this trend toward more severe regulation, there is a concurrent trend toward the use of fuels with higher nitrogen contents, especially coal and, in the not too distant future, shale oil.

The technology now used in the U.S. to meet NO$_x$ emissions standards is combustion modification, chiefly the two-stage combustion and flue-gas recirculation processes. The former involves burning the fuel with insufficient air, followed by

adding more air to complete the combustion. The latter involves diluting the combustion air with recirculated flue gas, thereby decreasing the peak flame temperature.

While both processes can provide substantial reductions of thermal NO$_x$, two-stage combustion has a very limited effectiveness against fuel NO$_x$, and flue gas recirculation is totally ineffective. Thus it seems probable that many areas in the U.S. will have more severe NO$_x$ emission standards that can be met with combustion modification technologies. In Japan, such severe standards are already law.

One control option

To meet this challenge, Exxon Research and Engineering Company (ERE, Linden, N.J.) has developed a new process based on the selective homogeneous gas-phase reduction of NO by NH$_3$ (ammonia). Practically, what this means is that no catalyst is required. One simply contacts the flue gas with NH$_3$ at the correct temperature. A rapid reaction occurs in the gas phase, and NO is reduced to N$_2$ and H$_2$O. Because this is a selective reaction, the amount of NH$_3$ needed is comparable to the amount of NO reduced.

Thus, the Exxon process is extraordinarily simple. One finds the location within a boiler/furnace at which the temperature of the flue gas is optimum for the process and, the internal configuration of the boiler/furnace permitting, installs the ammonia injection hardware in that location. However, use of the correct temperature is critical because the chemistry of the selective reduction of NO by NH$_3$ is highly temperature sensitive.

Laboratory data, obtained by ERE's Corporate Research Laboratory, illustrate this sensitivity. When the temperature is too low, NH$_3$ and NO tend to remain unreacted; when the temperature is too high the NH$_3$ tends to be consumed inefficiently, with relatively little net reduction of the NO. Thus, it is possible to achieve

Data. *Richard K. Lyon of the Corporate Research Labs of the Exxon Research and Engineering Company checks the temperature distribution within the boiler with changing load*

Reduction of NO_x up to 70% are possible...

...50% reduction was achieved under full load

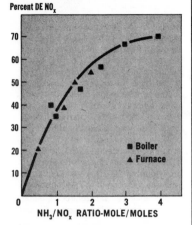

Percent DE NO_x

■ Boiler
▲ Furnace

NH_3/NO_x RATIO-MOLE/MOLES

Selective reduction is highly temperature sensitive...

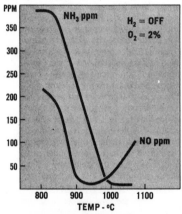

PPM

NH₃ ppm

$H_2 = OFF$
$O_2 = 2\%$

NO ppm

TEMP - °C

...but addition of hydrogen controls temperature for optimum results

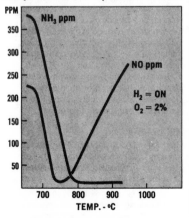

PPM

NH₃ ppm

NO ppm

$H_2 = ON$
$O_2 = 2\%$

TEMP. - °C

Source: Exxon Research & Engineering

Exxon thermal denox process, how it works[a]

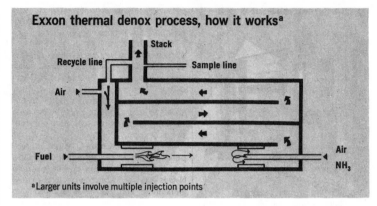

Stack

Recycle line

Sample line

Air

Fuel ▶

Air

NH_3

[a] Larger units involve multiple injection points

an efficient reduction of NO with little NH_3 remaining, but only within a narrow temperature range.

For furnaces and boilers that operate at varying loads, this would appear to pose a severe problem since the temperature distribution within the furnace/boiler varies with changing loads. However, the addition of a small amount of hydrogen (H_2) markedly decreases the temperature at which optimum results are achieved. The extent by which the temperature for optimum results is reduced depends on the amount of H_2 added; thus, by adding the correct amount of H_2, one may control the temperature for optimum results to match the variation of the temperature of the injection point caused by changing load. Laboratory data illustrated ERE's solution to this problem.

Field evaluation

Obviously, conditions cannot be as accurately controlled in practical combustion equipment as in laboratory apparatus. Consequently, the nearly quantative NO reductions achieved in laboratory experiments cannot be obtained in real furnaces and boiler.

This demonstration of the Exxon process on a real furnace and boiler was done at the Kawasaki refinery of Exxon's Japanese affiliate, and was an entirely retrofit operation: both the boiler and furnace were taken as found and the process installed within them without modification of any operating component of either unit. Reductions of NO_x of up to 70% were achieved by the Exxon Thermal Denox Process during its commercial demonstration (August 1974).

The installation was done by the regular operating staff of the refinery, and with no interruption or interference with normal refinery operation. The demonstration boiler is oil-fired, and operates at up to

140 million Btus per hour. It's a swing boiler, used when one of the other boilers is out of service. The furnace is gas-fired, operated at 500 million Btu/h, and is in continuous service.

Since this commercial demonstration the process has been installed in other units in the Kawasaki refinery with similarly good results. Further, one Japanese firm not associated with Exxon has taken a license for use of the process and installed it in a boiler. Although the internal configuration of the boiler was not entirely suited to the process, reductions of NO of 50% at full load were achieved.

Costs

While costs of the Thermal Denox Process are sensitive to the user's particular circumstances, for normal retrofit ERE estimates the range to be 7–15¢/10^6 Btus fired.

Naturally, considerable savings are possible if, when new equipment is designed, provision is made to accommodate the process. This is a consideration of importance to those now planning new boilers/furnaces for operation in regions where the ambient NO_x level is now considered marginally acceptable. In such areas, the standards for NO_x emissions may be made more stringent during the normal lifetime of combustion equipment. If provision is now made within the boiler/furnace design for the Thermal Denox Process, it will later be possible to install the process at reduced cost in the event that more severe standards are promulgated.

Secondary effects

There have been a number of cases, recently publicized, of pollution control processes which, despite their initial promise, were later found to solve one pollution problem only at the expense of creating a new and equally serious one. Recognizing the danger of other undesirable secondary effects, ERE has implemented an extended program of laboratory and field testing.

Special attention was given to the question of whether or not such substances as N_2O, CO, HCN, SO_3, and NH_4HSO_4 might be produced within the Thermal Denox Process and constitute new pollution problems.

Although the main products of the reduction of NO by NH_3 are N_2 and H_2O, N_2O (nitrous oxide) is also formed in trace amounts, about 1 or 2 moles of N_2O for every 100 moles of NO reduced. This amount of N_2O proved too small to be detected by gas chromatography but was readily measurable by long-path infrared spectroscopy.

This production of N_2O would not appear to represent a significant disadvantage to the Thermal Denox process, since N_2O is generally accepted as a harmless substance, and, harmless or not, it is naturally present in the environment at a worldwide ambient concentration of about 0.3 ppm. Thus, since N_2O is present in nature in massive amounts, its ambient concentration is not readily subject to change by human activity, and the production of traces of N_2O by the Thermal Denox process does not represent a problem.

Tests for CO (carbon monoxide) emissions show that the Thermal Denox reaction does not reduce CO_2 to CO. Thus, the reaction does not per se generate CO; however, it was also found that NH_3 does inhibit the oxidation of CO to CO_2. Thus, if any CO is left unburnt at the point where the flue gas reaches the NH_3 contactor, that CO may be left in the flue gas when it is discharged to the atmosphere. For normally operating gas- and oil-fired units, this is not a problem. In these units CO oxidation is complete long before the combustion gases reach the

ammonia injection point. At present, not enough information is available to determine whether or not incomplete CO oxidation will be a problem in coal-fired units.

It was found that HCN (hydrogen cyanide) can be formed if and only if hydrocarbons are present in the region in which the Thermal Denox reaction occurs. For a normally operating boiler, gaseous hydrocarbons will be present only if one injects them along with the NH_3. However, if one injects NH_3 with H_2 added for temperature compensation, HCN cannot be formed. If one injects NH_3 with CH_4 (methane) for temperature compensation, then a few ppm of HCN can be formed— less HCN than is found in automobile exhaust.

Whenever one burns a sulfur-containing fuel, the sulfur appears in the combustion gases chiefly as SO_2, but there are also traces of SO_3 (sulfur trioxide), normally 1–2% of the total sulfur. This small amount of SO_3 is a significant concern because at lower temperatures, SO_3 tends to form sulfuric acid mists that can cause corrosion problems at the back end of the boiler. Also, if the flue gas is not properly dispersed, the formation of sulfuric acid mist can be a severe local problem, although there is controversy about this in the technical literature. It appears that some of the SO_3 comes directly out of the flame, and some is produced by oxidation of SO_2 on the heat exchange tubes; that is, some SO_3 is produced by a homogeneous oxidation and some by a heterogeneous oxidation. Detailed laboratory experiments have been completed and the Thermal Denox process was found to cause neither additional homogeneous nor additional heterogeneous oxidation of SO_2 to SO_3.

However, although there is no change in the amount of SO_3, there is an effect. The thermal reduction of NO leaves some NH_3 unreacted, and downstream of the

Denox reaction zone the combustion gases are cool enough for the ammonia to react with SO_3 and H_2O to form ammonium sulfates. Whether or not these sulfates should be considered as an added source of particulate emission is a question of definition. Both SO_2 and SO_3, when discharged to the atmosphere, can form sulfate particulates. The effect of using the Thermal Denox process is that a small fraction of the total sulfur oxides are discharged as a particulate rather than forming a particulate at a later time.

While this change does not appear to be of environmental concern, it is of concern with respect to boiler/furnace operation. It is normal practice to avoid corrosion by sulfuric acid mist by running the back end of the furnace/boiler at a temperature above the acid dew point, generally around 180 °F. In contrast, ammonium bisulfate is a corrosive liquid in the temperature range 400–500 °F. Thus, one might fear that ammonium bisulfate formation might result in a new corrosion problem.

Prior to the first commercial demonstration, a laboratory simulation of a boiler was set up and the corrosion rate of steel in flue gas was measured in the presence and absence of NH_3. No change in this corrosion rate was observed up to the highest NH_3 concentration tested, 45 ppm. Moreover, short-term commercial demonstration has not disclosed difficulties with corrosion or fouling. Long-term experience is being accumulated.

Patent licenses

The Exxon Thermal Denox Process is covered by U.S. Patent 3,900,554. Those interested in purchasing licenses under this patent and technical assistance in using the process may contact Mr. S. Stahl, Patents, Licenses, and Technology Sales Division, Exxon Research and Engineering Co., 1600 Linden Ave., Linden, N.J. 07036.

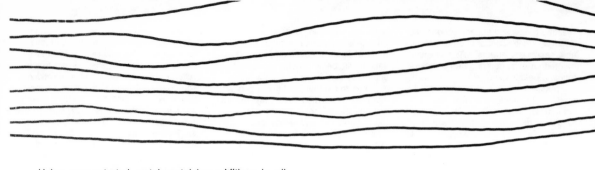

Using concentrated metal-containing additives in oil-fired furnaces is common practice today, particularly with boilers having high superheat and reheat temperatures and high heat release. Such additives improve boiler cleanliness, reduce corrosion of the firesides and air heaters, and reduce stack emissions and plumes.

In general, fuel additives should provide boiler cleanliness, high-temperature vanadium corrosion protection, prevention of loss of operating capacity by maintaining design steam temperatures, cold end (air heater) corrosion protection, reduction of stack emissions from hydrocarbon particulate matter and SO_3, and improvement in the handling characteristics of ash in the flue gas in oil-fired boilers equipped with precipitators and stack collectors. The significant fuel additives in use today usually contain MgO (with or without small amounts of aluminum oxide or hydrate), manganese, and MgO with manganese.

The application of additives to solve particular boiler requirements hopefully has evolved from an art to almost a science. Utility managers today will no longer consider seriously any fireside additive that does not contain a high metallic concentration. The choice of additive, however, depends upon the needs of the particular boiler, as well as environmental requirements.

Heavy duty additives: MgO vs. Mn

The addition of MgO-based products to the fuel oil or furnace will raise the fusion point of the fuel oil ash from an initial 950–1050°F to approximately 1350–1450°F at a Mg:V weight ratio of 1.5:1. A weight ratio of 1.5:1 is equal to a 3:1 molar ratio of Mg:V which corresponds to a dilution rate with an MgO slurry containing 50% MgO of approximately 1:2000 with a fuel oil having a vanadium content of 200 ppm. The fuel oil ash, consisting generally of magnesium vanadates, magnesium sodium vanadates, and magnesium sodium vanadyl sulfates, is less adherent to the superheat surface.

Depending upon the ratio of Mg:V used, the ash will range in texture from a soft but voluminous powder at high Mg:V ratios to a somewhat brittle "popcorn" configuration or even to a dense, layered, amorphous coating as the ratio of Mg:V is lowered. The "treated" slags are generally easier to remove by manual or air lancing; also, they are more water soluble which allows easier water or steam lancing, or preferably high-pressure "steam blasting" for maximum cleaning. The treated slags also have a higher pH, by 1–2 pH units.

There are several disadvantages, however, in using MgO-based products. Deposits are highly susceptible to rapid bridging of the superheater tubes because of the increased boiler ash loading (Figure 1 which compares a furnace treated with MgO to one treated with manganese).

With fuels of 1% sulfur, or less, and a vanadium content of less than 100 ppm, high-temperature superheat fouling and corrosion of tube supports are less likely to occur than with fuels of 2.0% sulfur with 200-300 or higher ppm vanadium. With the lower sulfur fuels, it has been possible to eliminate completely any interim boiler clean-

Additives can clean up oil-fired furnaces

Metal-containing additives improve boiler cleanliness, cut corrosion of firesides and air heaters, and reduce stack emissions and plumes

Ira Kukin

Apollo Chemical Corp., Clifton, N.J. 07014

Reprinted from ENVIRON. SCI. TECHNOL., **7,** 606 (July 1973)

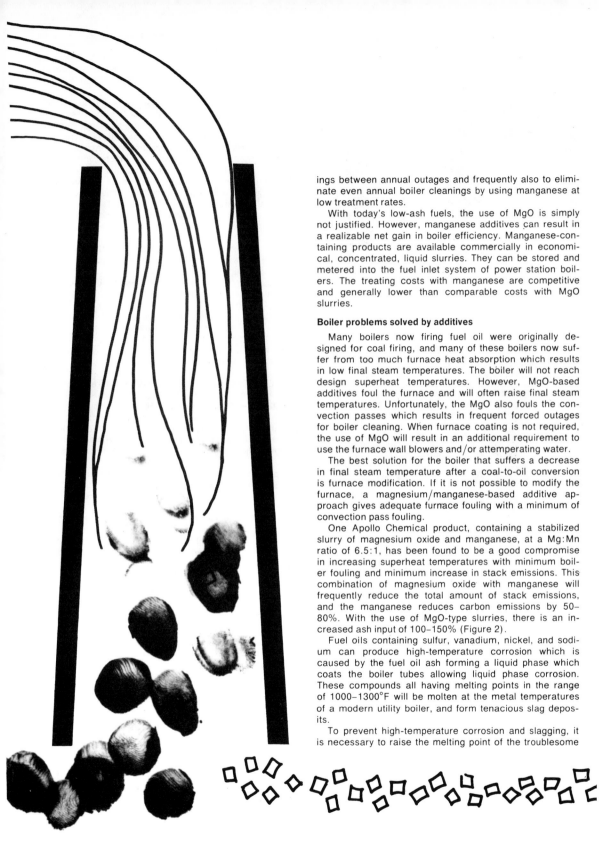

ings between annual outages and frequently also to eliminate even annual boiler cleanings by using manganese at low treatment rates.

With today's low-ash fuels, the use of MgO is simply not justified. However, manganese additives can result in a realizable net gain in boiler efficiency. Manganese-containing products are available commercially in economical, concentrated, liquid slurries. They can be stored and metered into the fuel inlet system of power station boilers. The treating costs with manganese are competitive and generally lower than comparable costs with MgO slurries.

Boiler problems solved by additives

Many boilers now firing fuel oil were originally designed for coal firing, and many of these boilers now suffer from too much furnace heat absorption which results in low final steam temperatures. The boiler will not reach design superheat temperatures. However, MgO-based additives foul the furnace and will often raise final steam temperatures. Unfortunately, the MgO also fouls the convection passes which results in frequent forced outages for boiler cleaning. When furnace coating is not required, the use of MgO will result in an additional requirement to use the furnace wall blowers and/or attempering water.

The best solution for the boiler that suffers a decrease in final steam temperature after a coal-to-oil conversion is furnace modification. If it is not possible to modify the furnace, a magnesium/manganese-based additive approach gives adequate furnace fouling with a minimum of convection pass fouling.

One Apollo Chemical product, containing a stabilized slurry of magnesium oxide and manganese, at a Mg:Mn ratio of 6.5:1, has been found to be a good compromise in increasing superheat temperatures with minimum boiler fouling and minimum increase in stack emissions. This combination of magnesium oxide with manganese will frequently reduce the total amount of stack emissions, and the manganese reduces carbon emissions by 50–80%. With the use of MgO-type slurries, there is an increased ash input of 100–150% (Figure 2).

Fuel oils containing sulfur, vanadium, nickel, and sodium can produce high-temperature corrosion which is caused by the fuel oil ash forming a liquid phase which coats the boiler tubes allowing liquid phase corrosion. These compounds all having melting points in the range of 1000–1300°F will be molten at the metal temperatures of a modern utility boiler, and form tenacious slag deposits.

To prevent high-temperature corrosion and slagging, it is necessary to raise the melting point of the troublesome

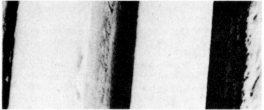

Figure 1. *The superheater section of a boiler using 2.3% sulfur fuel treated with MgO (top) shows deterioration while manganese-treated fuel has little effect (bottom)*

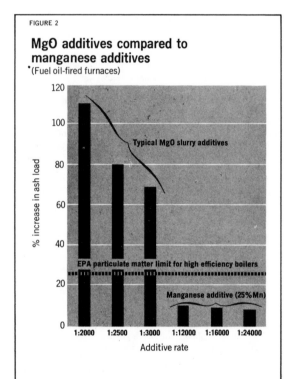

FIGURE 2

MgO additives compared to manganese additives
(Fuel oil-fired furnaces)

Chart — % increase in ash load vs. Additive rate

Typical MgO slurry additives

EPA particulate matter limit for high efficiency boilers

Manganese additive (25% Mn)

Additive rate: 1:2000, 1:2500, 1:3000, 1:12000, 1:16000, 1:24000

ash constituents to prevent formation of the corrosive liquid phase. This action produces a friable ash that normally responds to soot blowing. An alternate solution is to burn premium fuels containing low vanadium, sodium, nickel, and sulfur concentrations—i.e., a total ash content of 0.01–0.02%, or less.

Manganese-based additives will raise the fusion point of fuel oil ash. Where only a moderate increase in the ash fusion point is required, manganese treatment has the advantage because much less additive is required. The lower treatment rate possible with manganese results in improved boiler cleanliness which minimizes and often eliminates forced outages for boiler cleaning. Field experience has shown that manganese is superior to magnesium in maintaining boiler cleanliness, particularly with fuels of low ash content. Unlike MgO, manganese acts as a combustion improver and provides the potential for lower excess air firing.

Another problem is cold end corrosion, the result of SO_3 condensation, as sulfuric acid (H_2SO_4), on the air heater and gas duct surfaces. When the gas dew point is reached, a very sticky mist of sulfuric acid is formed and is extremely corrosive at the temperature conditions present in this boiler region.

This acidic stack fallout will also occur with fuels of lower sulfur content in the range of 2–3% and in many cases even with fuels having a sulfur content of 1 ± 0.5%. The most troublesome acid fallout problems occur under the following conditions:

● High excess air is required to obtain steam temperatures.

● Coal-fired furnace converted to oil-fired—generally such a furnace will require a high excess air to help maintain steam temperatures. Also, the exit duct work in these furnaces, whether or not the utility continues to use the precipitators present in coal firing (mechanical or electrostatic), often presents individual problems. In these systems, there generally occurs a high SO_3 concentration in the flue gas, and a tendency exists for the SO_3 to precipitate out with a rise in dew point temperature. The surfaces of the duct work cool rapidly because of the longer path lengths that must be traversed before the flue gases enter the atmosphere. Rapid corrosion of the air heaters, the fans, and the duct work will occur.

It is virtually impossible to experience low-temperature corrosion without cold end fouling. The sulfuric acid mist traps much of the particulate matter and fly ash carbon passing through the dew point area. Normally the cold end deposits are high in carbon which absorbs SO_3 and intensifies corrosive attack.

Use of excessive amounts of magnesium oxide slurries can actually accentuate the fouling and corrosion problem. The MgO resulting from use of the MgO additive slurries is dead burned as it passes through the boiler's flame zone, significantly reducing the reactivity of the MgO. This less reactive MgO then contributes to air heater fouling.

An alternate solution is to raise the unit's exit gas temperature. However, each 50°F increase in exit gas temperature costs 1% in fuel economy. With today's fuels costing in excess of $4.00 per barrel, it is usually more economical to control cold end problems with an additive program.

Generally, when a reduction of SO_3 of 25–40% is obtained by using fairly substantial quantities of MgO, it results from the MgO coating the iron tubes in the furnace so that the iron surfaces are no longer exposed. Otherwise, the iron acts as a catalyst to increase the percentage conversion of sulfur to sulfur trioxide:

$$S + O_2 \rightarrow SO_2$$

$$SO_2 + \tfrac{1}{2}O_2 \rightarrow SO_3$$

(Reaction catalyzed by iron and/or vanadium; extent of reaction controlled by high excess air.)

With the use of magnesium oxide, successive layers of this oxide and $MgSO_4$ build up on the heaters. As a consequence, air heaters must be water-washed at frequent outages to remove the insulating deposits.

Regarding air heater cleanliness, manganese is far superior to MgO slurries. Not only does the lower dosage rate with manganese reduce the unit ash loading, but the combustion catalytic activity of the manganese reduces the amount of unburned carbon and fuel oil residue by as much as 50–70%. Although this carbon residue represents only 0.2–0.4% of the fuel oil, it acts as a binder for air heater fouling and as an absorbing agent for acidic SO_3. Most boilers treated with manganese have required approximately one third the air heater washing frequency that they required when treated with MgO-based products.

Preventing cold end corrosion and stack fallout

Generally, application of MgO to reduce SO_3 is restricted to boilers that operate with excess air of 3% or less, but certainly no more than 5% excess air. Where it is not possible to restrict the amount of excess air to below 3–5%, then the use of MgO added to the fuel oil or furnace can actually increase the SO_3 content in the flue gas. A preferred method to reduce low-temperature corrosion and SO_3 emissions is the use of a supplementary cold end neutralizing additive.

English researchers, reporting on work carried out in England at the Marshwood generating station with boilers firing fuel oil in which back end injection with ammonia was attempted for reducing SO_3 in the flue gas, found that addition of ammonia at 20–33 lb/hr was required to control SO_3 attack in a boiler producing 550,000 lb of steam per hour at 915°F superheat temperature with a fuel containing 4.2% sulfur. At this feed rate, an increase of 3 in. of draft loss in the air heater occurred within 11 days. Further, the resultant deposits were coke-like mixtures of fused ammonium bisulfate and carbon, which were still acidic.

The economic losses to a utility because of corrosion of the air heaters can be substantial, particularly when a boiler is load-limited because of insufficient air. This situation occurs when the air heaters are plugged with iron sulfate deposits or have become corroded. In addition, utility managers report a worsening of the stack plume as air heaters become plugged or inoperative. Frequent outages to water-wash the air heaters then become necessary.

One leading process utilizing back end feed of a neutralizing complex, either as a solid or liquid injection, is offered to the utility industry by Apollo Chemical Corp. Worldwide patents have been applied for detailing this process which overcomes the problem where it occurs—for example, in the flue gas.

Summarizing, the preferred method to eliminate cold end corrosion in oil-fired furnaces is the use of a manganese, carbon-destroying catalyst in the fuel oil. This procedure often allows lower excess air firing, modifying the vanadium ash, coating the furnace iron tubes thus preventing the catalytic action of iron and vanadium in converting SO_2 to SO_3, and significantly reducing the flue gas carbon grain loading. Then, if air heater corrosion and fouling persists, or if the stack plume indicates SO_3 (acid) fallout as evidenced by a deep blue attached or detached plume, a supplemental flue gas–neutralizing treatment can be used, one that is injected into the economizer outlet.

At Union Electric Co.'s (St. Louis, Mo.) 500-MW coal-fired plant producing 3,290,000 lb/hr of 1000°F steam at supercritical pressures, there was a persistent buildup of deposits of iron sulfate in the tubular air heaters in the

Figure 3. *Untreated air heater (above) shows fouling, but back-end neutralizing additive decreases damage (below)*

bypass section. This bypass section, accounting for 10% of the exit flue gas, required forced outages at almost bimonthly intervals to water-wash and manually remove the deposits in the tubular air heaters.

Attempts to improve this condition by raising the steam temperatures through installing heating coils around the tubular air heaters were tried but were unsuccessful or otherwise economically prohibitive.

One of the twin tubular air heaters then was treated by tail end injection of a specially processed alkaline-containing 15–30-μ size complex. This product was over 90% active for removing SO_3 within a residence time of 4 sec. As shown in Figure 3, this method resulted in a dramatic improvement in the air heaters. No cleaning was required during the 12-week test trial. The untreated section had to be water-washed at monthly intervals.

Handling characteristics and air emissions

Flue gas supplemental feed has proved superior to the use of fuel oil treated with MgO with regard to electrostatic precipitator performance in oil-fired furnaces. The problem of cracking the porcelain insulators and shorting out the emitting wires can often be reduced by supplementary air cooling of the insulators, but this action will not entirely eliminate the problem.

Test period	Additive		Flue gas particulate matter, av concn. (mg/SCF)	Av % carbon (in particulate matter)
	Slurry of:	ppm Mn		
IA	None		0.17	72
IB	Manganese	45	0.04	31
IIA	None		0.14	58
IIB	Manganese	45	0.03	32

TABLE 1
Reduction of carbon particulate matter by manganese

The problem, it is thought, originates because of the accumulation on the insulators of carbon and acidic flue gas by-products. When subjected to high temperatures, these products destroy the insulation material.

This tail end neutralization has been useful not only for improving the overall efficiency of the precipitators, but it also often permits ash hoppers to be operated without interruption. The additive dries up the fuel ash in the back end by virtue of the acid-neutralizing properties of the injected additive. The ash no longer tends to clump, but rather flows freely. Indeed, this tail end feed will often eliminate the necessity to revert to the more expensive method of water sluicing of the ash to remove it from the hopper collectors.

Manganese reduces particulate matter

The use of manganese will result in significantly less particulate stack emissions, as much as 80% less (Table I). In this situation carried out at an oil company in the U.S., the fuel consisted of heavy asphaltic bottoms. This pitch was burned in a package boiler producing 200,000 lb of steam/hr at 900°F superheat temperatures of 900 psi. Measurement of the flue gas particulate matter showed a decrease from 0.17 to 0.04 Mg/SCF, with a reduction of carbon content of the emitting particulate matter from 72 to 31%.

Manganese also shows a significant effect in reducing particulate emissions when burning waste oils and bottoms in petrochemical and refinery heaters. In one exam-

ple, a Canadian refinery was burning a petrochemical waste pitch in a boiler. Because of the appearance of the stack, which exceeded a Ringelmann 2 value, the refinery could burn only 100 barrels per day of the pitch, rather than the 800 barrels produced daily and for which the heater originally was designed. Otherwise, the smoke number exceeded the local regulations that prohibited the stack having a Ringelmann value in excess of 1. Therefore, the refinery had to burn natural gas which was extremely limited in availability.

It was economically prohibitive to modify the furnace to meet the regulations. A quick solution was found in the use of an oxide slurry (25% by weight of manganese) at a feed rate of 1 gal/75 barrels of pitch. The required feed rate of manganese was determined with an in-line Bailey meter. From a visual point of view, the manganese kept the Ringelmann number below 1, permitting the refinery to burn all 800 barrels/day of pitch.

A chemical refinery on the Gulf Coast of the U.S. had a severe smoking problem since resultant plume opacity was not meeting local air pollution regulations. Three Combustion Engineering boilers, each rated for 460,000 lb of steam to be produced per hour, burned a mixture of natural gas and pitch (produced from the cracking of middle distillate fuel) which was used as feed stock to produce ethylene. The pitch had a low sulfur content (below 0.2%). Unfortunately, the stacks showed a noticeable carbon emissions plume having a Ringelmann 2+ value.

Manganese, in the form of a manganese oxide slurry containing 25% Mn was fed into one of the three units. The initial feed rate was 1 gal per 4000 gal of oil. The response was immediate and dramatic. The plume was reduced to essentially zero visibility, noted both by visible observation and recordings on a Shell Oil Co. light transmission meter which monitors stack plume.

The additive rate was reduced gradually to 1 gal per 8000 gal while still maintaining stack plume improvement. The exact feed rate required depended upon the amount of pitch burned and the excess air used. This relationship is not linear, but is easily controlled by the operators, resulting in minimum treatment costs. The use of the additive then was extended to the other two boilers. All three units are being treated at the above rate and maintain a stack plume reading below a Ringelmann 1. This dramatic response with the manganese was accomplished with a refinery pitch, one of the most troublesome fuels to burn.

Additional reading

M. Aramaki et al., *Tech. Rev.*, Mitsubishi Heavy Industries Ltd. (5) (1969).

Providence Evening Bulletin, October 6, 1970, pp 1, 11 (for reprint, write Apollo Chemical Corp., Clifton, N.J. 07014).

V. J. Cotz, *Power,* **114** (2), (1970).

I. Kukin, APCA-NES Spring Technical Session, April 15, 1971 (for reprint, write to Apollo Chemical Corp., Clifton, N.J. 07014).

R. J. Bender, *Power,* **116** (1), 42 (1972).

Ira Kukin *is president and founder of Apollo Chemical Corp. where he has pioneered the field of scientific application of chemical fuel additives. As a member of technical committees in the power industry, he has advanced the practical usefulness of chemical additives, and while employed by Gulf Research and Development Co. for six years, Dr. Kukin was in charge of research on fuel oils and fuel oil additives.*

Oil burning can be less "fuelish"

Sometimes its use is unavoidable. In that case, more efficient combustion is desired. Here are some methods than can reduce air pollution, as well

Reprinted from ENVIRON. SCI. TECHNOL., **11**, 954 (October 1977)

Whether one likes it or not, oil will continue to be a major fuel for the industrial economies of many countries, including this one, for the foreseeable future. Thus, it makes sense to try to find cleaner, more efficient ways of burning it. Indeed, it would be ideal if fuel oil could be made to burn as cleanly as natural gas does. That may be too much to expect, but perhaps one can come close. One approach was devised during the late 1960s by Sonic Development Corp. (Upper Saddle River, N.J.); that technique uses ultrasonics to atomize even very heavy oil. The oil burns with a clear, hot, blue flame, which resembles that of natural gas.

Another method with a touch of *savoir-faire* involves emulsification of fuel oil with water. It was developed by Société Nationale ELF Aquitaine (Courbevoie, France), and is being used at several hundred heating plants and industrial installations in Europe.

It's built to "soot"

The clean burning of fuel oils by emulsification with water is accomplished as follows: Light or heavy fuel oils are emulsified by a proprietary method developed by ELF. Then any mechanical pulverization or atomization burner can receive the emulsion and the necessary air. Regulation of emulsion flow can be automatic or manual. The emulsion is atomized by a pneumatic technique. This emulsion process can be installed in new or existing heaters, ovens, furnaces, or kilns, ELF's Claude Delatronchette told *ES&T*.

The proportion of water in the emulsion varies according to the application. It can reach 60% for furnace applications and 20% for boiler applications. The burning process is suitable for all mechanical atomization burners, and achieves a 90–95% reduction in soot and unburned solids, according to ELF. These lower unburned solids and soot deposits eliminate the need for soot collectors, cut operating costs, and help to produce a constant energy output. The oil flame is more uniform in temperature, more "filling", and more radiating, because of enhanced combustion.

Knocks down NO$_x$

Instead of 20–30% excess air, which oil burners normally use, the ELF burner concept uses 10% excess air. This use of less excess air leads to a slightly lower flame temperature, and attenuation of temperature gradients in and around the flame. These features, in turn, bring about a reduction in NO$_x$ emissions which, according to ELF, can be 10–50%. For example, when heavy fuel oil is burned in conventional systems, NO$_x$ can be 4200 mL/kg of oil burned, the company estimates. With a heavy oil/water emulsion fuel, NO$_x$ emissions could be 3800 mL/kg. With lighter oils, NO$_x$ emission could be markedly reduced.

Likewise, soot from heavy fuel oil alone could be 140 g/10 h of burning, while unburned solids are 7500 mg/kg of oil burned. With the emulsion, soot is reduced to 5 g/10 h, and unburned solids to 200 mg/kg.

Less SO$_3$

There will always be some small amounts of residual sulfur, even in low-sulfur, or well-desulfurized oils. When the oil is burned, the sulfur will become SO$_2$, upon which emulsification can have but little effect. Other means must be used to bring SO$_2$ within legal or regulatory tolerances, as necessary.

However, with conventional oil burning, some SO$_2$ will be "upgraded" to SO$_3$. This SO$_3$, generally 1–5% of the SO$_2$ generated, can cause severe flue and stack corrosion. According to ELF, the use of emulsions can reduce SO$_2$-to-SO$_3$ conversion by more than 50%, in many cases.

Thus, oil burners can be made to use fuel oils in a cleaner and more efficient manner. An example is the ELF fuel/air/water burner for furnace applications, which emulsifies the oil in the burner. Atomization by ultrasonics is another technique, and there might be more approaches extant, or under development. Be that as it may, oil will be needed in the economics of most countries and economic sectors for many years to come. For this reason, the optimization of its combustion efficiency serves the causes of fuel conservation and pollution abatement well. JJ

ELF's Delatronchette
clean fuel oil use with savings

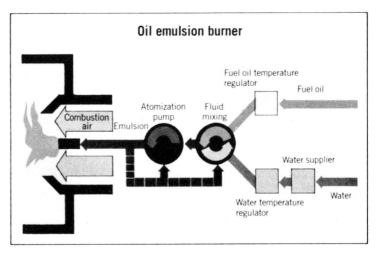

Sonic gas cooling systems cut...

Sonic Development Corp.'s units cool
hot effluent gases from smelters in Arizona and
asphalt roofing plants in Los Angeles

Reprinted from Environ. Sci. Technol., **8,** 212 (March 1974)

Optimum performance of air pollution control equipment with minimum maintenance cost requires more careful attention to preconditioning of the effluent gases. The use of ultrafine water sprays is helpful in pretreating gas stream to enhance the collection efficiency and reliability of the air pollution control system. Recently developed sonic atomizing nozzles and computerized analysis of gas conditioning problems permit the design of automated systems for cooling hot effluent gases. Such systems can be used to upgrade existing installations or to reduce the size and cost of new installations.

Sonic evaporative cooling equipment from Sonic Development Corp. (Upper Saddle River, N.J.) is now operating in numerous foundries, smelters, liquid waste incinerators, and various industrial furnaces with excellent results. In a typical application, sonic atomizing nozzles were installed in a cooling chamber at the Hayden, Ariz., facility of Kennecott Copper Corp. with the basic work handled by the plant engineering staff and operating personnel.

James Stocker, smelter operations superintendent at the Hayden operation, says that off-gases from the fluidized bed reactor are cooled from 1100–800°F before cleanup by an electrostatic precipitator. As originally designed, the cooling chamber used 30 conventional spray nozzles operating at hydraulic pressures up to 350 psig. Two sonic nozzles replaced the 30. The finer atomization delivered by these nozzles provided total water evaporation and eliminated the earlier experiences of wet chamber bottoms, sludge buildup, and unscheduled shutdowns. Stocker says the nozzles at the Hayden facility were installed in 1970, and in the past three years have saved many thousands of dollars by reducing plant downtime.

What is useful to smelter operators has been found useful by other manufacturing operations including asphalt roofing plants. For example, the Celotex Corp. recently installed a sonic gas cooling system in their Los Angeles asphalt roofing plant to cool hot gas fumes before cleaning by a Johns-Manville high-energy air filter (HEAF).

According to Lloyd Pfaff, senior project engineer for the Celotex Corp., the sonic system cooled 22,000 acfm from 150–130°F—the optimum temperature for efficient filtration by the glass fiber media. "Without the fine atomization of the sonic nozzles," Pfaff reports, "evaporative cooling could not have been accomplished in the existing ductwork within the available residence time of 1.6 sec. The sonic atomizing system enabled us to avoid costly shutdowns and satisfy EPA regulations."

Why spraying?

In evaporative cooling, water injected into a hot gas stream evaporates and absorbs heat from the gas. Virtually every problem associated with evaporative cooling systems is traceable in some way to the nozzles used to inject water into the gas stream. The low-cost pressure nozzle commonly used in such systems is inherently incapable of producing the ultrafine droplets required for evaporation and cooling. Large water droplets produced by such nozzles agglomerate with dust particles to form sludge or pass through the cooling chambers to foul collection equipment.

The large droplets also cause localized cooling resulting in refractory spalling and severe distortion of sheet metal ducts. The small orifices required for pressure nozzles clog frequently and wear rapidly causing high maintenance costs and frequent shutdowns. Lacking turndown capability, pressure nozzles cannot be modulated to suit varying inlet temperatures but must be actuated separately or in banks to achieve even crude flow modulation.

To compensate for the poor atomization, designers must often increase the required water volume by 50% to allow a "safety" factor, thus further compounding nozzle problems and creating a secondary water pollution problem. The combination of poor atomization, erratic cooling, and slow system response is often the cause of baghouse fires and precipitator malfunctions. These severe field problems clearly indicate the need for more sophisticated and reliable gas cooling equipment.

Preconditioning

In addition to cooling hot gases, fine water sprays also play a critical role in the preconditioning of the gas stream and significantly affect the performance and efficiency of the pollution control equipment. With electrostatic precipitators, efficiency of collection is a function of the electrical resistivity of dust particles.

Many dusts, such as glass and cement, have inherently high resistivi-

Kennecott's Stocker
*"Nozzles . . . have saved many
thousands of dollars . . ."*

20

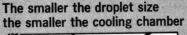

Why sonic cooling?

The smaller the droplet size the smaller the cooling chamber

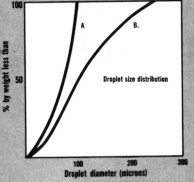

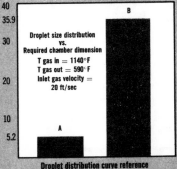

Use of dilution air for cooling increases baghouse requirement

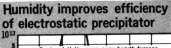

Humidity improves efficiency of electrostatic precipitator

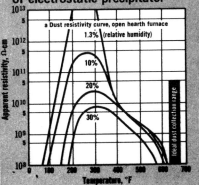

air pollution control costs

ties and, at normal collection temperatures of approximately 300°F, cannot be properly collected unless humidity levels approaching 30% are achieved. Hence, fine water sprays not only cool the hot gases but also condition the gas stream by raising the humidity level to enhance the collection efficiency of the precipitator.

With bag collectors, temperature and humidity controls are also critical. Precise control of cooling water is required to ensure temperature control below the limit of the materials and the prevention of condensation by maintaining the gas stream above the dew point.

Once it is recognized that water sprays play a critical role, not only in the cooling of hot gases, but also in the preconditioning of the gas stream to enhance collection efficiency, the need for improved design of gas conditioning systems becomes obvious.

Savings

Substantial cost savings can be realized by the use of improved gas cooling equipment in place of dilution cooling techniques. Dilution cooling involves the use of ambient outside air to dilute the gas stream and reduce its temperature. This technique increases the total cfm of gas to be treated with a corresponding increase in the size of collection equipment.

In a typical case, 100,000 cfm of dry air at 1500°F is increased to 185,000 cfm when cooled to 500°F by dilution with 100°F outside air. At $1.50/cfm cost of baghouse capacity using glass bags, the cost in excess baghouse capacity is $127,500. By contrast, the cost of a complete evaporative cooling system (including the cooling chamber) is $0.50/cfm or $50,000. The resulting savings in using evaporative cooling in place of dilution cooling is $77,500.

Smaller chamber size

The importance of small droplets to cooling chamber dimensions is

shown. A conventional pressure nozzle (B) producing maximum droplets to 250 μ requires a chamber of 35.9 ft in length. By contrast, Sonic Development's atomizing nozzles require only 5.12 ft for complete evaporation at the same water flow rate. Hence, to achieve dry bottom operation with smaller, lower cost cooling chambers, atomizing systems capable of producing extremely fine droplets are required.

Automating the cooling system

Equipped with accurate engineering data from computer studies together with nozzles capable of producing ultrafine atomization, we now are able to design a control system to modulate water flow to suit varying inlet gas conditions. For new installations, a complete system includes a cooling chamber, sonic atomizing nozzles, temperature sensors, automated flow controls, and an air compressor. In existing installations, the nozzle/control package can be retrofitted to the existing cooling chamber.

An automated control system takes full advantage of the modulation capabilities of the sonic nozzles and provides rapid system response. The temperature sensor at the inlet dictates the liquid flow rate to the nozzles and downstream sensor provides trimming feedback to correct for any system errors while providing fail-safe protection against excessive temperatures at the collector. The system is designed to use only the volume of water required to accomplish cooling thereby eliminating wetted walls, refractory spalling, sludge buildups and water wastage, carryover, and recycling problems.

Until recently, gas cooling has been the weakest link in most pollution control systems. The development of new evaporative cooling techniques using sonic atomizing nozzles and automated controls overcome the problems normally associated with wet systems while providing substantial cost savings.

21

Reprinted from ENVIRON. SCI. TECHNOL., **8**, 1062 (December 1974)

Stage charging reduces air emissions

U.S. Steel solves the coke oven charging problem at the world's largest coke plant near Pittsburgh, Pa.; its process is becoming the standard for the steel manufacturers

Deterioration of air and water quality is a matter of concern to the well-being of people everywhere. These problems are being attacked from many directions and many forces. The steel industry has made commitments of many hundreds of millions of dollars for controls, and additional millions are spent annually for operating costs of pollution control.

This industry has made substantial progress in its efforts to control effluents. Today, the steel companies in this country are removing 91% of all particulate matter from their emis-

USS' Munson
guided the development work

sions into the air, and effectively removing about 94% of the total pollutants from the water they use.

Nevertheless, controlling air emissions resulting from the coking of metallurgical coal required in making iron and steel is the one remaining troublesome environmental problem in this industry. These emissions are associated with the following three operations:

● charging of the ovens, when ports are opened to permit the entry of coal

● pushing, when coke is discharged from the ovens

● coking itself, through leaky doors and ports.

The largest of the three problems stems from the first operation, charg-

ing coal into the ovens. Without controls, such emissions represent about 70% of the total emissions resulting from the three operations.

John G. Munson, Jr., assistant to the executive vice-president, says, "U.S. Steel has solved the coke oven-charging problem through the development of stage charging." He continues, "What it has developed is rapidly becoming the standard for the

industry, and U.S. Steel is making the basics of their technology freely available to the industry."

Development work for the stage charging was performed at the USS Clairton Works—the world's largest coke plant—15 miles south of Pittsburgh, Pa., on the Monongahela River. This plant charges 30,000 net tons of coal per day and has highly advanced facilities for coke oven gas processing and coal chemical refining equipment.

Coke-making operations were begun here in 1916 and expanded over the next 30 years. In 1946, a complete rebuilding program for all its 1375 coke ovens was undertaken; the program was completed in the late 1950's. These ovens incorporat-

ed well-established design; all were equipped with double collector mains and steam aspiration for on-the-main charging. Now, stage-charging is in full operation on all batteries at Clairton. To date, there have been no serious problems resulting from the practice, and the lengthening of the machine cycle for charging has been accomplished without affecting overall operating schedules.

Stage charging: basic component changes

● Coal from the No. 1 and No. 4 hoppers is dropped simultaneously into the oven. Then No. 3 is dumped, and when empty, is followed by No. 2. The outside hoppers No. 1 and No. 4 hold approximately 67% of the coal charge; hopper No. 3 13%; and hopper No. 2 20%. All percentages are by volume and are only approximations. The height to which the coal peaks vary with bulk density, oil addition, moisture content, and grind of the coal.

● The angle of repose of the coal in the hopper as it is loaded from the bunker never exceeds 45°. This angle controls volume fluctuations due to flow

characteristics and was ascertained from a series of tests in which moisture, size, and bulk density were varied. The aprons, which control coal volume, are set at 45%, and extend from the base of the individual stationary sleeve to the perimeter of the hoppers. Deviations such as partial restriction of the bunker gate opening by the larry car operator can also affect this volume control system; therefore, it was necessary for the bunker gate opening to be automatically controlled.

● A specially designed smoke box was developed for the leveler bar to ensure a nearly perfect seal over the chuck door opening. This last change maximized aspiration during the leveling operation.

Backgrounder

During the 1960's a number of programs which improved charging procedures were put into practice at Clairton. Among these were:

● bulk density control of the coal charge

● improved design of volumetric cones in the charging hoppers

● development of the "pogo stick" to improve coal flow from gravity hoppers.

While these innovations improved efficiency of the charging practice, it was soon obvious that more work would have to be done to attain satisfactory control of charging emissions.

In 1969, an intensified research and development program was un-

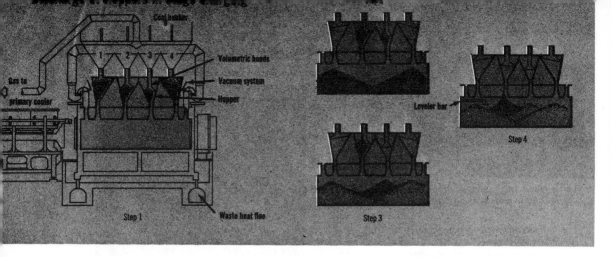

Coal bunker

Gas to primary cooler

1 2 3 4

Volumetric heads

Vacuum system

Hopper

Step 1 — Waste heat flue

Step 3

Leveler bar

Step 4

dertaken by personnel from Clairton Works and the U.S. Steel Research Laboratory (Monroeville, Pa.). The objective of this investigational program at U.S. Steel was to develop a specific charging system that would work effectively and reliably under the conditions existing at that location. The investigators aimed to find reliable methods for effective control of charging emissions that could be adopted without major alterations to the existing coke ovens. They recognized that a number of other items would have to be taken into account, including the effects on coal carryover into the collection mains, on tar quality, on excess air infiltration into the gas system, on lengthening of the machinery cycle, and on control of the oven coal line.

After carefully reviewing the work of others, this team concluded that charging emission control could best be achieved by containing the emissions in the ovens and exhausting them through the regular gas handling system. This achievement would require more effective aspira-

tion and a constantly unrestricted tunnel head in the oven (the space between the top of the coal charge and the oven roof) during the charging cycle. It was recognized that to maintain an open tunnel head, control of coal flow from the charging hoppers on the larry car into the ovens would be required. These principles were not new; other coke plant operators had previously experimented with various types of so-called sequential charging. As previously mentioned, Clairton Works experimented with this type of charging in the 1960's.

The old way—before stage charging

When the program began, Clairton Works was utilizing the following basic facilities and operating practices related to the charging operations. A typical battery had double collector mains equipped with steam aspiration for on-the-main charging. Ascension pipes had steam aspirators of various sizes, ranging from 7 ft 16 in. to $9/16$ in. in diameter, supplied at a header pressure of 60–70

psi. Ovens were from 10–14 ft high, with oven volumes from 521–837 ft^3. Each had four charging holes, serviced by a gravity-feed larry car.

The larry cars were of conventional design, with four hoppers of approximately equal capacity, equipped with volume control mechanisms and separate shear gate controls. The coal charge was a blend of high and low volatile coals pulverized to approximately 65% minus $1/8$ in., with the bulk density depending upon the type of coal charge and controlled by oil addition from 44.0–51.5 lb/ft^3 (ASTM cone test) at a natural moisture content. Coal moisture varied from approximately 4.5% to as high as 10%.

A first problem was to obtain adequate aspiration during charging. Initial studies of Clairton aspiration systems showed that they were not capable of aspirating all of the gases generated during charging. Standpipe and gooseneck size and configuration, type of damper valve, size of oven, plus other factors, all had to be taken into consideration in establish-

Stage charging: how it works

• After loading at the bunker, the larry car travels to the oven. The operator performs the required offtake cleaning and waits for the signal from the pusher operator to pour the charge. Upon receiving the signal which indicates that the pusher is in position to level, the larry car operator lowers the dropsleeves over the open charging holes. The lidman turns steam into the goosenecks, places the oven on the main being sure that the standpipe caps are seated properly, and signals the larry car operator when this step is completed. At this point, it is extremely important that the dropsleeves totally enclose the charging holes and that the standpipe caps are sealed securely to reduce the possibility of pulling air into the aspiration system. This would effectively increase the total volume of gas system's capabilities, forcing the excess into the air.

• The pouring procedure is of prime importance. At all times during the charging cycle it is necessary to provide a route for the gas to escape from the oven through the offtakes. This is necessary to attain smokeless charges. Therefore, in the stage-charging sequence, the No. 1 and No. 4 hoppers are discharged simultaneously. These hoppers must be emptied completely before the next hopper is charged.

The volume of these two hoppers is adjusted so that the coal peaked after discharge will be as close to the coal line as possible and still assure that a tunnel head opening will exist to the center of the oven prior to and during the leveling operation.

These hoppers can be emptied simultaneously since the coal charged from them does not significantly interact. Any disruption of the coal flow from one hopper will not substantially affect the

height of the coal peak under either hopper. As the hoppers empty, the rear gates are closed, the dropsleeves raised, and the lids replaced.

• Next, the No. 3 hopper is discharged. Since the No. 3 hopper is independently discharged and the volume is controlled, the peak of the coal is at the coal line and not blocking the tunnel head. Immediately after the hopper has emptied, the No. 3 lid is replaced.

• Last, the No. 2 hopper is discharged. After this hopper has stopped running and with sufficient coal remaining to assure a level charge, the signal is given to being the leveling operation. It is critical to the success of the operation that the chuck door remain closed until this time because premature opening or cracking will adversely affect the aspirating system.

Larry car. *This pollution-free unit carries coal to coke-making hoppers*

A larger size aspirator nozzle, increased nozzle pressure . . .

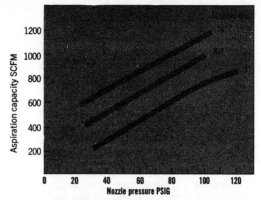

. . . a good cleaning program for the aspiration system

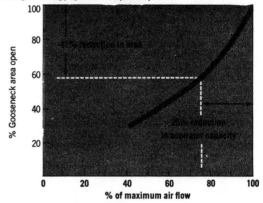

ing the aspiration needed to control emissions effectively.

Similar work in this area had been done under American Iron and Steel Institute sponsorship. The techniques developed in that program were adapted and modified for use in conducting test programs at Clairton. After extensive experiments, it was concluded that gas flows of at least 1700 scfm for the 12-ft (height) ovens and 2200 scfm for the 14-ft (height) ovens would be needed to assure a smokeless charging operation.

To obtain this gas flow, it was necessary not only to increase the size of the steam aspirating nozzles, but also the nozzle pressure which, in turn, required new and larger steam supply headers. These parameters are dependent on the standpipe-gooseneck configuration and the size of the oven.

Furthermore, it is essential that the aspirators be kept clean to be effective, and this requires a regular well-supervised cleaning program. The importance of clean goosenecks and standpipes cannot be overemphasized. Flow tests indicate that a relatively small reduction in the effective gooseneck opening due to carbon buildup can substantially reduce the volume of gas drawn through the off-takes and into the collector mains, resulting in gases escaping into the atmosphere. For example, a 1.5-in. thick buildup within a 13-in. diameter gooseneck results in an area reduction of 41% and causes a 25% reduction in aspiration capability.

These improvements in the steam aspiration system resulted in appreciable reduction of charging emissions; however, it was apparent that even with double collector mains, adequate steam aspiration and optimum conventional charging practices, consistent "smokeless" charging could not be realized with gravity-feed larry cars, because of frequent tunnel head blockage brought about by erratic flow of coal from the hoppers.

Sequential charging

As a matter of interest, extensive tests were made using "sequential charging." Through the installation of a time relay system on the larry car, the control of the discharge of the coal was taken out of the hands of the larry car operator. The No. 1 (nearest the pusher machine) and No. 4 shear gates opened simultaneously and were automatically followed 9–12 sec later by No. 2 and No. 3 shear gates.

At times this innovation showed considerable improvement on emissions, but further study revealed that fluctuations in rate of coal flow from the individual hoppers had a major effect on distribution of the coal in the oven. As a result, tunnel head blockage occurred prior to leveling, and smoke was emitted at the drop-sleeves and from the hoppers. It was also found that on one of the battery units serviced by a screw feed larry, the variability of coal volume in the individual hoppers and variable screw speed resulted in similar problems.

As a result of these tests, the conclusion was reached that absolute control of coal volume in each hopper and a pouring practice that minimized the effect of variable rate of coal flow on the distribution of coal in the oven were mandatory for a smokeless charging system.

Based on extensive tests conducted on a scale model of the charging system, it was found that coal from the No. 1 and No. 4 hoppers could be dropped simultaneously into the oven with little or no interaction. After these hoppers emptied, No. 3 hopper was dumped and when that was empty, it was followed by No. 2 hopper. Thus, a coal hang-up or rate of flow from the hopper would not affect distribution in the oven.

Since the hoppers were of essentially equal capacity, the test work made it apparent that the amount of coal put into each of the four would have to be changed considerably for stage charging. To do this, a procedure was developed for measuring the actual angle of repose of coal after it was dropped into the operating ovens. This information was then used to calculate the coal volume to be placed in each of the hoppers.

Reprinted from ENVIRON. SCI. TECHNOL., **8**, 600 (July 1974)

Gravel bed filters clean industries' hot and abrasive dusts

Rexnord units clean air emissions on lime kilns in Arizona,
clinker coolers in Florida, and a cement plant in Texas
and is applicable for any dusts that tend to agglomerate

Despite the fact that there are only a few examples where the gravel bed filter has been applied and fully evaluated on an industrial scale, nevertheless there are 16 installations in the cement industry that will be using this device to clean up their industrial emissions by 1975. By the end of 1973, Rexnord had installed 116 units at 12 sites and will be installing another 127 units at another 11 sites. Hot and abrasive dusts from clinker cooler operations are one of the most difficult jobs for the gas cleaning equipment manufacturers, one reason being that the quickly changing gas temperatures and gas volumes make it difficult to control operations.

Rexnord Air Pollution Control Division (Louisville, Ky.) nevertheless, has eight units operating today. The first REX gravel bed filter on the North American Continent went into operation in July 1972. The unit cleans the gases from a lime kiln of the Kennedy Van Saun preheater type, at the Paul Lime Co. (Douglas, Ariz.)

Stack emission tests were conducted by the Engineering Testing Laboratories, Inc. (Phoenix, Ariz.) in July, September, and October of 1972. Actual stack particulate emissions were about one half of those allowed by the stringent process weight requirements in the state of Arizona. The performance was accomplished in spite of the actual gas volume being approximately 50% greater than what was originally designed for. The company then ordered a second unit which has been operating since July 1973.

For hot and abrasive dust control problems, these gravel bed filters achieve high efficiency, low operating costs, and no maintenance. They operate at high temperatures (up to

900°F) with minimum maintenance, no bags, no water, and no replacement filters. The filter medium itself is relatively unaffected by temperatures up to 2000°F. In general, the unit can be applied to any dust which agglomerates. And although they have been used or ordered for dust separation in cement plants and lime industries, they may soon be finding application in other industries.

Other installations

The control of abrasive dusts has been confirmed in another geographic location. For example, General Portland, Inc. installed three units on three clinker coolers at their Tampa, Fla., plant between June and August 1973. Their spokesman reaffirmed

General Portland's Johnson
Meeting Florida standards

earlier experience. Plant manager Bob Johnson says, "Applicable Florida standards were achieved."

Then in August 1973, the units were installed by the Universal Atlas Cement Division of U.S. Steel in their operation at Waco, Tex. Plant manager Tom Ryan reports that compliance testing was performed by Ecology Audits (Dallas, Tex.) and are in compliance with Texas air pollution standards.

Despite the fact that these units are being used extensively by the cement industry in the U.S., other studies are indicating the usefulness of these units to control emissions from sinter plants in the steel industry, glass furnaces, fly ash, as well as other applications where hot and abrasive gases have to be cleaned.

At the installations tested, effluent quality has been well within the new source performance standard for portland cement operations (ES&T, Oct. 1972, p 886).

The European connection

As you may be aware, the gravel bed filter on the North American Continent is built in accordance with an exclusive licensee agreement between Gesellschaft für Entstaubungsanlagen (GfE) (Munich, Ger.) and Rexnord, Inc. (Milwaukee, Wis.). Prior to this agreement of March 8, 1972, the German company, GfE had built more than 150 gravel bed filters inside and outside Europe. The largest number of these were installed to clean gases from cement clinker coolers, one of the hottest and most abrasive dusts needing controls. Based on European experience, the unit is capable of achieving high efficiency and for all intents and purposes requires no maintenance, even when the most difficult clinker cooler application is considered.

25

Controls. *Emissions are well within EPA new source performance standards for portland cement plants*

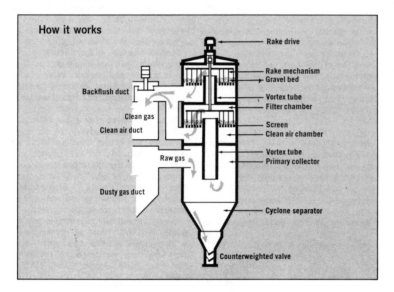

How it works

- Rake drive
- Rake mechanism
- Gravel bed
- Backflush duct
- Vortex tube
- Filter chamber
- Clean gas
- Clean air duct
- Screen
- Clean air chamber
- Raw gas
- Vortex tube
- Primary collector
- Dusty gas duct
- Cyclone separator
- Counterweighted valve

Why gravel beds?

Chief attributes of these units include:

- **High performance is achieved** without any necessity for maintenance
- **Unit accommodates quickly** changing gas temperatures and volumes and handles abrasive dusts. This combination is one of the most difficult applications of gas-cleaning equipment
- **Applicable standards are met at** the lowest possible operating costs
- **A wear-resisting filter medium** which will withstand much higher temperatures than fabric filters
- **Low maintenance costs because** of minimal moving parts and no necessity to replace the filter medium
- **Temporary surges of gas volume,** dust loading, or gas temperatures, such as from normal 500–900°F maximum, will not seriously affect equipment or performance
- **Relatively wide variations in dust** particle size will not significantly affect filter performance
- **No dilution air required for very** high temperatures. Almost complete independence from the dust particle structure
- **Occasional red-hot sparks will** not damage the filter medium

Reprinted from ENVIRON. SCI. TECHNOL., **7,** 494 (June 1973)

World-wide strategies for clean air rely heavily on the
burning of low-sulfur residual fuel oil whose demand
strongly points to the need for . . .

Desulfurization refinery capacities

In some places on the East Coast this past winter was the mildest in recent history. In others, it was the severest on record. In any case, the East Coast relies heavily on burning residual fuel for heating buildings and apartments and for operation of manufacturing industries and electric utilities. Clean air strategies along the East Coast rely primarily on the burning of low-sulfur fuel oil to meet clean air deadlines mandated by the Clean Air Act of 1970. Much of the fuel is low-sulfur residual fuel oil, and its availability is quite vague at this point. Desulfurization technology (i.e., removing sulfur from residual oil) is feasible but results in a significant cost increase for the "clean" fuel variety.

The public's attention to the supply and demand of low-sulfur fuel oil is focused none too early, at least considering these facts:

• the average annual increase for the demand of residual oil between 1970–1985 is projected to be 4.5%

• there has been a virtual stagnation in the growth of petroleum refinery capacity in the U.S. in relation to the U.S. demand for petroleum products

• over past years, U.S. refiners have maximized high-octane gasoline and distillate production, to the detriment of residual fuel production

• with the knowledge that it takes at least 2.0 to 2.5 years or longer to complete a refinery from contract to actual start-up, and with the position of the oil industry today, virtually nothing in terms of new capacity for low-sulfur residual fuel could be realized until sometime in 1976, after the mid-decade deadline for clean air.

Between now and 1976 the plaguing question is whether the supply of low-sulfur residual fuel will be available to meet the clean air deadline or will exceptions have to be granted? These questions simply cannot be answered at this time.

Of course, other fuels are burned to meet the clean air deadlines too. Natural gas is available on the West Coast, coming from Canada and Texas, and a certain amount will be available on the East Coast in accordance with the recent agreement to

import LNG from Algeria. But the main supply of natural gas for the East Coast comes from the Gulf Coast area; Canada doesn't plan to increase its exports to the U.S., and the Texas production is at capacity now.

What it is

Residual fuel oil is identified by two other terms—No. 6 fuel oil and Bunker "C." There are also No. 5 and No. 4 fuel oils (as the numbers decrease so too does the oil's viscosity); No. 4 contains an appreciable amount of distillate products. There is no No. 3. No. 2 is the light oil used for home heating, and No. 1 is kerosine which can be burned in home appliances.

Basically, there are two types of crude oil, sweet and sour. Sweet crudes are low in sulfur content, usually containing up to a maximum of 1% sulfur, and in general, most U.S. crudes are sweet. The sour crudes contain from 1–4% sulfur, and the object of desulfurization is to lower the sulfur content of the products obtained from sour crude to 1.0–0.3% sulfur.

Of course, the sweet crudes produced in the U.S. could be used to make low-sulfur residual fuel, and

some are, although the supply is limited. But U.S. refiners have chosen to make coke and distillate fuel oil and naphtha as the products from their operations. The coke is used primarily to make anodes for the manufacture of aluminum. Naphtha, of course, winds up as gasoline; and after passing through a catalytic cracker, the distillate too can produce gasoline.

There is a growing concern for availability of low-sulfur residual oil. Some sources indicate that there soon will not be enough such fuel oil available and that certain exceptions will have to be granted to the clean air deadline of mid-1975.

The supply

In 1972, the supply of residual fuel oil in the U.S. was 925.6 million bbl; the supply is made up from U.S. producers, imports, and inventory changes. Official tabulations of the Department of the Interior Bureau of Mines reveal that in 1972 a total of 591.7 million bbl of residual fuel oil were imported into the U.S. In the same year, the U.S. produced 292.5 million bbl.

The big user of residual oil is the East Coast, or PAD-1 (Petroleum Administration District-1), an area ex-

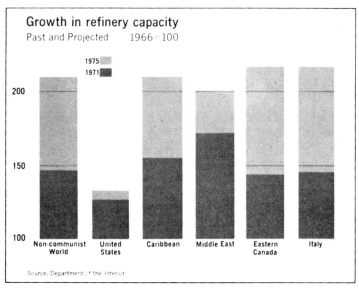

Growth in refinery capacity
Past and Projected   1966 = 100

1975
1971

200

150

100

Non-communist World · United States · Caribbean · Middle East · Eastern Canada · Italy

Source: Department of the Interior

New refinery. Two-2.5 years from contract to start-up

tending from Maine to Florida and including Pennsylvania and West Virginia. Various state and local air pollution control strategies require the burning of low-sulfur fuel oil, down to a minimum of 0.3% by weight, in high-density metropolitan areas such as New York City.

In 1972, 68.1 million bbl were used in PAD-1. (PAD-1 refineries produced 37.6 million bbl; most of the difference came from PAD-3, on the Gulf Coast.) In the same year, 99% of the imported oil was used in PAD-1. Incidentally, PAD-5, which includes California, is the next largest user of the residual oil produced in the U.S.

How to desulfurize

At present, many U.S. refineries have some capacity for removing sulfur from light and distillate oils but have little capacity for the production of low-sulfur residual fuel oil from sour crudes.

Low-sulfur residual fuel oil can be made from sour crudes by either a direct or indirect process. In the direct method, the heavy gas oil and residual are desulfurized at the same time. The indirect process splits the heavy gas-oil portion, desulfurizes it, and blends it back with the undesulfurized residual (see diagram). Sweet crudes are amenable to the direct desulfurization process or can produce low-sulfur residual directly without desulfurization. In fact, most of the U.S. production of low-sulfur residual oil is made from sweet crudes without any desulfurization process.

In the indirect process, the atmospheric residue is vacuum distilled to give a vacuum residual and a vacuum distillate. The distillate is passed through a hydrodesulfurizer which re-

moves the sulfur. The desulfurized distillate is then blended back with the vacuum residual.

The necessary catalyst is usually either a cobalt-molybdenum oxide or a palladium catalyst. However, some Arabian crudes can be desulfurized by the direct process also.

The sour crudes, especially those from Venezuela, contain nitrogen materials and high metals content—vanadium and nickel—which play havoc with the lifetime of the catalyst used in the desulfurization process. In general, Venezuelan crudes have high metals content; Middle Eastern crudes have medium metals content, and African crudes are low in metals. By vacuum distilling (the indirect route), heavy metals and nitrogenous materials are left in the residue.

Additional processes are being developed to recover sulfur—flue gas desulfurization, fuel gasification, fluid bed combustion. However, the lack of well-documented and publicly available information on these processes makes it difficult to judge their costs and technological merits.

Building

A number of well-known construction firms can handle the engineering and construction of desulfurization facilities—Badger, Bechtel, Fluor, Foster-Wheeler, Lummus, and Procon. A desulfurization plant costs on the order of $50–100 million. But it is important to remember that only plants that either have been announced or are in fact under actual construction at this time can produce a low-sulfur residual fuel oil commodity by 1976.

Who they are

Two units—Texaco Trinidad and Exxon (Creole refinery at Amuay, Venezuela) go on stream this year; the Bahamas Oil refinery at Freeport, Grand Bahama, will go on stream later (see box).

All of the desulfurization facilities which serve the U.S. market are in the Caribbean with the exception of Shaheen's Isomax plant in Canada. In 1972, 92% of the U.S. imported low-sulfur fuel oil came from six countries—Bahamas, the Caribbean area, Italy, the Netherlands Antilles, Trinidad, and Venezuela. The remaining 8% came from several other countries.

Of course, many of the refineries were attracted to these six countries originally by the proximity to large producing fields and in some cases the practically limitless amounts of cheap natural gas for plant fuel. More recently, large refineries have begun to develop in the island areas where the location incentives include

circumvention of certain import quotas (which have recently been lifted), low taxes or tax exemptions, the availability of deep water unloading sites, and less stringent environmental requirements than in the U.S.

The installed and planned facilities in the Caribbean area use the indirect process for desulfurization. Perhaps, the newest desulfurization plant to go on stream is the Texaco Trinidad (a subsidiary of Texaco, Inc.) Point-a-Pierre refinery on the island of Trinidad, off the coast of Venezuela. The refinery's total capacity is 350,000 bbl/day, about 25% of which goes for the production of low-sulfur residual fuel oil (some 90,000 bbl/day).

Gulf's Bayamon refinery (San Juan, Puerto Rico) can produce 12,000–16,000 bbl/day of low-sulfur residual fuel oil. Its capacity for this commodity is not significant, how-

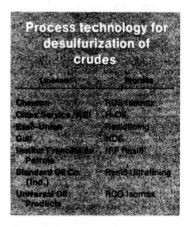

ever. In this case, the residual is prepared by blending distillates with the residual (the refinery apparently desulfurizes light distillates).

Of course, the construction of other low-sulfur residual fuel plants have either been announced or are under way today. If on stream by 1975, then they may help alleviate the shortage of low-sulfur oil. Such expansion announcements include:

• by the end of 1973, the Canadian firm—Shaheen Natural Resources—plans a new refinery for Newfoundland & Refinery, Ltd. The throughput of the refinery is 100,000 bbl/day, and it has been estimated that some 35,000 bbl/day of low-sulfur residual fuel oil will be produced

• also by the end of this year, Creole, a subsidiary of Exxon, will expand its capacity by 50,000 bbl/day for the low-sulfur residual fuel oil commodity. Creole's refinery at Amuay, Venezuela, uses the indirect desulfurization process (referred to as the ERE process)

• another desulfurization plant

scheduled for start-up in 1973 is Bahamas Oil and Refining. Owned by New England Petroleum Co. and Standard Oil & Refining Co., it will expand its capacity from 250,000 bbl/day to 400,000 bbl/day. What percentage of this goes for low-sulfur residual fuel oil is unknown. In this case, the $50 million expansion is being built by the Italian firm, SNAM Progetti

● also this year, Amerada-Hess in the Virgin Islands will expand its existing facility for the production of low-sulfur residual fuel oil

● another desulfurization plant is going into the Netherlands Antilles. Lago Oil and Transport (at Aruba, N.W.I.), a subsidiary of Exxon, plans to increase the capacity of its distillate hydrodesulfurization unit by 45,000 bbl/day by mid-1974

● by 1975, a new Virgin Islands HDS plant will come on stream. The Virgin Islands Refining Co. will produce low-sulfur residual fuel by an indirect process; this plant is currently under construction by Procon, a subsidiary of UOP

● Shaheen Natural Resources plans other desulfurization plants. A $223 million refinery is to be built in Nova Scotia at Port Hawkensberry on the Strait of Conso. Although no contract had been let at press time, the plans included a 200,000-bbl/day refinery, of which 80,000 bbl/day would be low-sulfur residual fuel oil

● Shaheen also plans a third refinery, but this refinery will not figure into the 1975 supply figure for the low-sulfur residual fuel oil commodity. A 300,000-bbl/day refinery will be built at Come by Chance, Newfoundland. Although the percentage of low-sulfur residual oil has not been revealed, there is press mention of use of the Isomax and Hydrobon processes, both of which are used for desulfurization

● there is also conjecture on other desulfurization plants. For example, Shell is building a new distil-

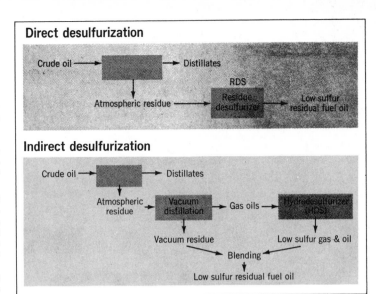

Direct desulfurization

Crude oil → [ ] → Distillates

Atmospheric residue → RDS Residue desulfurizer → Low sulfur residual fuel oil

Indirect desulfurization

Crude oil → [ ] → Distillates

Atmospheric residue → Vacuum distillation → Gas oils → Hydrodesulfurizer (HDS)

Vacuum residue → Blending ← Low sulfur gas & oil

Low sulfur residual fuel oil

lation plant in Curacao. Some sources say it's an additional refining facility; others say it's a replacement unit.

Foreign sources

Not only is the U.S. relying heavily on burning low-sulfur residual fuel to meet clean air strategies but so too are European countries. However, European sulfur fuel levels are projected to continue well above U.S. and Japanese standards with the exception of congested areas.

According to the CEC (Commission of the European Countries) report of December 1972, Japan is projected to be a big producer (and consumer) of low-sulfur residual fuel oil. According to this report, Japan will almost triple its production by the indirect desulfurization process and more than quadruple its production by the direct process. Prior to 1972, Japan produced 364,000 bbl/day by the indirect process but will be producing 911,000 bbl/day in 1975. Similarly, prior to 1972, Japanese

production by the direct route was 153,000 bbl/day which will increase to 645,000 bbl/day in 1975.

According to the Committee on Conservation of Air and Water, Western Europe, the published cost data for desulfurization only included the minimal cost, and the Committee concluded that the costs for adding desulfurization capacities were consistently higher than those usually quoted.

Foreign countries, including Australia, Belgium, Denmark, Finland, France, Germany, Greece/Italy, The Netherlands, Norway, Portugal, Spain, Sweden, Switzerland, and the United Kingdom, have each considered placing limits on the sulfur contents of fuels, and some have already done so. All of which points to a European need for additional desulfurization capacities and the availability of low-sulfur residual fuel oil.

Only a few new facilities are known:

● in 1973, a Caltex company at Bahrain (in the Persian Gulf) plans to bring on stream a 50,000-bbl/day plant

● in 1972, Petroleos Mexicanos at Salamanco, Mexico, started up a 18,500-bbl/day desulfurization plant

● also in 1972, Shell/Koppartrans at Bothenburg, Sweden, brought on line a 2500 direct residual plant.

In the short term, the availability of the low-sulfur residual oil will not meet the onslaught of clean air strategies. New desulfurization plants will be an area for expansion by the oil industry. In the long term, new and additional desulfurization plants will be added, and this "clean" air fuel will cost significantly more than its source. The big question is whether clean air is worth the price.   SSM

Fuel oil hydrodesulfurizer feed capacity

Company	Refinery	Capacity (thousand bbl/day) Estimated 1972	Estimated 1975
Borco	(Bahamas)	0	60
Exxon	Creole-Lago (Amuay, Venezuela-Aruba)	200	300
Hess	Amerada (St. Croix, V.I.)	70	140
Shell	Cardon (Curacao)	60	60
Texaco	(Trinidad)	0	90
	TOTAL	330	650

Rendering plants produce a variety of odorous emissions, depending on the types of raw material and process equipment used. Although both batch and continuous rendering processes are in use, the latter are becoming more popular for economic reasons, especially for new installations. Modern continuous rendering plants generally emit less odor than established installations because they feature enclosed material handling systems. In addition, odor abatement equipment is frequently an integral part of the total plant package.

Without controls, rendering plant odors have been reported discernible at distances up to 20 miles from the source. Unfavorable atmospheric conditions serve to magnify the problem so that a serious public nuisance develops, as evidenced by the number of complaints received by air pollution control offices both in the U.S. and abroad.

As of January 1969, there were about 850 domestic rendering plants owned by some 772 firms (Table 1). Approximately 460 of the total were operated by independent animal renders, 330 were controlled by the meat packing and poultry dressing industries, and the remaining 60 were owned by concerns with a variety of other interests. Some 275 of the plants controlled by meat packers also produce edible rendering products in separate locations at the same plant. There were 93 plants processing fish products and 11 feather mills not included in the above totals drawn from the most recent data available.

It is estimated that 20 new rendering plants will be installed in the U.S. each year. Virtually all will be replacement plants, and most will probably be of the continuous process type rather than of the older batch process technology.

All rendering plants, inedible grease and tallow producing works, and manufacturers and processors of animal and fish oil and by-product meal are in standard industrial classification (SIC) 2094. Processing of poultry wastes, including feather drying, logically falls in this category. Meat packing plants and captive or contract slaughtering houses (small game, on-premises canning and curing, and sausage and lard production excluded) are in SIC 2011. Although this includes abattoirs, blood meal production, and by-product hides, its primary group is food for human consumption and prepared feeds for animals and poultry.

The U.S. Environmental Protection Agency (EPA) is considering the development of standards of performance for new rendering plants in compliance with the Clean Air Act of 1970. The Act requires any control regulations to represent the best demonstrated technology, taking costs into consideration. One of the probable standards would be to limit the concentration of undiluted odors emitted from certain process equipment in any new or substantially modified rendering plant to no more than 200 odor units per standard cubic foot (ou/scf). An odor unit is defined as the quantity of any single odorous substance or a combination of substances which, when completely dispersed in 1 ft^3 of odor-free air, is detectable by a median number of observers in a panel of eight or more persons. The odor concentration is determined by a modification of the ASTM D1391-57 observer panel technique which, in its present form, allows optional approaches to sampling and dilution.

A standard of 200 ou/scf allows for increased odor levels which may result from substitution of distillate fuel oil when natural gas is unavailable for firing afterburners. Odor levels from oil combustion are usually slightly greater than those from burning natural gas. Because odor strength is so intrinsically related to the olfactory sensitivity of the "smeller," it is difficult to define the intensity of the proposed standard for any single individual. The vari-

Odor controls for rendering plants

Methods for curbing odorous emissions are used with varying degrees of success; without controls, rendering plant odors can be discernible at distances up to 20 miles from the source

Robert M. Bethea
Texas Tech University
Lubbock, Tex. 79409

Belur N. Murthy
Environmental Protection Agency
Research Triangle Park, N.C. 27711

Donald F. Carey
Environmental Protection Agency
Washington, D.C. 20460

Reprinted from ENVIRON. SCI. TECHNOL., **7**, 504 (June 1973)

TABLE 1.
Rendering operators in United States

State	Independent renderers	Livestock slaughter renderers	Poultry slaughter renderers	Marine renderers	Feathermill	Blender
Ala.	11 (6)[a]	12 (6)	2 (2)			
Ariz.	5 (1)	1				
Ark.	5 (2)	3	6 (6)			
Calif.	32 (15)	11 (3)	1	8 (3)	4 (4)	
Colo.	5 (3)	9 (5)				
Conn.	2 (1)		1 (1)			
Del.	1					
Fla.	7 (4)	6 (4)	1 (1)	7 (4)		
Ga.	9 (6)	14 (9)	4 (4)	1		3 (1)
Idaho	4 (2)	4 (4)			1 (1)	
Ill.	21 (16)	17 (8)				4 (2)
Ind.	15 (13)	11 (7)				
Iowa	21 (14)	29 (12)		1 (1)		
Kan.	7 (3)	2 (1)				1 (1)
Ky.	4 (3)	9 (3)				
La.	5 (2)	7 (4)	2 (2)	33 (33)		
Maine	1 (1)		3 (1)	4 (4)		
Md.	8 (6)	2	1 (1)	2 (2)		
Mass.	5 (4)			2 (1)		1 (1)
Mich.	8 (4)	6 (4)				
Minn.	15 (13)	12 (5)				
Miss.	5 (2)	9 (6)	5 (5)	3 (3)		
Mo.	14 (5)	17 (7)	2 (2)			
Mont.	4 (3)	9 (3)				
Neb.	8 (6)	3 (1)				
Nev.	2 (1)	1 (1)				
N.H.		1				
N.J.	10 (5)	4 (2)		3 (3)		1
N.M.	2 (1)	4 (2)				
N.Y.	8 (3)	5 (1)		2 (2)		
N.C.	7 (4)					
N.D.	3 (2)	2 (2)	2	2		
Ohio	14 (8)	11 (4)			1 (1)	1
Okla.	5 (2)	11 (6)				
Ore.	6 (3)	6 (2)		1 (1)	1	
Pa.	37 (27)	12 (6)	2 (2)			
R.I.	5 (5)					
S.C.	6 (4)		1 (1)			
S.D.	3 (2)	6 (3)				
Tenn.	7 (3)	18 (6)				1
Tex.	17 (6)	37 (4)	2	1 (1)		
Utah	4 (2)	4 (2)			2 (2)	
Va.	5	6 (2)	1 (1)	11 (8)		
Wash.	10 (5)	11 (6)		7 (5)		
W. Va.			1 (1)			
Wis.	11 (7)	12 (8)		1 (1)	2 (2)	1
Wyo.	2 (2)	2 (1)				
Alaska	1			1		
Hawaii	2 (1)					
P.R.				3		
TOTALS	389 (228)	346 (150)	37 (30)	93 (72)	11 (10)	13 (5)

[a] Numbers in () are installations in towns below 25,000 population.

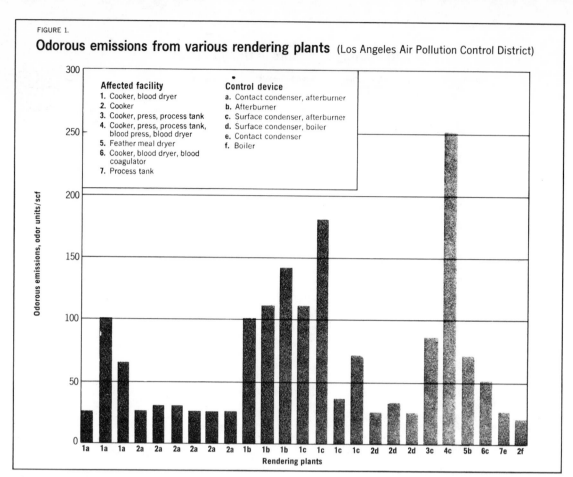

FIGURE 1.

Odorous emissions from various rendering plants (Los Angeles Air Pollution Control District)

Affected facility
1. Cooker, blood dryer
2. Cooker
3. Cooker, press, process tank
4. Cooker, press, process tank, blood press, blood dryer
5. Feather meal dryer
6. Cooker, blood dryer, blood coagulator
7. Process tank

Control device
a. Contact condenser, afterburner
b. Afterburner
c. Surface condenser, afterburner
d. Surface condenser, boiler
e. Contact condenser
f. Boiler

Odorous emissions, odor units/scf

Rendering plants

ation in sensitivity from person to person, although partially compensated by the use of a comparatively large panel in the test method, results in wide confidence limits in the test results. However, it would be reasonable to expect an odor concentration of 200 odor units from a rendering plant to be noticeable to most people. Coffee roasting or bread baking odors, with which the reader may be familiar, emanate at about 2000 ou/scf. In comparison, combustion of natural gas produces an odor level which has been measured at an average of 25 ou/scf.

The effect at ground level after atmospheric dilution of stack emissions of 200 odor units is less easy to define. By atmospheric dilution formula, the distance from the stack to the point at which the typical smeller would no longer detect an odor from the affected facilities can be estimated. Assuming the worst possible circumstances, the minimum distance would be less than 1000 ft from a typical rendering plant which incinerates 500–1500 scfm. This estimate does not allow for the contribution of odors from other sources in the plant which should be controlled by good housekeeping, covering or enclosing feedstock areas, and treatment of plant ventilation air.

Incinerator stack gas results from Los Angeles (Calif.) area were used as a data base. The standard, however, does not require the use of any specific odor abatement method. Tests by the Los Angeles Air Pollution Control District, averaging 67 ou/scf (Figure 1), reveal the following emissions when controlled by:

A contact condenser-afterburner system
   3 plants with cookers and blood dryers, avg 63 ou/scf

   6 plants with cookers, avg 27 ou/scf
Surface condenser–afterburner system
   4 plants with cookers and blood dryers, avg 99 ou/scf
   3 plants with cookers using boiler afterburner (appreciable dilution in boiler), avg 27 ou/scf
   1 plant with cooker, press, and processing tank, avg 85 ou/scf
Eight combinations of various control devices used on process equipment. Seven support the standard, and all eight average about 96 ou/scf.

Data indicate that odors from specified facilities can be held below the probable standard only if incineration is employed. In almost all cases, combinations of incineration and condensation reduced emission concentrations to less than 100 odor units. The combination is usually more economical as well as more effective because of the 20-fold reduction in gas volume which must be incinerated. With combination systems, an odor level of 200 ou/scf usually can be achieved at an incineration temperature of 1200°F. When no condensers are employed, it may be necessary to operate the afterburner at 1400–1600°F to achieve the required odor abatement.

EPA has not yet established air quality criteria for odors. The provisions of the Clean Air Act require that states develop regulations for control of odorous emissions from all existing rendering plant facilities covered under any federal new rendering plant performance standards.

The agency may be expected to suggest minimum odor control criteria to the states, probably under the new source performance standards. These standards will, of course, apply uniformly to all new rendering plants. The states will then set their own standards which may be more stringent than those recommended by EPA; the states will set their own control requirements for existing rendering operations.

Poultry rendering emissions are predominantly aldehydes and small amounts of dialkyl disulfides. The odors associated with dry rendering of dead stock, beef offal, slaughterhouse trimmings, to name a few, all contain a variety of carbonyl components with traces of sulfurous and nitrogenous compounds. Fish meal processing produces mainly amino compounds from trimethyl amine to putrescine. In nearly all cases, the odors are accompanied by particulate matter and fatty mists.

Odor control systems commonly used at this time are condensation followed by incineration, or scrubbing, or a combination of these. Incineration has been the most effective method to date, especially for low-volume, high-concentration odors. Recent advances in the design and utilization of scrubber systems have significantly increased their odor-destroying ability until they nearly equal the performance of incinerators. Scrubbers are especially useful in controlling high-volume, low-concentration odors. The renderer frequently uses both systems for full plant control of all odors.

Other odor control methods, used with varying degrees of success by the renderer, include catalytic combustion, adsorption, and ozonation. However, no odor control system can be truly effective unless general cleanliness, housekeeping, raw materials handling and storage, and spill prevention throughout the rendering plant are properly managed by the operator.

Odor control by incineration

Combustion or thermal incineration is the most effective technique for controlling high-concentration odors in low volumes of air. The economics of this technique are enhanced if the odoriferous materials provide a significant fuel source. Unfortunately, rendering off-gas has no appreciable fuel value. Careful design is necessary to achieve complete oxidation of these gases. Otherwise, the partial oxidation products may be more odoriferous than the original pollutants.

Many renderers in Los Angeles County have used thermal incineration for more than 10 years (Figure 2). The malodorous gases emanating from tallow presses, processing tanks, dryers, and related equipment have also been controlled by thermal incineration.

Carefully designed hoods and ducting systems are used to minimize entrainment of ambient air and fuel costs. Some renderers duct the low-volume, high-concentration odorous emissions to the boiler firebox where they are burned. This method is attractive since no additional fuel is required to destroy the odors. Users of this method, however, have found that the incineration system must be well designed and constructed to prevent both odor blowback and corrosion of the boiler tubes. In addition, provisions should be maintained to control the odorous emissions when the boiler is at "low fire" during slack periods of plant operation.

The volume of odorous gases to be incinerated (following any condensation for fatty particulate matter removal) ranges from 500–5000 cfm, depending on the size of the plant and the material being processed. A large plant rendering chiefly beef scrap and bone may incinerate about 2000 cfm while a smaller plant processing feathers may handle about 5000 cfm. Waste heat recovery equipment may be used in the latter case to reduce odor control costs. Such equipment is commonly used to preheat dryer air or to generate steam.

EPA has calculated the cost of odor control for typical new and existing rendering plants which use the conden-

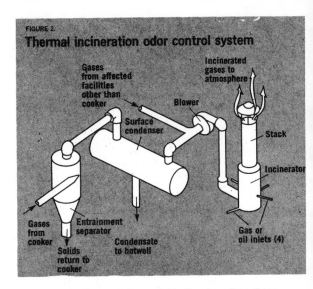

FIGURE 2.

Thermal incineration odor control system

sation-incineration technique. Table 2 summarizes these control costs, which include capital investment and operating costs of the abatement equipment required.

Feather cooking and subsequent drying in the rendering process generate large quantities of acrolein, acetaldehyde, methyl mercaptan, diethylamine, n-propylamine, ammonia, and hydrogen sulfide. Pyrolysis of the off-gases effectively controls odor. Reactive scrubbing with 2,4-dinitrophenylhydrazine is a second-best choice based on the economics of rendering operations.

If a renderer desired to incinerate all of his process and ventilation air, the costs would be excessive. For example, direct flame incineration of 100,000 cfm of air containing 5 ppm of 1-butanol would cost $1.15/hr/1000 cfm, without heat recovery. Also the costs would drop to $0.77/hr/1000 cfm if half of the waste heat were recovered and if the odor-laden gases in both cases contained no fuel value. With no waste heat recovery, the operating

TABLE 2

Control costs for typical rendering plants

Plant type	Continuous (new) 7500 lb/ hr product	Batch-1 cooker (existing) 780 lb/ hr product	Batch-3 cookers (existing) 2600 lb/hr product
Emission control	200 ou/scf	200 ou/scf	200 ou/scf
Required control equipment	Surface condenser cooling tower incinerator	Contact condenser incinerator	Surface condenser cooling tower incinerator
Production, T/yr	12,000	835	4,160
Control investment, $	43,000	6,050	35,200
Annualized control cost, $/year	12,000	2,460	8,900
Annual cost per unit of product, $/ton	1.00	2.95	2.10

Major assumptions: All control equipment depreciated over 20-year (straight-line) life. Product yield, 50% tallow, 50% protein meal.

cost of direct-flame afterburners for controlling rendering plant odors was $0.64/hr/1000 cfm. However, this amount could be decreased to $0.27/hr/1000 cfm with proper heat exchange equipment.

Odor control by catalytic oxidation

Catalytic oxidation has the advantage of operating at lower temperatures and thus with a lower fuel cost than thermal incineration, but neither process is economical for heating large volumes of essentially zero Btu of air. In addition, catalytic oxidation requires a suitable catalyst which will be unaffected by either particulate matter or sulfur compounds. For efficient and economical control of odors by incineration, odor sources should be isolated; their emissions should be treated separately and not mixed with ventilation air.

Catalytic oxidation has not become popular in the rendering industry despite the obvious advantage of using

Some rendering plants adequately controlling odor

Plant	Odor control method
Baker Commodities, Inc. Los Angeles, Calif.	Incineration
Calif. Protein Products Co. Los Angeles, Calif.	Incineration
Illini Beef Co. Joslin, Ill.	Incineration and scrubber
Longueuil Meat Exporting Co., Ltd. Quebec, Canada	Scrubber
Minneapolis Hide and Tallow New Brighton, Minn.	Incineration and scrubber
Peterson Mfg. Co., Inc. Los Angeles, Calif.	Incineration
Pine State By-Products, Inc. South Portland, Me.	Scrubber
Pride Packing Co., Inc. Los Angeles, Calif.	Incineration
Reliable Grease Los Angeles, Calif.	Incineration
Stovall Tallow Co. Wylie, Tex.	Scrubber
Van Hoven Co. South St. Paul, Minn.	Scrubber
West Coast Rendering Co. Los Angeles, Calif.	Incineration

Note: The assessment of the degree of control has been made by the authors, not by EPA. The mention of company or products is not to be considered an endorsement or recommendation by EPA.

less fuel. The higher investment cost over a simple thermal incinerator is discouraging. The major deterrents, however, have been rapid catalyst poisoning with subsequent loss of activity and the inability to destroy rendering odors at temperatures lower than 1200°F. These conditions have made installations of catalytic afterburners in the Los Angeles area unsuccessful. Within a relatively short time, such installations have had to be operated like direct-fired afterburners at 1200°F or more. The Los Angeles County Air Pollution Control District states that direct-fired afterburners are ordinarily installed to incinerate smoke and organic aerosols that are difficult to burn. Emissions of this type are typical of rendering plants.

Wet scrubbing

Odor removal by gas scrubbing with water has problems as a control technique. Too many of the components already identified in rendering odors are essentially insoluble in water. Even if a suitable universally effective scrubbing solution could be found, the odor would merely be concentrated in the liquid phase and thus become a potential water pollution problem. Odor removal by wet scrubbing is a function of the residence time of the odor-laden air in the scrubber, the available contact area, the solubility of the odorant in the water, and the concentration of the odor in the inlet gas stream.

Reactive scrubbing, as all types of chemical removal, requires that the contacting operation either yield inert precipitates upon reaction or form easily removable, nonodorous complexes. All such reaction products must then be amenable to the usual waste water treatment methods for proper disposal. Other problems are corrosion of the equipment, the necessity for maintaining the minimum effective concentration in the scrubbing fluid at all times, and considerable reagent costs. Examples of potential scrubbing reagents are $KMnO_4$, $NaClO$, $NaHSO_3$, lime water, Cl_2, and ClO_2; however, no single scrubbing reagent can effectively remove all the odorous compounds from meat or poultry rendering plants.

Rationale and criteria associated with designing effective reactive scrubbing systems for controlling rendering plant odors include preconditioning of air to remove particulate matter, removal of insoluble components by condensation or adsorption, and other steps for operational improvement of odor control scrubbers. There is also a need for low- to medium-pressure drop venturi scrubbers for gross particulate matter removal. (Removal of fine particles carrying odors requires high-energy (pressure drop) scrubbers, which are usually the venturi type.) Particulate matter can also be removed by a wide variety of wet collectors and impingement devices. Demisters prevent contact tower plugging by particulate-laden gas.

For five scrubber systems evaluated, the total operating costs were approximately $0.10/hr/1000 acfm; the air flows involved were 21,000–55,000 acfm. Approximately a 100-fold reduction in odor levels was achieved.

A Canadian rendering plant equipped with such a system was tested by EPA using modified ASTM D-1391-57 odor concentration determination methods. The scrubber inlet odor concentration ranged from 50,000–77,000 ou/scf, while the outlet concentration was 10–40 ou/scf. Sodium hypochlorite solution was used in a packed tower following a once-through venturi scrubber supplied with river water. An overall pressure drop of 10 in. of water was observed. About 10,000 cfm of the total 32,000 cfm handled was process air, largely from the blood and feather dryers; the remainder was ventilation air.

This particulate installation would be able to meet the EPA-proposed rendering plant standard of 200 ou/scf, which requires that any dilution air must be taken into account. In this case the equivalent odor concentration would be about 128 ou/scf.

The initial operation of this scrubber used potassium permanganate as the oxidizing agent in the packed tower. The subsequent substitution of the sodium hypochlorite reduced chemical costs and achieved slightly greater odor abatement. In addition, the earlier manganese dioxide residue removal problem was eliminated.

A comparison of several aqueous reagents for scrubbing indicated that, while potassium permanganate was uniformly effective in chemically oxidizing all the odoriferous substances, specific compounds were more efficiently removed by reagents depending on functional group reactions. The most effective reagents for amines, mercaptans, and aldehydes were, respectively, hydrochloric (or sulfuric) acid, sodium (or calcium) hydroxide, and sodium (or calcium) bisulfite.

In reducing the odors of aldehydes, ketones, organic acids, and organic nitrogen and sulfur compounds produced during rendering and food processing by 80–98%,

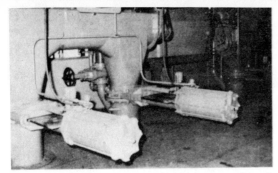

Fresher. *Modern continuous rendering plants usually emit less odor because they feature enclosed material handling*

the efficiency of odor removal can be increased by raising the pH of the scrubbing solution. Here, as with all reactive odor control systems, the ultimate disposal of the reaction products and spent scrubbing solution presents a significant problem.

Waste gases from a combined slaughtering and rendering operation have been deodorized by a two-step process; the gases are first passed through a filter and sent to a tower where they are scrubbed with a weak aqueous chlorine solution. A seawater scrubber operating at 70 gpm/1000 scfm of gas gave 50–80% reduction in the odors from a fish meal dryer. Odor removal efficiency could be increased to 90% by adding 20 ± 5 ppm of chlorine to the seawater.

Ozonation

Used successfully for the treatment of sewage digestion odors in Nagoya, Japan, ozonation should be adaptable to enclosed operations for the control of noxious gases. The Japanese system, handling 56,400 acfm, required 0.37 lb of ozone per hour to maintain a 1-ppm O_3 concentration; it had (in 1968) an annual operating cost of $290/year. With a 4-sec residence time allowed for gas mixing and reaction, ammonia concentration decreased 60%. None of the normal principal components of sewage odors (hydrogen sulfide, indole, and skatole) could be detected in the treated air effluent.

For implementation of the ozonation process, all odorous emissions from rendering process steps and the plant ventilation air should be collected by means of exhaust fans. This mixture should then be treated with a small concentration of ozone. Laboratory studies are needed to estimate the required ozone concentration and

contact time. An ozone concentration of 2–4 ppm with a minimum contact time of 4–6 sec could be used for initial studies. Ozone concentrations as high as 20 ppm may be needed even with contact times of 20–40 sec.

Adsorption and masking

Adsorption with activated carbons can be an effective and economical odor control method for emissions with low concentration of odorous compounds. The technique requires collection of both the ventilation air and the process effluent gases, and separation of particulate matter and fat droplets before adsorption to prevent plugging and inactivation of the bed. Adsorbent beds must be carefully designed and continuously monitored to avoid displacement and breakthrough of some odorants caused by the presence of more strongly adsorbed materials. Adsorption problems include batch to batch variation in adsorptive capacity, low capacity for some odorants, and, without regeneration, relatively short adsorbent life. These problems have prevented this technique from achieving popularity with renderers.

Although activated carbons are the most common adsorbents in use for odor control, their application is limited to the control of low-temperature gases or low-odor concentration. Rendering plant emissions should be cooled to 120°F or less before adsorption can be used for effective odor control.

Activated carbon could be used to control odor emissions from rendering plant operations. The control system, consisting of a combination condenser-deodorizer, reduces blowovers from the cooker. The abatement cost (in 1962) was $550/year. Activated carbon was chosen because of its low cost, high capacity for organic compounds, and simplicity of operation.

Masking and counteraction techniques have not proved suitable for rendering operations. Not only are too many different chemical species present, but the added chemicals merely compound the problem by becoming air pollutants themselves. The great disadvantage of these procedures is that the odorous compounds are not removed from the rendering plant effluent air.

R&D needs

Air pollution control regulations in some states call for either incineration of the cooker and expeller off-gases at 1400–1600°F for not less than 0.3 sec or other treatment to give equivalent results. Thus, there is a great need for the development of low-cost catalytic afterburners, which are not affected by catalyst poisons, to meet the odor control regulations both effectively and economically.

Robert M. Bethea *is on the faculty of Texas Tech University where his primary research interests are in design of control systems for agricultural particulate matter, in situ analysis of reactive gases, and controlling odors from agricultural operations.*

Belur N. Murthy *is a research chemical engineer with Control Systems Laboratory, National Environmental Research Center. His interests have evolved from odors and hydrocarbons to the general area of control of pollutants from stationary sources. Address inquiries to Dr. Murthy.*

Donald F. Carey *is a chemical engineer recently transferred to EPA's Stationary Source Enforcement Division, Technical Support Branch. Formerly his responsibilities included developing new stationary source performance standards for sulfuric acid and rendering plants.*

Another approach to the solution of rendering odor problems is the development and demonstration of regenerable package sorption systems. The advantages of such systems are simplicity of operation, applicability to a wide variety of installations, and economic acceptability to small processors. Engineering and economic feasibilities of using the combined adsorption/regeneration-incineration system for rendering plant odor control require detailed investigation. Of particular importance for the successful implementation of sorption systems is the removal of fatty particulate matter from the air and process streams prior to the package adsorption unit. This procedure is necessary to avoid plugging of the bed, thereby lengthening the time between required bed regenerations.

A coordinated research effort is also needed to find a scrubbing chemical or mixture which can be used for the wide variety of odorous compounds present. Sodium bisulfite or calcium perchlorate can be used to control the odors associated with aldehydes; lime water can be used to remove ammonia and mercaptans; hydrogen sulfide can be removed by potassium permanganate or sodium hypochlorite which are effective oxidants for many other organics. This research should center on finding a scrubbing mixture which is highly effective, inexpensive, regenerable or recyclable, and amenable to standard waste water treatment operations. Concentration(s), pH value, residence time, mass transfer area, and contacting method (packed, spray, or tray tower) should be investigated. A preliminary study of these factors has been included in a research effort, supported by EPA and the Fats and Proteins Research Foundation, which is currently under way at the Research Institute of the Illinois Institute of Technology. The goal of this research is to develop an effective and economical method for rendering plant odor control.

Other possible research areas with potential for achieving economical control of odors from rendering operations are:

- investigation and development of economical, poison-resistant catalysts for application in more attractive catalytic incineration processes
- development of analytical techniques and instrumentation for identification and quantitive evaluation of the level of odorous pollutants in process streams and in effluent streams
- investigation of refrigeration or freezing as a means of minimizing decomposition and odor formation in materials prior to rendering
- study of solvent extraction of fats or other low-temperature processes as possible alternates to high-temperature cooking operations
- better housekeeping in plant operations.

Additional reading

Anderson, C. E., Adolf, H. R., "Odor Control in the Rendering and Food Processing Industries," Paper No. 71-22, 64th Annual Meeting, Air Pollution Control Association, Atlantic City, N.J., 1971.

Lundgren, D. A., Rees, L. W., Lehmann, L. D., "Odor Control by Chemical Oxidation: Cost, Efficiency, and Basis for Selection," Paper No. 72-116, 65th Annual Meeting, Air Pollution Control Association, Miami Beach, Fla., 1972.

Mills, J. L., Danielson, J. A., Smith, L. K., "Control of Odors from Inedible Rendering and Fish Meal Reduction in Los Angeles County," Paper No. 67-10, 60th Annual Meeting, Air Pollution Control Association, Cleveland, Ohio, 1967.

Murthy, B. N., "Odor Control by Wet Scrubbing: Selection of Reagents," Paper No. 73-275, 66th Annual Meeting, Air Pollution Control Association, Chicago, Ill., 1973.

Yocom, J. E., Duffee, R. A., "Controlling Industrial Odors," *Chem. Eng.*, **77** (13), 160-8 (1970).

Reprinted from Environ. Sci. Technol., **11**, 333 (April 1977)

Upgrading electrostatic precipitators

Research-Cottrell and Standard Oil (Calif.) show that imaginative
engineering for meeting today's air pollution regulations can
keep within refinery space limitations as well as budget

Many companies that installed air quality control equipment before the current regulations were established now face a serious problem—bringing this equipment up to standards. Having made a substantial investment in air control equipment, they now find it obsolete, unable to meet current regulations. They have a "white elephant" on their hands that must somehow be incorporated into current or future air control plans or discarded at a total loss of investment.

More than 20 years ago, Standard Oil (El Segundo, Calif.) installed a Research-Cottrell electrostatic precipitator as the final stage of a dust recovery system to capture "fines" from a fluid catalytic-cracking unit. With normal maintenance and parts replacement the unit continued to perform satisfactorily. Since that time, however, more stringent air quality control has become one of the conditions for doing business, so the original unit could no longer meet the regulations.

Faced with a seven-figure price tag for replacing the precipitator in its entirety, Standard Oil instead solved the problem by "capping" the original precipitator shell with a 4-ft extension. This allowed Research-Cottrell to design new internals that increased precipitator treatment capacity by more than 13 000 ft²—at far below the cost of new construction. The actual cost to Standard Oil was less than 10% of the cost for new precipitators.

The choices

In most cases where precipitator efficiency must be increased, companies usually select one of two basic solutions. One is to add on the required treatment capability by building another precipitator either in series or parallel with the existing precipitator. The other is new construction, either by tearing down and replacing the old precipitator with a larger one or by building in a new location.

To Standard Oil engineers, both of the foregoing solutions were impractical because of cost and available space limitations. Also, the existing precipitator shell, though undersized, was still in good condition and they were reluctant to scrap it.

Their solution

The solution was to increase the capacity of that shell by increasing its height. In the original installation, 31 collection plates in each unit treated flue gas in three separate fields—inlet, middle, outlet—in the precipitator. Each plate measured 9 ft wide by 20 ft high for a total collection area of 66 960 ft². By replacing each 20-ft plate with a 24-ft plate for a new collection area total of 80 352 ft². Standard Oil gained 13 392 ft² of treatment area for roughly the construction cost of a four-foot high precipitator shell.

The steel cap itself, 4 ft high × 16 ft wide × 18 ft long, was constructed on the

Capping. *This 4-ft extension cost 10% that of new units*

ground in an area near the old precipitator shell and lowered into position by a crane. The cap was then welded to the old shell and the new internals installed.

The precipitator, as designed and built by Research-Cottrell, now has a normal rated gas capacity of 150 000 acfm at 650 °F. It is two chambers wide and three fields long in the direction of the gas flow. Each field has 30 9-in. each wide ducts, formed by 31 Opzel-type collecting plates, each 24 ft high by 9 ft long.

These collection electrode plates are cleaned by magnetic impulse, gravity impact rappers. The 2160 copper Bessemer discharge electrode wires are cleaned by electric-type vibrators. Electrical energization to the precipitator is supplied by four 25 kVa, 75 kV vacuum tube rectifiers.

Six-stage collection unit

The particulate collection system is installed downstream of the fluid catalytic cracking unit to control emissions of silica alumina catalyst fines. The cracking operation generates a large quantity of these fine particles. Under current regulations Standard Oil must collection all but 30 lb/h, and they do.

To collect these highly abrasive fines, Standard Oil has three stages of cyclone separators and the three-field Research-Cottrell electrostatic precipitator. The cyclone separators filter out the larger catalyst fines, which are collected and recycled into the cracking process. Those small fines, which cannot be collected by the cyclones, pass on to the electrostatic precipitator where they are removed from the gas stream before it is released to the atmosphere. Catalyst particles collected by the precipitator that are too fine for re-use are hauled away.

Double savings by remodeling

With proven technology and some imaginative engineering design, Standard Oil has avoided a potential air pollution problem on a critical process plant. What's more, by not building a new precipitator housing shell, which would have meant an extended shutdown of the cracking unit, they saved, in addition to construction costs, literally tens of thousands of barrels of lost product.

Electric utilities seriously look at fabric filters

With increased use of low-sulfur western coal, baghouses may capture 35–50% of the particulate control market for this sector in 5–7 years

Reprinted from ENVIRON. SCI. TECHNOL., **11**, 856 (September 1977)

In 1970, fabric filter sales to electric utilities for fly ash control were nonexistent. Last year sales totaled about $14 million. This year sales will mount to $15–20 million. And by 1982 sales should hover near the $90–100 million mark. These figures were supplied to *ES&T* by Ken Aken, chairman of the Industrial Gas Cleaning Institute's (IGCI) fabric filter division, who cautions that IGCI tracks only bare hardward sales figures that exclude the cost of auxiliary equipment and erection.

To say the least, this penetration of the utility market has been very gratifying to the companies designing, engineering and constructing baghouses. And the future indeed looks promising. Today, 10% of utility air pollution control devices sold are fabric filter collectors. Should the economic climate for utilities remain good, fabric filter suppliers anticipate a surge in baghouse sales over the next two years. In 5–7 years these suppliers speculate that fabric filters will capture 35% (and possibly 50%) of the electric utility particulate control market.

Why the interest?

Fabric filters have been around for at least 80 years, and have found application in solving industrial control problems. The earliest use of fabric filters, in the late 1880's, was to control emissions from nonferrous metal smelters; today, these collectors are being widely used to clean the waste gas from steel-making furnaces and to collect dust in the cement industry. Utilities—free to burn oil and/or coal with no restriction on sulfur content—found electrostatic precipitators technically and economically viable systems for particulate matter control.

But 1970 saw enactment of the Clean Air Act with its tighter strictures on particulate and SO_2 emissions, and 1977 heard President Jimmy Carter's call for utilities to switch to coal to fire their boilers. Though widely spaced in time, both events force increased reliance on low-sulfur western coal. And therein lies the problem.

Because of the chemical characteristics of low-sulfur coal, the fly ash produced from the burning of this coal is a poor conductor of electricity. The fly ash is difficult to charge, which is a necessary precondition for the use of electrostatic precipitators, and once charged, the collected material on the plates of the precipitator is difficult to discharge. Engineers call this high resistivity.

Fabric filter baghouses, which act essentially like huge vacuum cleaners, require no preconditioning of the fly ash, and are able to efficiently filter the residue produced from the combustion of low-sulfur coal. In fact, baghouse collection efficiencies are usually rated at better than 99%.

Because precipitators are size and cost sensitive to the type of coal being burned, the size and cost of a precipitator designed to control emissions from low-sulfur coal could negate against these

IGCI's Aken
chairman, fabric filter division

systems as the pollution control device of choice if high collection efficiencies are required. However, it is risky to generalize about economic and technical tradeoffs between fabric filters and precipitators because the pollution control device of choice is dependent on what engineers term "site specific factors."

Generally speaking, Sidney Orem, technical director of IGCI says that "Fabric filters will find application in new electric utility plants dedicated to the burning of western coal." But he cautions that utilities, traditionally conservative, are waiting to judge the ultimate bag life. If the bag life is only one year, then fabric filters become costly alternative systems; however, if bag life is five years, baghouses become very acceptable options.

Western Precipitation Division of Joy Manufacturing Co. supplied the first baghouses to four of six coal-fired utility boilers (175 MW total) at the Sunbury Station of Pennsylvania Power and Light, about 5 years ago. Bags from three baghouses were changed after two years' operation, but the fourth baghouse with the original bags has been operating well for about 4 years. However, this is considered a small unit, and utilities are waiting to see the performance of fabric filters on several larger units—Texas Utilities' two 575-MW Monticello Units 1 and 2 (about 80% of the flue gas will go through bag filters and the remaining 20% will pass through the existing electrostatic precipitator), and Southwestern Public Service's 350-MW Amarillo, Tex., unit—scheduled for start-up in 1978. At present, engineers calculate bag life at a minimum of two years.

New directions

Despite the fact that baghouses can efficiently collect the fly ash from coals of varying sulfur content, these units, as presently designed, can control only particulate matter. The problem of SO_2 control still remains unless the utility has flue-gas desulfurization units . . . or the bag can be treated to simultaneously collect particulate matter while absorbing SO_2 emissions.

The most promising treatment to date that uses baghouses is called dry sorption of SO_2 in which the bags are "seeded" with a naturally occurring dry sodium bicarbonate additive called nahcolite. Several companies have conducted laboratory-scale tests of this and other sodium carbonates, but only one company — Wheelabrator-Frye — has completed pilot testing of the process on a utility boiler. Wheelabrator conducted the test in conjunction with a consortium of 7–8 electric utilities (the Coyote Group), Bechtel and Superior Oil.

Wheelabrator coated its bags with nahcolite supplied by Superior Oil. The baghouse, operating on the Leland Olds Station of Basin Electric Cooperative at Stanton, N.D., was tested for six months.

According to Dick Adams, vice president for systems and technology, the tests were successful. A report of the test results will be available from Wheelabrator later this month or next.

Adams feels that this nahcolite conditioning process is commercially feasible and that its introduction to the marketplace is dependent only on the availability of the carbonate. But that availability depends on commitments to mine the nahcolite. Three companies—Superior Oil, Industrial Resources and Rock School Corp.—are seeking mining rights. However, Larry Thaxton, vice president, fabric filter, Buell-Envirotech, points out that $10–14 million is involved in sinking a mine shaft and that the mining of this material is a serious drawback to the ultimate commercialization of the process.

Buell-Envirotech is supplying a baghouse to Colorado Springs Municipal Utility. But, in addition to the full-scale baghouse, Buell-Envirotech is also building a fully instrumented pilot-size baghouse on the 85 MW Martin Drake No. 6 boiler. A portion of the flue gas from the boiler will be diverted to the pilot baghouse for SO_2 removal. According to Thaxton, variously designed tests will be conducted; for example, the flue gas stream itself will be continuously seeded with different physical configurations of nahcolite (and other sodium salt compounds) and the bags will also be coated with nahcolite (and other sodium salt compounds) after a compartment of the baghouse has been cleaned. Test trials are scheduled to begin in May–June 1978.

Neither Wheelabrator-Frye nor Buell-Envirotech has received federal funds for their seeded fabric filter studies. EPA is now funding a feasibility study of the process and, depending on study results, may fund a demonstration project with a Southwestern utility in fiscal 1979.

Some problems

Because of the mining problems, Wheelabrator's reaction process with nahcolite will not be commercially available for at least another year, according to Adams. In the meantime, Atomics International Division of Rockwell International has developed a regenerative aqueous carbonate process that uses commercially available sodium carbonate as the SO_2 scrubbing agent.

In the Atomics International process, the sodium carbonate is dissolved and then spray-dried. The spray-drying unit, through which the boiler's gas stream passes, acts as a chemical contractor for SO_2 removal. The products formed are sodium sulfite, sodium sulfate and untreated sodium carbonate, which are entrained in the flue gas as fine particulate matter. The gas stream from which SO_2 is removed is then passed through a particulate matter control device—either a

fabric filter or an electrostatic precipitator. The collected particulates are then treated to recover elemental sulfur and regenerate sodium carbonate, which can be reused in the spray-drier.

A demonstration of this process is planned at Niagara Mohawk's 100 MW Huntley Station; possible startup date, January 1980. The project is to be funded by a consortium of New York electric utilities—the Empire State Electric Energy

Research Corp., of which Niagara Mohawk is a member, and the EPA.

Detractors of the SO_2 dry sorption process point to a solid materials-handling problem created by the formation of sodium sulfate, a soluble solid. If disposed of in the ground, this reaction product could easily leach into groundwater.

Adams counters by saying that the disposal problems are site specific and besides, any disposal problem may be offset by the fact that dry sorption of SO_2 is generally less costly than the purchase of a scrubber or low-sulfur western coals.

Another counter-argument can be found. In addition to Atomics International's regenerative aqueous sodium carbonate process, American Air Filter has a patented process for the regeneration of nahcolite.

Needless to say, the fabric filter people are excited by the potential to simultaneously control flue gases for particulate matter and SO_2 emissions.

Competition, fast and fierce

Fabric filters have successfully penetrated the electric utility market. Their share of the utility particulate control market, at the expense of electrostatic precipitators, may grow to 50%, although many company spokesmen felt this unlikely. More realistically, they estimated the share of the market at 35%.

There was general consensus, however, that over the next 5–7 years, fabric filters will establish a very firm place in the utility market, but they will not take over the primacy established by electrostatic precipitators.

The spokesman all agree that the market is there to be captured. But Bill Henke of Western Precipitation adds this note of caution: "Too many inexperienced companies are racing to get into the field and competition will be fierce."   LRE

Baghouse installations on utility boilers

Utility	Location	Equiv. MW	Startup Date
Board of Public Utilities	Kansas City, Kans.	44	1979
City of Colorado Springs	Colorado Springs, Colo.	200	1980
Colorado Ute Electric Assn.	Nucla, Colo.	13	1978
	Montrose, Colo.	12	1977
Crisp County Power Co.	Cordele, Ga.	10	1975
Minnesota Power & Light Co.	Cohasset, Minn.	75	1978
Nebraska Public Power	Bellevue, Nebr.	125	1976
New South Wales Electric Commission	Tallawarra Station	39	1976
	Wangi Station	180	1976
Pennsylvania Power & Light Co.	Holtwood, Pa.	44	1975
	Sunbury, Pa.	175	1973
Public Service of Colorado	Palisade, Colo., Unit 1	22	1977
	Palisade, Colo., Unit 2	70	1978
Sierra Pacific	North Valmey, Nev.	250	1981
Southwestern Public Service	Amarillo, Tex.	350	1979
Texas Utilities	Monticello, Tex.	880	1978

Source: Power Magazine, Jan. 1977

Manufacturers of fabric filters

American Air Filter Company, Inc.
Buffalo Forge Co.
Carborundum Co./Pollution Control Division
C-E Air Preheater Co.
Combustion Equipment Associates, Inc.
DCE Vokes, Inc.
Donaldson Company, Inc./Torit Division
Dustex Division/American Precision Industries, Inc.
Environmental Elements Corp./Subs. Koppers Company
Envirotech Corp./Buell Emission Control Division
Fisher-Klosterman, Inc.
Fläkt, Inc.
Fuller Company/Dracco Products
Industrial Clean Air, Inc.
Lear Siegler, Inc.
MikroPul Corp.
Peabody Air Resources Group
Research-Cottrell, Inc.
W.W. Sly Manufacturing Co.
Standard Havens, Inc.
Western Precipitation Division/Joy Manufacturing Co.
Wheelabrator-Frye, Inc.
Young Industries, Inc.
Zurn Air Systems

Source: IGCI

Reprinted from ENVIRON. SCI. TECHNOL., **8**, 508 (June 1974)

Future bright for fabric filters

With efficiencies in the range of 99.9% and favorable cost comparisons, fabric filters are the way to go for many source operators

With certain caveats, the most promising technology for controlling small, or submicron, particulate matter is fabric filters, according to most people working in the field. For one thing, these filters are extremely efficient. Removal efficiencies in the range of 99.9% easily can be achieved. Although other kinds of filters—fibrous and aggregate, for example—also are used successfully, results indicate they are effective only under specialized circumstances.

A fine particle usually is defined according to the range most detrimental for breathing, 0.1–1.0 micron. Related health problems, however, also can arise because of exposure to particles ranging all the way from 0.001 to 10 microns. John K. Burchard, director of the U.S. Environmental Protection Agency's Control System Laboratory at Research Triangle Park, N.C., points out that fine particles not only cause haze and smog, but can penetrate deep into the lungs, sometimes causing serious respiratory conditions, and can act as transport elements for other pollutants.

Burchard told a Symposium on the Use of Fabric Filters for the Control of Submicron Particulates in Boston, Mass., recently that 50% of the particulates of the magnitude 0.01–0.1 micron will remain once they reach the lungs. (The symposium was sponsored by GCA Technology, Inc. and drew about 200 attendees.) Silicosis, bronchitis, and other lung conditions can result. Larger particulates, of a magnitude of 5 microns, are deposited on the mucus linings of the mouth and nasal passages, and can cause stomach and other health problems when swallowed, he said.

Burchard pointed out that the Council on Environmental Quality's Fourth Annual Report revealed that a reduction in particulates has been achieved in six of 10 major U.S. cities. Still, a massive effort is needed to meet the nation's clean air needs. Burchard said improved fabric filtration will be an important step toward accomplishing this task.

Basically, fabric filters work inside a "baghouse," or overall structure, where the dirty air steam—from a metal processing or fuel-burning plant—is filtered by cloth tube-like bags. The process is a mechanical one where the particulates are trapped on the cloth inside the bags. Removal of the particulates is accomplished by shaking the bags, reversing the air flow, quick jets of air, or by rapidly expanding the bags with compressed air.

A variety of fabrics is available. The choice depends upon the particular needs of the air-cleaning operation. Temperature, moisture, chemical composition of the gas, and the physical and chemical composition of the particulates all affect the choice. While a cotton fabric would be acceptable for a low-temperature gas stream, a glass fiber fabric would be more suitable for a high-temperature gas stream. The chemical composition of both the gas and particulates, as well as such physical properties as weight, affect the rates at which the fabric filters will wear out and need to be replaced.

Alternative methods

Fabric filters compared well with such other particulate control techniques as scrubbers or electrostatic precipitators, according to John D. McKenna, vice president of Enviro-Systems & Research, Inc., Roanoke, Va. In a comparison with the Venturi scrubber, he said although the initial, or capital, cost for the fabric filter would be slightly higher, the operating costs would be less. Thus, the overall annual cost would be less for the fabric filter, he said.

In most cases, the scrubbing system—where the gas stream travels through a narrowed "throat" portion of an air duct into which water is forced to impact the particulates—is not as efficient as the fabric filter, McKenna said. He also pointed out that in some cases of fly ash scrubbing or where there is acid in the gas stream, even the cost of equipment would be higher than for fabric filters. This is because special metals and materials would be required in the construction of the scrubber.

McKenna said electrostatic precipitators are cost competitive with fabric filters in such large projects as power plants. But, he added, the precipitators are not competitive on a smaller scale, such as would be suitable for an industrial boiler. Also, the cost of electrostatic precipitators increases greatly as additional equipment is added to achieve efficiencies of 99.9%. As codes and standards tighten up, efficiency will more and more be at a premium. "With this in mind, fabric filters become the way to go," McKenna said.

Needed research

On a final note, Knowlton J. Caplan, University of Minnesota, outlined four areas where additional research is needed to help perfect fabric filtration. The first concerned the mechanism of "bleeding" or "seepage" where after a period of time, particles begin collecting on the clean side, or outside, of the filter bags in defiance of classical filtration theory.

Caplan also said in-stack sampling, shaking speeds and bag accelerations, and the electrostatic phenomenon, all need further study. Electrostatics is an especially important area of consideration because explosions can result on a large enough scale to destroy entire plants, he said. The problem arises from the fact electrostatics is difficult to measure and the bags are difficult to ground, Caplan told the symposium. Such research is increasingly of interest to manufacturers, Caplan said, because the market for such equipment is growing at a rapid rate.   WSF

Baghouses. *New baghouses greatly reduced particulates at this cement plant*

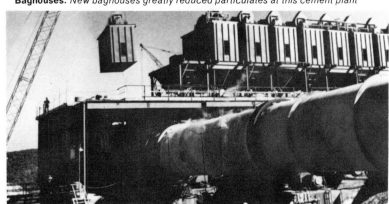

Reprinted from ENVIRON. SCI. TECHNOL., **8**, 127 (February 1974)

Fabric filters abate air emissions

Frank R. Culhane

Wheelabrator-Frye Inc., Pittsburgh, Pa. 15222

Lower in energy requirements and cost in certain applications, they are preferred over other air pollution control devices

The highly efficient fabric filter is one of three air pollution control devices which was, and is, broadly and successfully applied throughout industry. Design and application engineers have incorporated engineering features into the fabric filter, enabling it to be successfully utilized in areas routinely considered the realm of the electrostatic precipitator and scrubber. It is a device which has grown from a method of product recovery in the late 1800's to a sophisticated machine widely applied today to meet a variety of air pollution control problems.

Tracing its development

Stack losses—smoke, dust and/or fume—consisting of either semifinished or finished product values created the proper economic climate for the adaptation and development of the baghouse as a means of effectively separating finely divided solids from the carrier gas stream. The nonferrous metallurgical and carbon black industries are cases in point.

The carbon black industry went through a 5-year period in the late 40's and early 50's during which considerable pilot plant and prototype baghouse work established the necessary application and design parameters required for successful operation of fabric filters in a very severe and harsh environment. After this period, the baghouse became the exclusive means of gas cleaning throughout the carbon black industry, achieving virtually 100% recovery of fine particulate carbon black from stack gases.

Much earlier, at the turn of the century, the same economic reasoning and justification by the nonferrous discipline of the metallurgical industry led to the adaptation and development of the baghouse for recovery of stack metal losses. The magnitude of these losses range between 5 and 6% of the furnace feed for most nonferrous pyrometallurgical processes. Large tubular bags, fabricated of wool, mounted in a hopper, connected to a flue, and enclosed in a brick house, best recovered these stack values.

As community pressure increased the demand to control air pollution to maintain the improve the quality of life, the function of the baghouse was changed from a device used for product recovery to an effective means of stack gas cleaning. Commercially, the baghouse became known as a fabric filter or fabric collector.

Figure 1
Woven-type low-energy systems

Most established manufacturers of industrial gas-cleaning equipment have added fabric filters to their existing line of equipment, providing a greater degree of equipment selectivity, and offering the optimum and best means of solving a specific emission problem. Today the process, plant, and consulting engineers have a broad choice between a commerical baghouse, the high-energy scrubber, and the electrostatic precipitator in determining the most effective and best means of solid particulate control.

The prime growth and development of the fabric filter has followed a corresponding technical growth of synthetic fibers. The early and traditional barriers to baghouse development have been operating limitations of temperature, fiber resistance to chemical attack, and fiber dimensional stability and sustained tensile and flex strength. These limitations have dictated the design of the system.

Initially, the only available fibers—cotton and wool—limited operating temperatures to either 180 or 220°F, although their stability formed ideal grid systems. These systems support and hold a sufficient quantity of the gas-separated particulate as a filter cake, preventing bleeding during filtration and puffing after periodic cleaning. The first man-made fiber—rayon—became. available in the twenties. This enabled experimenting with a wool–rayon combination which provided greater tensile strength and increased the abrasion resistance qualities of the wool. Nylon was introduced in the thirties, and offered greater abrasion resistance than either cotton or wool but did not increase the operating temperature significantly. These fibers added the filament fiber form to the fabric, thereby increasing the ease of cake removal during periods of cloth cleaning.

The greatest stride in fiber development came in the late 40's with the advent of acrylic, polyester, and mineral fibers. In 1946, the upper temperature operating limits were increased from 220° to 275°F through the application of acrylic fibers. This increase of 55° represented a major application breakthrough for the baghouse, allowing the use of spray towers for thermal control, with resulting dew points in the 150–165°F range. In 1953, through improved finishing of fiber glass fabric, the upper operating temperature limit was further increased from 275° to 550°F.

Research and development continue on fiber improvement with an emphasis toward the increase of operating temperature limits beyond 550°F and the reduction of thermal control requirements. However, regardless of higher operating temperatures of the baghouse, the design engineer must consider the trade-off between the savings of gas cooling and the increase in gas-cleaning cost because of the increase in gas volume. If, on the other hand, gas cooling is carried to an extreme to reduce the gas volume, increased chemical attack is experienced as the acid dew point is approached.

Uses and limitations

The characteristics of most submicron particles to agglomerate have resulted in the ability of filter fabrics to remove 99.9% by weight of submicron solid particles. This high order of fine particulate efficiency is routinely being achieved in separating:

- carbon, as fine as 320 Å, from process reactors
- silicon dioxide, in 0.02–0.25-μ range, from submerged arc furnaces
- 100% of particles below 5 μ and 95% below 2 μ, in ferrous fume emissions

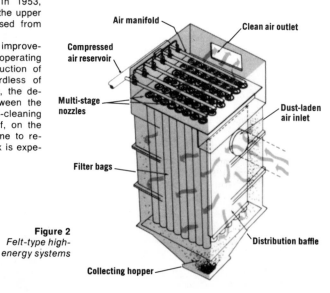

Figure 2
Felt-type high-energy systems

Air manifold · Clean air outlet · Compressed air reservoir · Multi-stage nozzles · Dust-laden air inlet · Filter bags · Distribution baffle · Collecting hopper

Figure 3. *Primary aluminum reduction plant uses fabric filters*

Submicron fume is molecular in size and the ability of the baghouse to remove it is due to the filter cake formed by proportional number of like agglomerated particles. The finer particles cause a tighter filter cake, resulting in greater head loss per unit of gas flow in comparison with a cake formed from coarser dust particles.

A critical factor in sizing a baghouse is the resultant pressure drop across the filter fabric. Within reasonable cloth resistance, the filtration efficiency for all practical purposes is constant. The selected filtering velocity varies from application to application, depending upon the filtration characteristics of the dust and fume particulate burden. In general, the pressure drop across the cloth will vary directly with the time between cloth-cleaning periods, exponentially with the filtering velocity, and directly with the filterability constant.

The filterability of any given quantity of particulate depends on the particle size and distribution, shape, surface properties, and electrostatic forces. Experience and application know-how are the final determining factors regarding sizing of a baghouse for a given application. The combination of the variables affecting the sizing of cloth collectors are many, and it is necessary to work from known results in like applications, and to allow for unknowns when direct experience is not available.

For new applications, a 500–2500 cfm pilot unit will provide specific answers to filtration characteristics of the dust and fume in question. Most new baghouse applications have developed through pilot test units. Current work being done on coal-fired boilers in the power plant industry for the collection of fly ash has been preceded by extensive pilot plant work.

It is broadly accepted within the industry that the baghouse will not work for gas below the dew point. If, however, other factors eliminate a precipitator or scrubber, and if a high degree of dust and fume recovery is critical, an external source of heat may be indirectly or directly

Characteristics of baghouse design

Cloth-cleaning classification	Low-energy systems	High-energy systems
General description	The filter fabric is formed into tubes, opened at one end and closed at the other. The open end is mounted into a cell plate or thimble floor which is located over a trough-type hopper, in a single compartment. Several compartments are manifolded into a single baghouse. The gas flows through a dirty gas inlet manifold into the hoppers and floats upward into the bottom of the tube. Preseparation of coarse particles in the hoppers is thus achieved, beneficially reducing the load to the fabric. The flow pattern is usually from the inside of the bag outward, depositing the dust and fume inside the bag, and enabling inspection and maintenance from the clean-gas side of the baghouse	The filter fabric is formed into tubes, opened at one end and closed at the other. The open end is mounted in a thimble ceiling with the closed end hanging vertically over a hopper, in a single casing. The gas flow is from outside to inside the bag and a metal cage is used to keep the bag from collapsing during filtration. Inlet designs provide preseparation of particles within the casing, reducing the load to the fabric
Filter type	Surface filter	Depth filter
Media	Woven fabric	Felt fabric
Cloth-cleaning description	Filament fiber fabric is usually cleaned by repressuring. Staple fiber fabric is usually cleaned by mechanical shaking. A combination of both repressuring and shaking is used for difficult-to-remove particulates (Figure 1)	High pressure compressed air is used to instantaneously and frequently pulse the bag, snapping the particulate off the fabric surface (Figure 2)
Filtration stops	Filtration is interrupted during periods of cloth cleaning, achieved by use of compartment dampers	Filtration is interrupted during periods of cloth cleaning, achieved by a pulse of counterflow air
Spacing requirements	Sufficient centerline spacing between tubes is necessary to prevent rubbing and to provide unrestricted gas flow from cloth to compartment outlet duct. For example, 6-in. centers on 5-in. diameter bags, 9-in. centers on 8-in. diameter bags, and 14-in. centers on $11^{1}/_{2}$-in. diameter bags are recommended	Sufficient centerline spacing between tubes is necessary to prevent rubbing during periods of cloth cleaning. For example, 6-in. centers on 4-in. diameter bags, 8-in. centers on 6-in. diameter bags are recommended

applied to the gas, raising the baghouse operating temperature 50–75°F above the dew point. Calcining gypsum in a kettle is a case in point.

In this process, a fire box below the kettle indirectly heats the gypsum, and the resultant gases created by the release of the water of crystallization carry a copious quantity of rock dust and partially calcined gypsum dust into the atmosphere in a steamy, dust condition. These untreated saturated gases could never be directly handled with a baghouse. However, by use of indirect or direct heat exchangers, the waste heat from the fire box can readily increase the dry bulb temperature 50–60°F. This heating enables the dry operation of the fully insulated baghouse and system.

The hazard of hygroscopicity will generally rule out a baghouse from consideration. However, special adaptation of light chains, suspended inside the bags, will break encrusted dust so that strip heaters within the hoppers will keep the unit hot and dry during periods of downtime. This adaptation permits use of a baghouse. In addition, an all-filament fabric will reduce cloth-cleaning problems usually associated with hygroscopic materials.

Space factors often prohibit consideration of a baghouse. As a result of this problem, most manufacturers offer two or three bag diameters and a corresponding range of bag lengths to obtain the required filtering surface area within the confines of the available space.

Incandescent sparks in the gas stream are also reasons for eliminating the use of a baghouse. Basically, sparks could be prevented from contacting the filter fabric by using spark arresters in the form of mechanical collectors or screens. Another solution would be to precoat the bags with an inert mineral before start-up. This precoating would provide a noncombustible surface on the surface of the bags.

If metallic fume is not completely oxidized, the collected metallic values will be pyrophoric and care should be exercised to avoid accumulating a large quantity of this kind of material within the baghouse hoppers.

Low vs. high energy

The fabric collector's design generally falls into two broad categories classified by the energy level—low vs. high—used to clean the filter fabric (see box).

Design parameters common to both types of cloth collectors are as follows:

• Fabric selection is based on gas temperature and chemistry, dimensional and chemical stability, tensile and flex strength, and permeability.

• Unit design is based on pressure or suction. A suction unit will prevent erosion and accumulation of dust burden on the fan scroll and impellers, but is higher in first costs than a pressure unit.

• Materials of construction are based on economics and corrosion resistance. Typical casings are fabricated of carbon steel, but stainless steel, aluminum, tile, and concrete are not unusual.

• Trough-type, live-bottom hoppers are based on design with conveyors for continual dust removal.

Fabric filters have been designed and installed from 100 cfm to 4.5 million cfm in a single installation. A well-designed baghouse is capable of handling the entire range of gas-borne stack emission particles—fine or coarse in size, light or heavy in grain loading. It also can withstand the most abrasive and corrosive conditions. Its efficiency is constant, regardless of varying moisture, temperature, particle size, or gas flow.

On applications wherein the dust resistivity is high, the electrostatic precipitator becomes large and expensive, and first cost of the fabric filters becomes comparatively attractive. A typical example is the filtration of iron oxide fume from the melting operations of ferrous scrap in the steel and foundry industries. Because of the lack of moisture, fume resistivity is high, making it difficult to separate particles through use of electrostatic precipitators. Because of the extremely small particle size, the energy requirements of the scrubber (approximately 60-in. water gauge), is high, resulting in a preference for the fabric filter.

Industrial use

The cement industry employs both the electrostatic precipitator and the fabric filter to collect and recover material in a dry, reusable state. On wet cement kilns, the high moisture content of kiln gases favors electrostatic precipitators use. On dry kiln emissions, a conditioning tower is required to improve dust resistivity through the addition of moisture. Here, the combined first cost of the conditioning tower plus the precipitator will often result in a higher first cost compared to the baghouse. Wet scrubbers have been applied to kilns only in a few cases, mainly because of the disadvantages of additional slurry-handling equipment required.

Controlled emissions from clinker coolers have been achieved predominantly through use of fabric filters. Cloth collectors are preferred over wet scrubbers since the collected clinker product is returned to the dry process without further treatment. The gas is bone dry and high in resistivity, thus requiring a large precipitator, and making the fabric collector more attractive in first cost.

An interesting development is the use of precoated fabric filters to clean large volumes of exhaust gases generated in primary aluminum reduction plants. In this case, the fabric is not only used to remove the fine solid particles from the gas stream, but also the gaseous fluorides. The filter tubes are precoated automatically at regular intervals by a reagent that chemically reacts with the gaseous fluorides. The discharge gases are dry and hot, resulting in a plume-free discharge with excellent buoyancy and providing good dispersion and atmospheric dilution of the remaining small amount of fluorides in the gas stream. In a typical plant (Figure 3), four dirty-gas manifolds convey the off-gases from the reduction cells into four suction baghouses just to the right of the stack. Four double-inlet, double-width exhaust fans move approximately 1,330,000 cfm of filtered gas into a common manifold and breeching for final discharge through a single stack.

Selection criteria

Current energy crisis and changing economic values make the correct choice of air pollution equipment vital. Power requirements for high energy scrubbing is critical today and pales the demand for a new set of installed bags every two or three years for a baghouse. The pending closed-loop water pollution control requirements of waste water demands sophisticated treatment of scrubber effluent. Overtones resulting from a reduction in the currently accepted bleed-off allowance can appreciably shift an evaluation from the scrubber to a baghouse. If proposed fine particulate codes are made law, a technological reassessment will be required.

Frank R. Culhane, *presently vice president APC Products, Wheelabrator-Frye Inc., is a graduate mechanical engineer with over 24 years experience in the field of air pollution control.*

The largest U.S. wet scrubber system

At the Four Corners plant, Arizona Public Service Co.
is not leaving any corner unturned in its search for
stack gas controls for power production

Reprinted from Environ. Sci. Technol., **8,** 516 (June 1974)

Out in the desert in the southwest portion of the U.S. near where four states come together is the Four Corners power plant. The actual plant is located on the Navajo Indian Reservation, about 20 miles west of Farmington, N.M. All told there are five power generating units at this location with a generating capacity of nearly 2.1 million kW. All units are coal fired, the coal being supplied by Utah International, Inc., in its surface mining operation on the Navajo Indian Reservation.

What makes this plant unique is the fact that emission control systems have been installed on all the units, but more controls—costing an estimated $158 million—will be added to meet the "moving target" of newer air quality standards proposed by the State of New Mexico and the Environmental Protection Agency. The experience of Arizona Public Service Co. (APS), the plant's operator, with so-called reliable and available technology is noteworthy. Despite considerable setbacks, APS continues to get the bugs out of such reliable and available controls. They haven't given up yet. Their experience serves as an example to others.

Controls for particulate matter have been installed on the larger units 4 and 5; yet others are to be added. For example, Research-Cottrell electrostatic precipitators were installed on unit 4 in July 1969 and on unit 5 by July 1970 when the units were constructed. These precipitators cost nearly $6 million; they were designed for 97% removal of fly ash. Their removal efficiency was verified by Truesdail Lab (Los Angeles, Calif.).

Since the installation of these controls, which are in actual operation on a day-to-day basis, there have been only two violations of New Mexico air pollution codes, once with both units on July 18 and 19, 1972, and again with one unit in November 1973.

Electrostatic precipitators were also considered for units 1, 2, and 3, but removal efficiencies would not satisfy existing codes even as late as 1971. Chemico wet scrubbers were installed on units 1, 2, and 3; they became operational in December 1971. While designed only for fly ash removal, test results of York Research, the EPA, and the State of New Mexico confirmed that the units were also removing 30–35% of the SO_2 emissions from the coal-burning units. The scrubbers met efficiency guarantees and met all codes in existence for fly ash removal.

Lyman Mundth, vice-president—power production of APS, told *ES&T* that the utility has experienced difficulties with the scrubber control devices but in no way does APS intend to give up on controls.

The moving target

But neither these control devices nor the removal efficiencies are good enough in this continuing age of moving targets. The federal EPA late in March said that by July 31, 1977, 70% of the SO_2 emissions from the plant's four stacks (units 1 and 2 have a common stack) must be removed.

A new regulation on particulate matter has been issued by the State of New Mexico to become effective December 31, 1974. This new requirement means supplementing present controls on units 4 and 5. For example, particulate emission

control for units 4 and 5 would have to be increased from 97% fly ash removal efficiency to 99.74% efficiency. And then in order to comply with the federal EPA requirement of 70% SO_2 removal efficiency, an SO_2 absorption removal system would have to be installed downside of the precipitators—a scrubber system the owners intend installing on units 4 and 5. Present estimated cost of the scrubbers and SO_2 removal system is $140 million. Of course, scrubber SO_2 removal efficiency in units 1, 2, and 3 would have to be upgraded to 70% too, from 30–35%. Estimated cost of this upgrading is $18 million.

Lyman Mundth said that the additional controls for units 4 and 5 were recommended by Bechtel Corp., the San Francisco-based consulting firm with enormous electric utilities experience. APS is going full steam with these controls.

If limestone is used in the SO_2 removal process by APS, an estimated 400–500 tons of limestone would be needed each day. At present, the consulting firm of Dames & Moore is making a survey of lime and limestone availability for APS. There are six limestone deposits within a few hundred miles of the plant. The utility needs a 35-year supply!

Lyman Mundth said that APS plans to be in the business of power production at the Four Corners plant for a long time. He explained that the utility has a 35-year agreement with Utah International for the coal supply. Bill Grant, spokesman for Utah International says that its area of coal mining today would last 125 years at the present rate of coal consumption at the Four Corners plant. The plant burns about 25,000 tons per day.

Nevertheless, the plans and actual experience do not go hand in hand. Plant manager Walker Ekstrom of the Four Corners plant explained that the plant is a base load power plant with a high-capacity factor, meaning that the fuel—the surface-mined coal—is cheap, and the goal of the plant is maintenance of the highest availability of power production with limited outages.

However, the maintenance costs are high. Ekstrom explained that the ash content of the coal is 22%—one

Scrubber experience at Four Corners

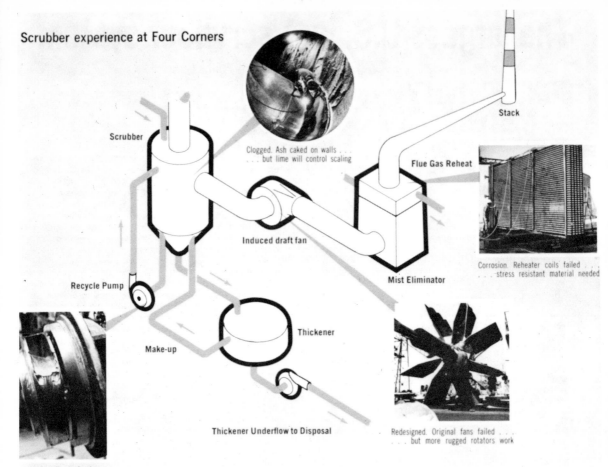

Scrubber

Clogged. Ash caked on walls . . .
. . . but lime will control scaling

Stack

Flue Gas Reheat

Induced draft fan

Mist Eliminator

Corrosion. Reheater coils failed . . .
. . . stress resistant material needed

Recycle Pump

Make-up

Thickener

Thickener Underflow to Disposal

Redesigned. Original fans failed . . .
. . . but more rugged rotators work

rn out. Impeller broke down . . .
but rubber lined pumps are in

of the highest ash contents of any coal in the U.S.—and that the ash is very abrasive on plant equipment.

Each unit at Four Corners is shut down for scheduled outage once a year. During this outage normal and routine repair and replacement of equipment are performed. But, he pointed out that the operation of the different wet scrubbers has caused more delays and maintenance of equipment. Ekstrom said that in 1973, the first full year of scrubber operation on units 1, 2, and 3, it was necessary to add 15 to 20 people to the maintenance department which in prescrubber days operated at slightly more than 100 people.

Getting the bugs out

The scrubbers have been plagued by numerous problems which have created other than routinely scheduled outages. "It must be remembered that this system reflects 1969 technology, was the first of this magnitude, and that a lot of lessons have been learned and are being applied," said Robert Blinckmann, vice-president and general manager of Chemico Air Pollution Control Co. "Since the initial start-up, Chemico has worked with APS to solve the problems encountered."

Some of the original recycle pumps had to be replaced after only 700 hr of operation because they wore out. APS tried using six different pumps made of as many types of alloys in order to overcome the abrasion which was damaging the pumps. None of the six pumps worked well, so APS has been testing a rubber-lined pump which now has operated successfully for more than 4500 hr. Originally, rubber-lined pumps for this application simply were not available.

Scaling of the scrubber led to clogging problems and other downtimes. All six of the induced draft fans failed and had to be replaced. The original fans were installed December 1971 and cost about $1 million. The replacements were of a new, more rugged design and utilized Inconel for greater strength.

But the most serious and perhaps most expensive experience to date is the problem of stack gas reheat coils. Normally, the stack gas is heated with steam coils to raise the temperature of the flue gas above condensation temperature to avoid condensation in the stack and breeching. Heating of the gas also expands the gas, giving it more buoyance so it rises faster and higher and is more readily dispersed in the air.

Ekstrom said it would take several million dollars to replace these reheat coils if a material could be found with greater resistance to stress corrosion, but none has been found yet. The coils were out of service when *ES&T* visited. The two stacks for units 1, 2, and 3 have been coated with a protective material, Stak Fas. These units now operate with wet stacks. Custodis Construction Co. (Chicago) completely removed the brick lining from the stacks, washed the chimneys with acid and sprayed on a $3/16$-in. layer of the protective material. Chemico and APS are studying other alternatives such as the possibility of reheating stack gases directly with oil burners.

It all adds up to a basic lesson in economics. In 1972, the capacity factor of units 1, 2, and 3 declined considerably from prescrubber days. APS vice-president of engineering Thomas G. Woods, Jr., said before the New Mexico Environmental Improvement Board last December, "Capacity factors for each of the units dropped 13%, 8%, and 29%, respectively, compared to prescrubber experience."

Much of this reduction is attributable directly to the induced draft fans. Since they were replaced in 1972, there has been a marked improvement in scrubber availability and in plant capacity factor. Unit 1 had the same capacity factor in 1973 as in 1972. Unit 2 showed a 3% increase and Unit 3 was up 34%, a net gain of +5% for Unit 3. Woods said that Chemico has been most cooperative in trying to find solutions to the problem.

To add to all the materials and equipment problems, it cost APS an estimated $1 million a year to operate the scrubbing controls on the three units. Despite all the troubles and delays, APS has not thrown in the towel. For example, in 1974, additional, correctional actions will be taken, including:

• replacing the remaining original recycle pumps with rubber-lined pumps

• adding a permanent lime-feeding device to control the degree of scaling and buildup of sulfates and sulfites throughout the scrubber system.

In 1974, APS will begin sharing the output of the Navajo Power Plant (Page, Ariz.), another coal-fired station. It will be equipped with "hot side" electrostatic precipitators of Western Precipitation, a division of Joy Mfg. Co.

What about monitoring

Just 10 miles north of the Four Corners plant is the San Juan power plant which is jointly owned by the Public Service Co. of New Mexico and Tucson Gas and Electric Co., which are cooperating with the Four Corners plant owners in an air-monitoring program for the area.

All told, the plants' owners have invested about $300,000 in that program. They have eight monitoring stations in the area. The companies intend keeping emissions at a level which meets the national air quality standard and keeping a record of emissions. Four ambient air monitors were installed in 1972, another four a year later, in the spring of 1973. A weather tower was added at Four

APS vice-president Mundth
The company has not given up on power plant emission controls

Corners in May 1973. The cost breakdown shows a $200,000 investment for the eight monitors (about $25,000 each) and $60,000 for the weather station. The companies spend about $65,000 on operating expenses for the monitoring.

Recently, the companies elected to go the contract route for the evaluation and reporting of their monitoring and meteorological data. Western Scientific Service (Fort Collins, Colo.) maintains and collects the data from the weather tower. Sierra Research (Boulder, Colo.) does likewise for the monitoring data from the eight monitors. Loren Crow, a consultant from Denver, advised the firms on the placement of the monitors. They are scattered in a circle around the two power plants. Intercomp (Houston, Tex.) has a diffusion model; the monitoring data are fed into the model to see if any violations of standards occur.

In addition, the Four Corners plant owners added an on-site environmental laboratory at an expense of $65,000, budgeted $50,000 in 1974 for testing which required a continuous 2–3 people effort, and has instrument repair personnel on hand. A full test of an electrostatic precipitator may run as much as $50,000; normally it takes four people three days to test a unit.

In 1974, continuous stack monitors for SO_2 and NO_x emissions will be installed on units 4 and 5, APS says. The installation of such stack gas-monitoring systems entails an additional investment of $93,000.

Again, in this age of environmental moving targets it is useful to point out facts. In its implementation plan,

which was submitted to the federal EPA, the State of New Mexico required that the Four Corners plant install a continuous stack gas monitoring system but that portion of the plan was disapproved by EPA!

Keeping up their environmental guard on another area, the owners of Four Corners and those then planning San Juan contracted for a study of potential vegetation damage study with Dr. Clyde Hill of the University of Utah. The study has been continuing since 1971. In data for the past three years, Dr. Hill has found neither damage to alfalfa, the prime concern since the material is of commercial value, nor to other native species. His investigation involves more than 40 test plots; his main concern during the growing season is observation of young plants since they are potentially more susceptible to air pollution damage. The study is funded at the $50,000/year level by the owners of the Four Corners and San Juan power plants.

Sludge problem unresolved

APS has a scrubber sludge disposal problem that is common with other scrubber-equipped power plants. At the Four Corners plant, the total ash production has been estimated to be 1.5 million tons per year; about 75% of which is returned to the mine by Utah International in its reclamation practice. But for the scrubber sludge, there is no apparent and immediate solution.

Mundth mentioned that APS is looking at the process of International Utilities/Conversion Systems, Inc. (*ES&T,* Oct. 1972, p 874) as well as the sludge disposal process of the Radian Corp. (Dallas, Tex.). Furthermore, APS has contracted for a study on how to dispose of this sludge with Stearns-Roger Corp. (Denver, Colo.). Results are forthcoming.

Arizona Public Service Co. represents a $1,003,218,000 investment. Total business for the company was $224,956,000, according to 1973 corporate figures (year ending December 1973). By the end of that period, APS had invested and authorized $133 million in environmental controls at its Four Corners and Cholla plants and spent in excess of $1 million a year on the operation of these controls.

During all of this costly testing and reduced capacity production, APS has requested and been granted two rate increases since 1970; a 6.3% rate increase became effective May 1973, and a 4.65% increase became effective in October 1973. Despite the troublesome experience, APS continues to seek reliable, efficient, and available controls.   SSM

Removing three air pollutants at once

Reprinted from Environ. Sci. Technol., **8,** 788 (September 1974)

The Modine Manufacturing Co., Racine, Wis., produces aluminum fin- and tube-type condensers for automotive air conditioning units. Their manufacturing operation begins with the assembly of preshaped fins and tubes temporarily clamped together and drenched with a slurry mix. Then the assemblies are passed through gas-fired ovens where they are subjected to a high temperature, often in excess of 1000°F, to weld the fins and tubes into a single unit. Subsequent operations include quenching, drying, assembly of additional components, hydrostatic testing, and painting.

Both gaseous pollutants and particulate matter result from this operation. For example, the aluminum bonding ovens release a corrosive combination of hydrogen chloride gas and aluminum hydroxide and aluminum chloride particles. Of Modine's 14 operating plants, the one at Clinton, Tenn., has been equipped with an air and water pollution control system costing approximately $285,000.

But not all the 13 other Modine plants will be equipped with scrubbers because they are not required. Operations vary from plant to plant. For example, a plant at McHenry, Ill., also has a Ceilcote scrubber for a completely different set of reasons. A small prototype unit also was installed recently at Racine, Wis.

The Ceilcote Co., Berea, Ohio, designed and manufactured the venturi scrubbers which are unique in that the venturi diverging section is located within the main scrubber chamber and passes vertically through the packed entrainment separator. The principal advantage of this design, in addition to saving considerable space, is the fact that water and gas are discharged directly into a separator sump at the bottom of the scrubber, where most of the larger entrained liquid particles from the venturi throat are removed.

While the discharge from the ovens is reasonably mild, the Tennessee Department of Air Pollution required the installation of an air pollution system.

Before the Ceilcote scrubber was installed at Clinton, hydrogen chloride was being emitted at the rate of 50 lb/hr. After the installation, the amount was virtually undetectable. Also, before the scrubber installation, particulate matter was emitted at the rate of 50 lb/hr. After installation, particulate matter was being emitted at the rate of 3 lb/hr.

Stephen Schwartz, Modine's manager of environmental protection, says, "The combined water and air pollution control system at the Clinton, Tenn., plant incorporates almost total water reuse. The system goes a long way in allowing the company to comply with state effluent limits.

He continues, "Another important feature of the system is that the entire oven emission is vented through fiberglass-reinforced resins (with the exceptions of the fans), which should require almost no replacement or repair due to corrosion."

Schwartz says, "The plant's two operating ovens release approximately 65 lb/hr of aluminum hydroxide and aluminum chloride. Particulate matter averages about 0.3 μ in diameter. "The selection of high energy venturi units corrected the emissions problem."

Schwartz says that air and water pollution control is being solved on a continuous basis and most Modine plants do not present any major problems. While the Clinton, Tenn., plant processes aluminum, most other plants are working with copper and brass. He said that by the end of 1974, the company will have achieved 85–90% control at all plants.

How the scrubber works

In operation, gas effluent from the aluminum bonding oven enters the top of the scrubber at the rate of approximately 35,000 acfm while water is injected at 200 gpm. Both water and gas speed through the highly constricted venturi throat and are thoroughly mixed as the water passes over a small turbulent producing "knee" projection around the periphery of the venturi throat.

Each 16-ft high scrubber is fabricated of 3/8-in. thick fiberglass-reinforced plastic using a bisphenol-type polyester with Dynel veiling. A 12-in. thick entrainment separator is set approximately 6 ft above the base of the scrubber and is packed with a 1-in. diameter Tellerette plastic packing. The entrainment separator serves as a second stage scrubbing operation.

Modine's Schwartz
Complying with effluent limitations

A circular spray header located below the separator plays a constant overlapping pattern of fresh water up into the packing to keep the packing clean of lime scale and to give further absorption of HCl gas. Passage through the packing provides additional contact time between the water and the gas stream, and the unique shape of the filamentous Tellerettes provides the final absorption of the last traces of hydrogen chloride and removal of entrained liquid particles down to those submicron and larger in size.

Another spray header located

Ceilcote venturi scrubbers clean emissions from an automotive air conditioning unit manufacturing plant in Tennessee; they control gaseous, liquid, and solid emissions

above the packing section can be used to spray acid, when the scrubber is not in operation, to dissolve accumulated scale.

The inlet section of the venturi utilizes an underflow weir for liquid distribution, an arrangement which eliminates the possibility of dry spots leading to solids buildup, which overflow weir or sprays may induce.

When water and gas drop to the bottom of the scrubber, the water falls into a sump area while the air, containing some entrained liquid droplets, makes a 180° turn and is drawn back up through the entrainment separator.

This complete turnaround of the air stream causes most of the entrained liquid particles to drop out in the sump. Cleansed air is pulled through two 32-in. diameter exhaust pipes (at 16,000 acfm each) at the top of the scrubber, then to the base of a 40-ft high exhaust stack.

Exhausted air is approximately 99% free of pollutants and produces

a barely visible steam plume that dissipates itself within about 20 ft of the stack, depending primarily on the difference between ambient temperature and the heat of the stack discharge.

Passage of the gas through the scrubber system is handled by fans downstream of the scrubbers which are powered by 250-hp motors. Each fan develops 25,000 scfm at 34 in. w.g. of pressure. Fan blades are 4 ft in diameter and are of mild steel construction with a baked phenolic coating.

In-plant water reuse

At this Modine plant, water does double duty. The same water used in the manufacturing process is also used in the pollution control system. At Clinton, approximately 4.5 million gal of water are maintained at pH 7.5 in a 2.5-acre, man-made pond located about 100 yd from the plant.

From the pond, the water is pumped to the plant where it is utilized first as a quench for cleaning the fin and tube assemblies after the bonding process. The water is then pumped to the oven hoods where it serves as a coolant and lowers discharge gas temperature from 800°F to 140°F.

From the oven hoods, the water (which by now has picked up considerable acid) is gravity fed to a rapid mix tank where a lime slurry is automatically added to bring the pH of the water up to 9. From there, the water is pumped to the top of the scrubbers where it begins its work as a gas cleaning agent.

Heavily laden with gas and solid pollutants, the water drops to the bottom of the scrubbers.

A 6-in. diameter drain set 16 in. from the bottom of the scrubber creates a 16-in. deep water sump. The velocity of the air stream in the system keeps the sump water in a continuous state of agitation, which precludes particulate matter from settling in the bottom of the scrubber. Agglomerated particles and water overflow from the sump, through the drain, and flow by gravity to a 16-ft diameter concrete clarifier, where a sludge raking blade rotates at $\frac{1}{8}$ rpm and helps move sludge, settled along the sloped floor, to the center of the clarifier overflow. They then drain back into the pond at a slightly alkaline pH of 8.0 virtually free of any pollutants, except some solids that settle in the pond, and long before reaching the water intake for recycle.

Breakdown of pollutants and control costs

Aluminum hydroxide } Aluminum chloride }	65 lb/hr
Hydrogen chloride	50 lb/hr
Particulate matter	50 lb/hr

	Costs
Scrubbers[a]	$ 30,000
Other major equipment[b]	50,000
Installation of all equipment	205,000
TOTAL	$285,000

[a] Including fans, motors, starters, lime system, clarifier drive, pumps, pH instrumentation, and stack.
[b] Including concrete pads, piping, electrical and structural supports.

How water is recycled

pH 9.0

Lime for pH control

Scrubber

pH 8.0

Clarifier

400 GPM

Pond

To river less than 50 GPM

Rapid mix tank

Oven 800°F

Quench for product cooling

49

Reprinted from ENVIRON. SCI. TECHNOL., **7**, 988 (1973)

The business of air pollution control

ES&T's Stan Miller reports on

activities in this growing field

The time is good and ripe for the air pollution control industry. It certainly could be much better, but obviously, it couldn't be worse. There is no question but that federal legislation has had an impact on sales. Each year that new air pollution control laws came into being—both in 1967 with the Air Quality Act and again in 1970 with the Clean Air Act Amendments—federal laws triggered sales to new plateaus.

This special report cites the companies in the business and how their spokesmen view each company's role; the equipment suppliers and their trade association, the Industrial Gas Cleaning Institute (IGCI); and the technical experts and their professional association, the Air Pollution Control Association (APCA).

Together, this team stands ready to make progress on cleanup of a national vital resource—air. Several indicators point to the fact that business in this field will definitely increase in the near future:

• Legislation has been a very strong forcing factor.

• In 10 years (1971–1980), an expenditure of more than $106 billion, the largest expenditure for any environmental category, is needed for air pollution control alone, according to the Council on Environmental Quality annual report.

• In 1973, American business will spend $3.6 billion to control air pollution, up considerably from previous years, according to a survey released this May by the McGraw-Hill Publications Co.

• The renewed challenge: an unrelenting deadline for clean air by the middle of this decade with achievement of the national air quality standards (protective of human health) by July 1, 1975.

But what perhaps can be misleading to the casual air pollution control watcher is that business opportunities in this industry lie somewhere between the optimistic, pie-in-the-sky

CEQ figures, the hard-fact booking figures of the IGCI, and actual shipments of control equipment documented by the U.S. Department of Commerce from Census Bureau data. Obviously, somewhere within these limits, industry leaders view a future.

Total environmental companies—such as Envirotech, Combustion Engineering, Research-Cottrell, Peabody-Galion, Wheelabrator-Frye, and Zurn Industries—have capabilities in all segments of environmental control including, of course, air pollution control. On the other hand, a company such as American Air Filter deals only in air. Each of these companies can handle all aspects of engineering design, equipment, and construction and erection of control systems for virtually any air pollution emissions problems.

In contrast with the water pollution control business (*ES&T*, Nov. 1972, p 974) which is in both industrial and municipal markets, the business of air pollution control (apc) is mainly industrial, the exception, of course, being municipal incinerators.

In 1972, apc markets were ranked in the following decreasing order: electric utilities, rock products (including cement), steel, chemical, and petroleum refining. Utilities accounted for approximately 40% of the total market. And Envirotech's vice-president Hugh Mullen noted that there does not seem to be much reason to predict a change in ratings, only an increase in all markets.

What these companies provide—the four traditional apc devices, electrostatic precipitators, fabric filters, mechanical collectors, and wet scrubbers—is what industries need to control emissions. Basically, this equipment is used for particulate matter (soot and fly ash) removal. Some companies handle all four types of equipment. Gaseous pollution control devices—SO$_2$ removal systems for stationary source and

catalytic devices for mobile sources —are in their infancy. Control of the five gaseous pollutants, along with particulate matter, is required by the national air quality standards.

Other companies provide ancillary equipment such as fans, blowers, duct work, instrumentation, motors and cooling towers, all of which are necessary to ready the control devices for operation.

Without question, electric utilities in this country are a main target for SO$_2$ and particulate matter control systems. Conservatively, there are about 2000 power plants in the U.S. and about half of them, including industrial boilers, must control emissions.

But we soon learn that SO$_2$ control systems are simply not available, at least the variety that has a performance guarantee for continuous operation, 24 hr per day, 7 days per week, for a full year. Without this type of guarantee, utilities simply are not rushing in to purchase such control systems which, incidentally, are quite expensive. One economist estimated that stack gas cleanup for utilities will cost from two-thirds to one-and-one-half times the original cost of the plant, not including operational costs.

The situation with regard to SO$_2$ control was put in perspective by Research-Cottrell's chief executive in an interview (*ES&T*, Jan. 1972, p 16). "Everything has to be done quickly these days," John Schork told *ES&T*. "We had 50 years to de-

velop controls for particulates; we've had only four years for SO₂ controls!"

Nevertheless, the utilities and apc companies have been evaluating promising SO_2 control systems for a number of years. There are well over 100 so-called joint demonstration projects under way. In any case, it doesn't look as if SO_2 control systems will be commercially available till after the clean air deadline.

J. H. Oxley and associates at the Battelle Columbus Laboratories surveyed manufacturers, government officials, and utility owners. Performed for the American Electric Power Service Corp., their 1973 survey found that there are more than 100 stack gas treatment processes in various phases of development; several are being tested at the prototype level. The potential systems were discussed in detail by TVA's chief chemical engineer A. V. Slack earlier (*ES&T*, Feb. 1973, p 110). None of the systems has demonstrated proved reliability in the sense that it has met environmental regulations and has operated on a coal-fired plant of 100 MW or greater in the U.S. for a period of one year with an on-stream factor of even 50%; in other words, half the time.

Another look at SO_2 control device availability was taken by the federal interagency committee, the Sulfur Oxide Control Technology Assessment Panel. Its final report stated that processes will be available within the next five years. This panel

reaffirmed the expense of such control systems and noted that if stack cleaning costs are passed on, consumer cost for electricity could increase by as much as 17%. Nevertheless, SO_2 removal from stacks is mandatory.

In the business

AAF, **American Air Filter** (Louisville, Ky.), has been in the business of air pollution control for 48 years. Sales in 1972 hit a new high, $124 million, up from $116 million the previous year. During the next few years, AAF anticipates an annual average increase of approximately 15% in sales.

AAF offers all four types of air pollution control—electrostatic precipitators, , wet scrubbers, mechanical collectors, and fabric filters. Systems are engineered to accommodate practically any air cleaning need. However, to say the company has on-the-shelf equipment that can simply be shipped to a customer and installed would be a great oversimplification of the situation.

AAF specializes in the engineering, manufacture, and installation of complete air cleaning and handling systems. In such turn-key projects, company engineers perform a consulting service; however, the company does not engage in consultative engineering only.

Examples of air cleaning systems that AAF has engineered throughout the U.S. are numerous. Probably the most important, in terms of the na-

tional air quality standards, is the system now under way for Kentucky Utilities Co. The system is a mobile bed contactor scrubber system for removal of SO_2 and particulate matter from power plant effluent gases. Based on its achievement in a two-year pilot operation, AAF recently announced a $2.6-million contract for the full-scale system at the Green River Generating Plant in western Kentucky.

AAF's new system will permit the utility to use high-sulfur content coal from nearby western Kentucky fields and will go on stream in April 1975.

Another recently completed project involved the engineering, manufacture, and installation of an AAF Elex electrostatic precipitator for the Ideal Cement Company at its Portland, Colo., plant. These precipitators are being used by several cement plants to remove dust. Ideal, for example, with these precipitators, collects and returns to production 350 tons of cement per day, making air cleaning economically practical.

Other AAF projects include control of iron machining dust at a Caterpillar Tractor plant (Peoria, Ill.); emissions from an alumina kiln of the Reynolds Metals Co. (Corpus Christi, Tex.); and emissions from U.S. Steel Co.'s Duquesne Works. AAF also has provided air filtration products for the twin towers of New York's World Trade Center and Chicago's Sears Tower and Standard Oil Buildings. In municipal incineration, AAF has engineered several systems that will

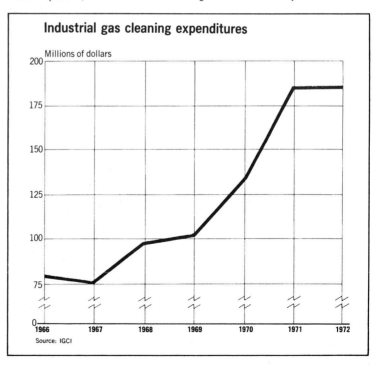

Industrial gas cleaning expenditures

Millions of dollars

Source: IGCI

meet or exceed local and EPA standards.

Also, the company maintains a complete research facility at its corporate headquarters in Louisville. Representative of AAF developments in the field of air cleaning is a nuclear containment system for nuclear power generating stations. AAF filters trap and hold any airborne radioactivity released during a system malfunction. AAF has also developed the only nuclear carbon filtering media currently meeting Atomic Energy Commission standards.

Summing up, AAF supplies systems for most applications, from small dust collectors to complete turn-key installations. It's not difficult to understand why the company's slogan has long been "Better Air is Our Business."

Aerodyne Machinery Corp. (Hopkins, Minn.) has a subsidiary, the Air Purification Methods, Inc., for the manufacturing, sales, product design and marketing of air pollution control equipment and systems. Among Aerodyne's products and services, the company lists fabric filters, centrifugal collectors, and turn-key.

Buffalo Forge Co. (Buffalo, N.Y.) produces and sells air-handling equipment—fans, blowers, and other ancillary equipment—with total sales of $54 million in 1972. Buffalo offers an Aeroturn Fabric dust collector, for example, which removes wood dust from a 150,000-cfm sanding operation in a board manufacturing plant. Another Buffalo Forge fan draws 300,000 cfm through a wet scrubber which cleans air from a sintering process in a steel mill. Blowers are used to boost the suction required to exhaust industrial process air which in another particular case has been cleaned in a high-energy venturi scrubber in a chemical plant. Perhaps not a true apc company, Buffalo nevertheless produces products necessary to the erection of apc systems.

Combustion Engineering, Inc. (C-E) (Stamford, Conn.) had sales of $1.18 billion in 1972, up from $1.073 billion the previous year. It, too, is a supplier of environmental control equipment, systems, and services in all areas of environmental controls. Somewhere between 5–10% of these sales could be characterized as environmental equipment.

C-E is a manufacturer of heavy equipment for many markets. Principal among them are the utility industry; paper and pulp industry; steel industry; stone, clay, and glass industry; and petrochemical industry. The company also has architect engineering capabilities in its C-E Tec organization and industrial design and construction capabilities in its C-E

Air pollution control equipment

Bookings have more than doubled over a 6-yr. period . . .

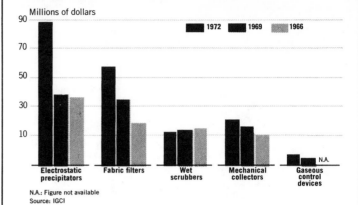

N.A.: Figure not available
Source: IGCI

. . . but actual shipments only show 50% increase over a 4-yr. period . . .

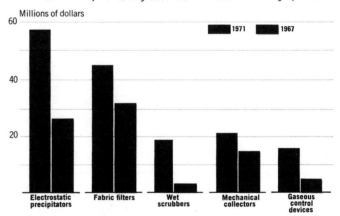

Source: Dept. of Commerce, Social and Economic Statistics Adm.

. . . as backlog of orders hits $139 million as of December 31, 1971

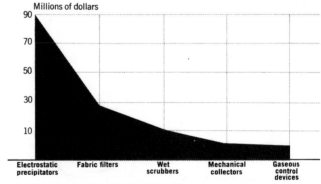

Source: Dept. of Commerce, Social and Economic Statistics Adm.

Lummus organization. C-E's Combustion Division developed an SO_2 removal system for the utility market, and elsewhere within the corporation, the Air Preheater Co. makes and sells fabric filters, wet scrubbers, mechanical collectors, fume incinerators, and small incinerators to the industrial market, whereas C-E Bauer and C-E Raymond make equipment for the waste water treatment market.

C-E's top volume sales for the year 1972 in the environmental area was in the field of large industrial boilers that incinerate waste products and produce steam for the related industrial complex. The area of incinerators of one type or another is the equipment that had the greatest sales in 1972. However, C-E architect engineering firms have many large dollar value contracts in waste water treatment plants, and the Combustion Division bookings of flue gas desulfurization systems were top money items.

The St. Louis/Union Electric refuse burning in a utility boiler demonstration incorporates basically a C-E refuse disposal system; the boiler innovations are entirely C-E, although within the responsibility of the architect engineer. C-E has many other contracts with cities for studies that project steam generation from prepared refuse. C-E's Combustopak, a modular 3-ton per hour municipal incinerator, has been installed in the city of Reading, Mass., and has met all the air quality regulations, both Federal and State.

C-E has a SO_2 removal process too. In fact, its Combustion Division had an order for its 10th SO_2 removal system at press time. It is a tail-end line/limestone wet scrubbing system for SO_2 removal. It has been installed at a 65-MW coal fired steam generator at the Louisville Gas & Electric Co.'s Paddy's Run No. 6 unit. The system will remove more than 80% of the SO_2 from the flue gases of the steam generator that fires bituminous coal, 3.5% sulfur, 11.5% ash content, with a heating value of 11,500 Btu/lb. Calcium hydroxide is the additive being used to remove the SO_2 in the wet scrubbing.

C-E is also providing an $8 million SO_2 particulate removal system for Southwestern Public Service Co. (Amarillo, Tex.). The new Nichols Station B, a 350-MW coal-fired controlled circulation steam generator (also ordered from C-E) is scheduled for operation in September 1975. A slurry of lime will be used to remove SO_2 from the steam generator's flue gas by wet scrubbing.

The company has a research staff consisting of 1196 professional R&D people but does not have a testing

subsidiary devoted to actively verifying unit performance as such. (In many instances, an outside testing concern is required by law to determine a unit's compliance with legislative regulations.)

The **Chemico Air Pollution Control Co.** was announced this June. It is an outgrowth of the Air Pollution Control Division of the Chemical Construction Corp. (New York, N.Y.), a wholly owned subsidiary of the Aerojet-General Corp. (El Monte, Calif.). Total sales for 1972 were approximately $35 million for engineering, design, and equipment supply (excluding construction). A major portion of the sales was for fly ash and SO_2 control in the electric utility industry.

The company's lime-scrubbing technology for SO_2 control has been demonstrated in a full-scale system that has been providing trouble-free operation for more than one year at

the Mitsui Aluminum Co. Ltd's Miike Power Station in Ohmuta City, Japan. The Chemico process, a regenerative, recycle SO_2 process, is now progressing from proved process chemistry to development commercial criteria, such as reliability, availability, and economics emphasizing cost reductions.

Basically, the Chemico method uses magnesium oxide as the flue gas scrubbing agent, recovering the magnesium for reuse and also producing commercially useful by-products such as sulfuric acid and sulfur. The process started in the Chemico R&D center as a bench-scale project in 1967. It was field tested in 1969 and placed in full-scale service in 1972. The strength of its R&D center varies from 20–50 people, depending on the specific active projects.

Chemico has several units in operation or under construction for major utilities including Boston Edison Co., Penn Power and Light, Arizona Pub-

lic Service Co., Duquesne Light Co., Potomac Electric Power Co., and the Penn Power Co.

Virtually all of Chemico's sales go to the industrial market, and although Chemico did provide an emission control system for a municipal incinerator, the company does not see heavy involvement in this area in the near future.

In the past 20 years, the company has developed sophisticated gas scrubbing systems to control emissions from the various operations in the steel industry. More recently, Chemico participated in development of process techniques to remove particulate matter and recover carbon monoxide as a usable fuel (O-G System). This system also has the capability of increasing the yield and productivity of a basic oxygen furnace.

In 1973 (year ending March 31) sales for **Combustion Equipment Associates** (CEA) (White Plains, N.Y.) were $66 million, up from $50 million the previous year. Environmental products accounted for $19 million (29% of the total) and environmental energy systems $35 million (53%).

Within CEA's Environmental Systems Division, an automation and instrumentation group reduces pollution by regulating the process. An emission control devices group concentrates on scrubbers, fabric filters, and mechanical collectors to remove the pollutants from the combustion effluents.

CEA utility stack gas cleaning systems under contract represent a total capacity of 1285 MW and a value of $27 million. SO_2 fly ash removal systems are slated for the Nevada Power Co., the Gulf Power Co. (40 MW), and joint ventures of Montana Power Co. and Puget Sound L&P.

Envirotech Corp. (Menlo Park, Calif.), one of the total environmental companies, recorded sales of nearly $176 million for 1973 (year ending

March 31), up 21% from the previous year. Its Air Pollution Control Group accounted for $30.4 million in industrial sales, up 77% from the previous year for this activity and representing 17% of the corporate total.

Envirotech's apc group, formed in 1972, consists of the Buell, Norblo, and Arco Divisions. It manufactures all the major types of equipment used to control air pollution from stationary sources. Buell manufactures electrostatic precipitators and cyclones to remove particulate matter from air streams. Arco manufactures gas scrubbing systems, and Norblo produces a full line of fabric filters with one exception—Envirotech is not involved in the pulse jet fabric filter business.

Envirotech is also in the great SO_2 apc race; the company announced its double alkali scrubbing system at the American Power Conference in Chicago this May. Since early 1972, this system has been operating on a 3000-acfm gas stream (equivalent to less than a 10-MW power plant) 24 hr per day, 5 days a week, on the Gadsby station of Utah Power & Light Co. (Salt Lake City). No evidence of scaling has been observed.

Envirotech is ready to offer its double alkali process to electric utilities for application on boilers burning low-to-medium sulfur content coal. And since the process is also applicable to industrial boilers, the company is looking for an opportunity to demonstrate the system in an industrial situation.

The electrostatic precipitator, Envirotech's leading apc product, will continue to be so for several years. The company anticipates a high dollar volume of precipitator sales for at least five years. Fabric filter sales also are growing rapidly; the company projects that among its equipment lines, this product will grow fastest. In another few years, its SO_2 system will become a major item, the company anticipates.

At press time the company had acquired The Bahnson Co. (Winston-Salem, N.C.), whose specialty is in-plant control of air in textile, tobacco, food processing, and tire and rubber manufacturing plants, among others.

A subsidiary of the Riley Co., **Environeering Inc.** (Skokie, Ill.) manufactures wet scrubbers and anticipates being in all major markets for these control devices including utilities and foundries. The company offers a limestone scrubbing system for SO_2 removal and venturi scrubbers for small particles. With some 1200 scrubber installations in the U.S., this company anticipates major innovations in the field of wet scrubbers.

An equipment company, **Joy Man-**

ufacturing Co. (Denver, Colo.), has 99% of its sales going to industrial markets—power (SO_2 controls), steel, chemical processing industries, and pulp and paper industries. The company also provides spiral degritters for controls on municipal incinerators. Its top dollar volume items for 1972 were pumps and clarifiers which are used on SO_2 systems on power plants and scrubbers on cement, sand, and gravel plants.

Showcase examples include a lead oxide removal system on a battery line of the Gates Rubber Co.; pumps and clarifiers for an SO_2 pilot plant of Southern California Edison; and pumps for SO_2 control systems for Arizona Public Service, Detroit Edison, and Nevada Power. The company does no research, nor consulting, but does provide testing services. Joy's Western Precipitation Division is busy with SO_2 control systems for electric utilities.

Peabody-Galion Corp. (New York, N.Y.) is another total environmental company. Peabody's President John McConnaughy told *ES&T* in an interview (*ES&T*, Feb. 1973, p 97) that 70% of Peabody's revenue is derived from a balanced position in all segments of pollution control—air, water, and solid waste, and testing services. Sales in 1972 were $134 million, 25% of which relates to its Air Pollution Control Group. Peabody has more than a thousand employees in its air group which accounts for the largest percentage of group sales, $33 million in 1972, up 25%.

In practically every line of air pollution controls with a notable exception of electrostatic precipitators, the Peabody air group entered fiscal 1973 with a $25-million backlog. Activities of this group concentrate in eight apc markets including SO_2 controls and particulate removal from power plant stack gases, incineration systems for liquid and solid waste generated by industry and municipal-

ities; combustion equipment conversions, in-plant cleaning equipment, dry dust collection systems, gas-inerting systems for oil tankers, process cleaning for industries such as pulp and paper, steel, chemicals, and other metals and food; and burner management and fuel preparation.

Operating within the air group, Peabody Engineering (Stamford, Conn.) has 52 years of experience in the field; a main activity is in the scrubber area. Peabody also has designed and engineered fly ash–SO_2 removal systems. Peabody's limestone scrubbing process has been installed at Detroit Edison's St. Clair generating station and started operation this fall. Under a joint agreement between Pfizer, A. D. Little, and Peabody, the company is checking out the commercial feasibility of the Bu-Mines citrate process for SO_2 removal at Pfizer's chemical plant (Terre Haute, Ind.).

Peabody is also building and installing one of the largest wet-scrubbing systems in the U.S. for a Maryland utility, Potomac Electric's Chalk Point plant. The company also has scrubbers work under way for other industries including paper, steel, and chemical plants.

Fabric filters, an area which Peabody entered in 1970, showed a strong increase in 1972. Now available, its reverse-air Lugar model has been installed atop the storage elevators of a large Northwestern grain grower to control dust emissions. Last year Peabody introduced the pulse-jet model and acquired American Brattice Cloth Corp., a manufacturer of filtration fabrics. The company also offers in-plant Ven-Kinetic air washers for the removal of airborne particles in manufacturing operations. A new aerosol model is said to be the only in-plant equipment capable of removing ultra-fine aerosols. **Research-Cottrell** (R-C) (Bound Brook, N.J.), another heavyweight in

the air pollution control business, is also a member of the total environmental pollution control club. R-C has been in the air pollution control business for more than 60 years. Sales in 1972 hit $133 million; its air pollution control division accounted for $67 million.

As in the past, electrostatic precipitators were by far the dominant equipment item in their sales, representing approximately 77% of the apc division sales; fabric filters (baghouses) ran second at 12% sales.

R-C has a complete line of equipment and products including, in addition to the four well-known devices for particulate matter controls, thermal incinerators, catalytic converters, and catalysts. What is more, the company performs air emission source testing, contracts for R&D, and has an air pollution consulting service subsidiary—Cottrell Environmental Sciences, Inc. With a staff of 43, this subsidiary is engaged in research and development for the operating divisions and performs contract research as well.

R-C typically has spent approximately 2% of sales for activities devoted to the evolution of new products and services. The company currently is in the process of introducing the hot precipitator and two SO_2 control systems; R-C sees nothing in the near future in terms of radically new concepts that would prove practical in commercial-scale operation. Development of pollution control techniques is largely evolutionary; R-C thinking is directed toward further development of known technologies.

Examples of R-C operations are numerous. Air pollution control watchers, for example, are keenly interested in its large-scale system for SO_2-particulate matter control which was due to be operational at press time. Installed on a 115-MW generating unit of Arizona Public Service Co., this system has separate provi-

sions for fly ash removal (by wet or dry means, depending on the individual circumstances) and SO_2 absorption by a slurry of lime, limestone, or other alkali. It is important to point out that this process is still a demonstration rather than a commercial installation.

Thus, R-C's major contribution to the technology of SO_2 scrubbing is a wetted-film absorption tower packing which has high specific surface, low-pressure drop, and high liquid rate availability. Little deposition of solids occurs within the absorption tower, thus minimizing the plugging problem that has plagued calcium-based SO_2 scrubbing systems.

In the area of small particulate matter collection, the latest R-C development is the hot precipitator. Performance of the hot precipitator is essentially independent of the coal sulfur content of the coal and the presence of sodium, magnesium, and other constituents in coal that must be considered when designing precipitators to operate at conventional inlet temperatures.

Thus, performance of these precipitators will remain high even though fuel changes may occur during the installation's lifetime. In this day of fuel conservation, this versatility is necessary since a utility cannot be sure at the time of specifying a precipitator what fuel will be burned during the lifetime of the installation.

Obviously, with a large company like R-C, there are many other examples of users. The R-C precipitator installation at Montreal's Des Carriers incinerator is believed to be the first—certainly it is the largest—for steam-generating incinerator furnaces on this side of the Atlantic.

At a large eastern steel plant, R-C was called upon to upgrade 18 older and lower-efficiency precipitators on nine open-hearth furnaces from which stack emissions exceeded current code requirements. Rather than replacing the existing units, R-C engineers took advantage of the older precipitator's capability and manifolded the effluent gas from them into a secondary system, two larger double-section precipitators and a new single stack.

Big as a water pollution control equipment company, **Rexnord Inc.** (Milwaukee, Wis.) entered the air pollution control equipment field in 1972. The company offers a gravel bed air filter unit which can be used under very dusty, abrasive, and such hot industrial conditions as in cement plants, lime kilns, refractories, and foundries. When the company entered the field, Rexnord Chairman William Messinger said, "More than 100 major installations in Europe and Asia have proved the effectiveness of

the system under a variety of difficult conditions. There are more than 500 cement plants in the U.S. and more than 80% of them could use the technology."

The Torit Corp. (St. Paul, Minn.) produces fabric filters and mechanical collectors principally for industrial markets. This company sells a full range of fabric filters—manual, semiautomatic, and automatic. Although Torit does not provide consulting services, it does maintain a small research staff and has the capability of testing installations for emissions.

In 1972, sales for the **United States Filter Corp.** (Newport Beach, Calif.) hit $123 million, up from $92 million the previous year. Its air pollution control group accounted for about one third ($38 million) of the total; backlog orders for the division were $12.8 million, up 30% from the previous year. Another one third of their business is in water treatment and pollution control, and a final one third is in energy and nuclear engineering.

The company's air pollution control group is made up of the old Mikropul Division of the Slick Corp., the Ducon Division of the old U.S. Filter, and that part of the Menardi-Souther fabric filter operation devoted to filters for dry applications. Their line includes virtually every apc device including fabric filter baghouses, mechanical collectors, cyclone separators, wet and dry electrostatic precipitators and wet scrubbers.

With 1972 sales of nearly $0.5 billion, UOP, **Universal Oil Products Co.** (Des Plaines, Ill.), announced plans late last year to build a multimillion dollar plant near Tulsa, Okla., to make catalysts to control exhaust emissions from automobiles. This March, Chrysler placed a contract with UOP for 100% of its catalyst requirements beginning with the 1975 model year vehicle. UOP construction is scheduled so that commercial quantities of the catalysts can be delivered by April 1974 in time to meet 1975 model production needs of automobile manufacturers.

UOP also has a Mini-Verter, a device for improving the quality of exhaust emission from vehicles already on the road; a number of states are testing and evaluating this device for possible use in retrofit programs.

Procon, Inc., the design engineering and construction arm of UOP, was selected to devise and recommend a complete SO_2 removal process for the Navajo Generating Station, a 2.31-million-kW station now under construction near Page, Ariz. This work is being performed by UOP's Air Correction Division (Darien, Conn.) and Corporate Research.

UOP is also designing and constructing a stack gas cleaning system for Commonwealth Edison Co. at its State Line Station (Hammond, Ind.). This station will use the UOP Sulfoxel process, a wet process, which will permit the utility to burn high-sulfur fuel. The process removes SO_2 from the flue gas and recovers elemental sulfur.

With the largest backlog in its history, the Air Correction Division is also building the world's largest device designed to mechanically remove fly ash from the flue gas of an oil-fired boiler at Virginia Electric and Power Co.'s station (Yorktown, Va.).

UOP also designed and built twin electrostatic precipitators to remove fly ash material from a Westvaco bleached board plant (Covington, Va.). It is said to be the first successful application of a precipitator to a coal–wood bark burning boiler. Particulate emissions were slashed from several tons per day to 200 lb. In addition, UOP has all the refining processes needed by the petroleum industry to produce leaded, low-lead, and lead-free gasolines.

Wheelabrator-Frye, Inc. (New York, N.Y.), another heavyweight in the air and total environmental controls club, recorded sales of $174 million in 1972, up 18% from the previous year. Wheelabrator International Inc. was formed in 1972. W-F, an old timer in air pollution control, installed its first air pollution control system on a lead smelter in 1913 to capture particulate matter. Today, its Environmental Systems Group (Pittsburgh, Pa.), with a team of more than 1000 engineers, technicians, and support personnel, accounted for $85 million of the 1972 corporate total. The company offers all four generic types of air pollution control devices as well as high-energy fans and blowers.

In 1972, the company obtained its first major electric utility contract valued at more than $8 million for a Wheelabrator Lurgi electrostatic precipitator system for a joint venture of Northwest Public Service, Otter Tail Power, and Montana-Dakota Utilities Co. at Big Stone, S.D.

Other examples of Wheelabrator Lurgi precipitators include:

● cement kiln protection at the Louisville Cement Co.'s Seed, Indiana, Brixment kiln, the first W-F system on a cement kiln operation

● dolomite lime kiln at St. Joe, Fla., magnesium oxide plant of Basic, Inc. collects 237 lb an hour of extremely fine fluffy powder from a dolomite lime kiln, again another example of making air pollution controls economical. The system cleans 100,000 cf of air per min.

● cleaning pitch from the air in Revere Copper & Brass anode manufacturing operation (Scottsboro, Ala.)

Wheelabrator also has showcase examples of fabric filter installations:

● One system of 2112 fabric filters collects 5 tons of particulate matter every 24 hr from the silicon–chromium electric furnaces at the Chichibu Works of Showa Denko Co. (Japan).

● Another is for removal of dust fibers from asbestos processing. At the Johns Manville's Jeffrey Mill (Quebec, Canada), a total of 75,000 filter tubes and 30 fans move 4.5 mcfm of air every minute. The air is clean enough to be recirculated through the mill.

● A third is for recovery of clay, some 7000 lb every hour at the Freeport-Kaolin plant in Gordon, Ga.

Zurn Industries, Inc. (Erie, Pa.) is a total environmental company whose sales hit $196 million in 1973 (fiscal year ending March 31), the best year in Zurn's history. Sales, earnings, and backlog hit record highs. Approximately 78% of sales consisted of orders for the company's environmental pollution control systems and services. Its air and land pollution control (which includes solid waste activities) accounted for $45 million total sales.

Zurn designs, markets, manufactures, and installs a wide range of air pollution control products and systems with emphasis on turn-key approaches. The company also provides air management services through its environmental engineering components ranging from initial feasibility studies to final monitoring of the installation.

Major products and systems include mechanical dust collectors, fabric filters, high-energy wet scrubbers, SO_2 scrubbers, waste heat energy recovery systems, blowers, fans, dampers, exhausters, ventilators, dryers, and controls. Approximately 80% of its sales are to industrial markets, the remainder going to commercial and institutional markets, municipal waste treatment, and solid waste disposal.

The company doesn't make electrostatic precipitators. As Frank Zurn told *ES&T* readers in an interview (*ES&T*, June 1972, p 496), "We are able to buy such equipment from a half-dozen sources." The company views the scrubber business as a better opportunity and projects that the top dollar volume sales items will be SO_2 scrubbing systems. Over the past three years, sales of these devices have shown excellent growth; the trend should continue, the company says. Zurn SO_2 scrubbing systems are installed at the Key West

Power & Light Co. (Key West, Fla.) and the Caterpillar Tractor Co. (Joliet, Ill.).

In 1972, a first Zurn baghouse filtration system, an important addition to Zurn line of products, went on stream at the Sorg Paper Co. (Middletown, Ohio). The control system was installed on three pulverized coal-fired steam generators of the paper company. In the past, baghouses were not used on pulverized coal systems, because until now they have not been rugged enough to operate for the required 18-month period without downtime. Nor were they used to operate in a gas stream where sulfuric acid is likely to be formed. In this case, the system cleans the exit flue gases from 60,000 lb of steam per hour pulverized coal-fired steam generators.

Other Zurn showcase systems include controls for Incineration, Inc. (Chicago, Ill.); San Diego Gas & Electric (Calif.); Mistersky Generating Station (Detroit, Mich.); Noranda Aluminum Co. (Madrid, Miss.); Kimberly-Clark (Coosa River, Ala.); and Public Service Co. (Platteville, Colo.).

In addition to electric utilities which are a market for all apc companies, Zurn is involved in air pollution controls for a variety of various industrial manufacturing plants including metals processing, paper, chemical, and petroleum plants.

The company also has an environmental testing installation, Enviro-Engineers, Inc., that checks actual installation and verifies emission results for state, local, and federal air pollution control inspectors. Looking ahead, the company has contracts totaling $25 million in negotiation; contracts being finalized are in the order of $5 million.

Associations

With a membership of 6500, the Air Pollution Control Association (Pittsburgh, Pa.) is a nonprofit technical and educational organization dedicated to advancing the science and art of air pollution control. Organized in 1907, APCA is divided into 18 geographical sections throughout the U.S. and Canada; members also reside in 40 other countries throughout the world. APCA works toward international adoption of reasonable engineering performance standards. It also seeks to establish definitions, methods, processes, procedures, and to recommend practical limits of air pollution emissions.

The national association of equipment manufacturers of industrial air pollution control equipment is the Industrial Gas Cleaning Institute (Stamford, Conn.); member firms sell more than 75% of all industrial air pollution control and dust control equipment purchased in the U.S. During 1972, its 29-member companies sold equipment totaling $175.6 million in the U.S. and $10.5 million in Canada.

It is important to point out that these sales figures do not include any field construction costs associated with the equipment. Nor do IGCI figures include costs of auxiliary equipment such as foundations, duct work, instrumentation, cooling towers, fans, and motors. Often, expenditures for these auxiliary items are included in the gross figures compiled by other sources.

IGCI spokesman and technical director Sidney Orem conservatively estimates that a good average multiplier (to the IGCI sales figure) might be three to arrive at the actual cost to an owner for ready-to-operate units.

For a list of individual companies that supply any one or more of the specific equipment, the reader is referred to the 1973–74 ES&T Pollution Control Directory, published this month and available from ACS Special Issues Sales.

Reprinted from ENVIRON. SCI. TECHNOL., **8,** 199 (March 1974)

American Air Filter's Jesse Shaver

What is the sales breakdown for AAF business? Since its formation in 1925, AAF's main thrust has always been toward designing and manufacturing products and systems for control of environmental air. Today, AAF's general activity encompasses air pollution control, air filtration, noise pollution control, and air conditioning, heating, and ventilation. The company keeps no separate figures for these activities because it frequently occurs that orders will include air filtration equipment and air pollution control equipment as well as heating, ventilating, and air conditioning equipment on the same job. Sometimes noise pollution control is also included. Because these activities are intertwined, sales are not broken down into separate categories by function. There is no meaningful method of separating orders by types of products.

What is the fastest growing segment of the business? All categories have had substantial growth and it appears they will continue to expand in the future. My guess is that what is defined as air pollution control probably is growing faster than the others. But it is not a runaway race, other areas are moving ahead as well. With a field like noise pollution control, which started from a small base, the rate of increase is quite good. In air conditioning, recent developments in energy-conserving systems have produced good growth.

As to air filtration, a little background might be helpful here. Air filtration has been on a growth curve of its own since the mid-1920's. At first, AAF predecessor companies had difficulty in establishing the concept of air filtration. Prospects found it difficult to understand the advantages in relation to the cost. In 1929 when the AAF name came into the field with the consolidation of four companies, there were two major classes of customers—automobile finishers and department stores. The former were interested because they had to apply as many as a dozen coats of finish. It was necessary to keep surfaces dust-free during the slow drying of each coat.

Department store owners were quick to recognize that air filtration reduced merchandise soilage, allowing it to be kept on shelves longer. In such stores air filtration systems could pay for themselves in a matter of months. Gradually others began to recognize the advantage of good air filtration systems.

Now, the basic advantages of air filtration are beginning to be more generally recognized. Today, air filtration systems are being used in food processing, textiles, clean rooms for electronics, and even motor rooms in steel mills. However, there are still large buildings being constructed today with inadequate air filtration systems. As you can see, AAF is involved in many facets of air quality control. However, recognizing the interest of your readers in air pollution control, I would like to channel my remarks toward the air cleaning segment of our business.

What is the rationale for industry to clean up in light of the current energy crisis and resources considerations? I believe there is a fallacy current with the public in relation to air pollution versus the energy crisis and resources considerations. People appear to believe that air pollution control is going to be relaxed and that standards will be abandoned. This is erroneous. The pressure for

and the needs to correct air pollution will not change. Although the pollution controls most familiar to the public, the automotive pollution controls, reduce efficiency in relation to the energy consumed, the same is not true of industrial pollution controls. In this latter area, there is relatively little sacrifice of operational efficiency by the installation of controls. Indeed, in some cases, you can recapture otherwise wasted resources which can then be recycled. As an example, AAF has a system allowing steel mills to collect gases generated by oxygen steel conversion, then recycle the gases as fuel.

I personally do not see much future relaxation in the codes. The possible exceptions are one or two areas where changes would have been required in any event. One is the previously mentioned automotive area. However, this is not of AAF interest.

Another area where there may be some modification, one which is of definite interest to AAF, is the SO_2 requirement for electric utilities and large industrial fossil-fuel burning plants. From the time the codes were published, many people in the air pollution control industry believed it would be virtually impossible to meet all the requirements under the established timetables. Consequently, it

Jesse M. Shaver is president and chairman of the board of American Air Filter, one of the world's largest suppliers of systems and technology for improving the quality of air. The 54-year-old executive tells ES&T's Stan Miller that AAF is an international company with 23 foreign subsidiaries, has been in business 49 years, and has doubled in sales since 1968. AAF's corporate motto, "Better Air Is Our Business," very well describes the company's basic and practically exclusive interest in improving air. AAF systems do nearly everything one can do to air— clean it, filter it, heat it, cool it, move it, silence it, and change its moisture content.

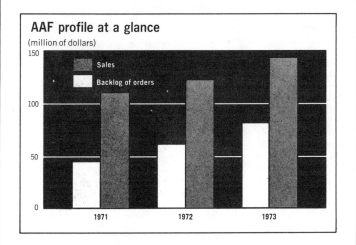

AAF profile at a glance
(million of dollars)

Legend: Sales, Backlog of orders

AAF milestones in air quality equipment

1973 Acquired Dynapure mist collectors from General Dynamics

Licensed for IRSID-CAFL system of gas collection for basic oxygen steelmaking

1972 Mobile bed contactors for SO_2 control

1971 Acquired Pulsco noise attenuation and pressure pulsation systems

1970 Acquired Elex electrostatic precipitators

1968 AMERpulse pulse cleaning cloth collector

1967 Arrestall self-contained dust collector

1962 Electro-Pak electronic air filter

1960 Environmental control systems for Minuteman missiles

1953 Introduced fabric collectors with AMER-jet reverse jet collector

1952 AMER-Clone centrifugal dust collector

1950 Acquired Herman Nelson air conditioning, heating, and ventilation systems

1939 Introduced electronic air cleaning with Electro-Matic self-cleaning electronic filter

1935 Introduced wet collection systems with Type W Roto-Clone

1932 Introduced dust collection in the U.S. with Type D Roto-Clone dust precipitator

1929 American Air Filter Co. formed from predecessor Reed and three other companies

1925 Reed Air Filter Co. established to manufacture air filtration equipment

---

appears there will be forthcoming modifications in the deadlines, allowing a more practical time schedule for accomplishing the requirements.

What sort of stretch-out do you foresee? I believe an orderly approach needs to be taken to the problem, one that is in keeping with the financial ability of the polluters and capacity of the air pollution control industry to provide the necessary systems. Preferably, we at AAF would like to see the market increase on the order of 15%, 20%, or 25% each year, rather than expanding at an explosive rate. Under the present timetables, the industry would have to expand immediately to something like an impossible 1000% or more each year just to meet the primary requirements. Therefore, AAF looks for a time lapse of at least the next decade or two before the principal air pollution problems are corrected.

CLEANUP PROJECTIONS

What then, is your prognosis for air pollution control cleanup progress over the six years remaining in this decade? In the past half dozen years many estimates have been made by research firms, trade publications, and associations, often based on imprecise figures. Frequently the same forecaster has doubled or tripled his estimate from one year to the next because of the very impreciseness of his previous figures. There still remain several problems when one attempts to estimate this market over the next few years.

One of the problems is that forecasters seldom take into consideration the amount of money required to do the job, as well as the source of those funds. Billions will be required. For example, in the area of SO_2 control alone, estimates of the

cost have ranged as high as $30 billion, and I do not recall seeing any less than $5 to $10 billion. In addition, one must also take into consideration potential materials shortages, and a very real long-term shortage of engineers and other competent people needed to handle this type of work. They have to be technically trained first, then gain on-the-job experience. I believe considerable progress will be accomplished over the remaining six years of this decade; however, the biggest part of the iceberg will remain to be done in the 1980's and beyond.

TECHNOLOGY

When will utilities begin to purchase SO_2 stack gas cleaning systems? This is an area in which disagreement has been rife for some time. The federal government is of the opinion that the technology exists, and it is eager to move the utilities rapidly toward installation of this equipment. And, of course, environmentalists are vocal in their support. On the other hand, electric utilities argue that proved, reliable systems do not yet exist. They point out that they are being pressed too hard and that the time is not yet right. There is truth in each position because, although there are exceptions, there is hardly any SO_2 system that presently operates on a continuous 24-hr basis, seven days a week. However, it is a fact that the technique for cleaning stack emission has been proved in the laboratory and in test installations.

Utilities are usually managed by prudent businessmen who are cognizant of their shareholders and are careful in their money management. They tend to require substantiated evidence that their expenditures are actually going to meet the needs for which the monies are allocated. All along I have felt that utilities would like to install a test unit and closely observe its progress. After they are confident that the unit can actually achieve its purpose they would proceed with further utilization, adding new units on a regular basis.

How well is the AAF full-scale unit proceeding at the Green River generating plant in western Kentucky? Have other AAF units been purchased by other utilities? The Green River project, a unit of Kentucky Utilities, is an outgrowth of the pilot plant test we operated at Louisville in 1971 and 1972 in cooperation with Louisville Gas & Electric and three other electric utility companies. The Green River project is to go operational in 1975. As for other utilities, there is considerable interest in plants with proved track records. Presently, we are in various stages of negotiation or discussion with more than 20 utilities. As I commented previously, the entire air pollution control industry should receive quite a few orders this year and an increasing number next year. We at AAF anticipate our fair share of these projects. We know what our system can achieve and are willing to contract on a guaranteed performance basis.

SHOWCASE

Can you cite a number of recent air pollution control projects in which the company is involved? AAF has many in our worldwide operations. The Soviet Union recently awarded AAF a $1.9 million contract for air pollution control equipment for what will be the world's largest foundry complex at the immense Kama River truck plant project in central Russia. We won the contract in competition with other international companies based on our experience, expertise and proven equipment.

During the past few months, we also received a $3.5 million project at TVA's Johnsonville steam plant. This involves fitting six large AAF electrostatic precipitators to remove fly ash from the exhaust of six boilers at that TVA plant. In other industries, we recently installed air pollution controls on a major steel mill, a large automotive plant, a major construction equipment fabricating facility, and in woodworking operations and cement mills. Because the need for better air is universal, AAF's systems virtually know no industry boundries.

RESEARCH

Does AAF have any new devices for control of air pollutants? Many of the air cleaning techniques now widely used by others were inventions of American Air Filter. All told, AAF has more than 1000 patents worldwide. Gradually there have been improvements in these techniques. Of course, AAF tries to keep ahead of its many competitors. However, rather than "new" inventions, the process is an on-going series of improvements in the techniques that exist.

Are there likely to be breakthroughs on air pollution control devices in the next few years? In the next six years, there will probably be no pure technological breakthrough. There will, however, be a steady progression of

improvements. In our industry, the target is continually to get increased efficiency at lower cost with improved reliability. It is in these areas where the improvements will come. It's a gradual activity which results in general overall improvement. It's unlikely that totally new techniques will be invented. AAF, as well as others, has done a great deal of research and experimental work on many different types of devices. However, one soon arrives at the question of economics. Although we developed some new techniques 25–30 years ago, economic considerations still do not favor their use, even today. Over the years others have done the same.

What makes a control device better today than 10 years ago? For example, what's new with electrostatic precipitators? In almost every instance, if you compare the performance of today's systems with those

"**AAF looks to at least the next decade or two before all the air pollution can be corrected.**"

AAF's Shaver

of 10 years ago you will find they will do better cleaning jobs in terms of efficiency, and they will continue to do their jobs over longer periods with considerably less maintenance. Today's market requires this sort of improvement; however, manufacturers like AAF are continually working to improve the efficiencies of their products in anticipation of market requirements.

As to the electrostatic precipitator, improvement has been along the same lines, increased efficiency and higher reliability. When AAF acquired the Elex precipitator in 1970, Elex already had an outstanding record of technological capability and reliability in the European market. In Europe, reliability has been far more important, and businessmen there are more insistent on performance guarantees than has been the case in the U.S. Consequently, there was more testing and guarantee of reliability there than was the practice in the U.S.

From that base, improvements have continued. There are two recent developments of significance. First is the unbreakable discharge electrode. By using a rigid, tubular assembly in place of the usual wire electrode, one of the major causes of downtime is eliminated—broken wire electrodes. The second major improve-

ment is the change to solid state controls with their inherent reliability and quick response.

What is happening in the area of gaseous emission controls? The principal gas receiving attention is sulfur dioxide. There are various devices, including afterburners and catalytic devices, for removing odors truly offensive to the public. Many noxious gases can be removed with wet systems. Some can be burned with catalytic converters, but converters represent a limited market. Carbon filters are also useful, but again, the market for these controls is in its infancy too. The use of fuel to handle odors is another method, but it is an expensive alternative. Too much fuel is required to heat up a large volume of air to cure a problem that may be caused by only a few parts per million.

How do you silence air? AAF is interested primarily in industrial noise pollution. There are many locations in chemical plants, for example, where the relief of pressures and vacuums causes intense screeching noises that must be eliminated. There are a number of techniques for eliminating them. Basically, these techniques cancel, contain, or absorb noise. Some work on the principle of diverting half the noise to a more circuitous path. That path is designed so that the portion of the noise traveling takes longer to reach the end point, arriving directly 180° out of phase with the other half. When the two rejoin, they cancel one another. Although it sounds somewhat simple, such systems are fairly sophisticated and require extensive knowledge in the field. There is also a market for such devices in the aviation business, where they are installed on the hydraulic landing gear systems to quiet the operation.

PERSONAL ASIDE

What is your background and how did you ascend to the top at AAF? After training as an engineer at Purdue University and as a MBA graduate from the University of Chicago, I worked for many years in the management consulting field with the firm of Booz, Allen & Hamilton in the area of electronics. I joined AAF in 1962. At first, I managed one of the groups and, after working on reorganization of that activity and bringing in another manager, I moved to another group. Over a period of time, I managed each of the major segments of the AAF business before I became president in 1967 and chairman in 1968.

Reprinted from ENVIRON. SCI. TECHNOL., **11**, 556 (June 1977)

Northeastern utilities are meeting the clean air challenge

By installing emission control equipment, building tall stacks,
burning lower sulfur fuels, optimizing air/fuel ratios
and reducing the operation of some units

**Joseph J. Cramer, Frank B. Kaylor, Edward J.
Schmidt, and Ernest R. Zabolotny**
*Stone & Webster Engineering Corp.
Boston, Mass. 02107*

The National Environmental Policy Act of 1969 and the Clean Air Act of 1970 have had major impacts on the utility industry, particularly in the highly populated and industrialized northeastern U.S. Yet despite the period of confusion that immediately followed the enactment of these laws, northeastern utilities have been able to achieve considerable control over emissions of combustion products, with significant improvement in the air quality of urban areas.

On a national scale, utilities, regulatory agencies, public interest groups and legislative bodies are now coming to grips with some of the more subtle consequences of the earlier legislation. Questions such as "significant deterioration" and location of new sources in "non-attainment" areas (where national ambient air quality standards have not been achieved) are now being debated. Considering the implications final decisions on these policies will have on national growth and public well-being, the delays are understandable. However, the effect of these delays on utilities has been to create considerable confusion.

The examples described in this article give a broad perspective of the type of environmental problems experienced by northeastern utilities during a period of constantly evolving legislative, judicial and regulatory action. Both urban and non-urban problems, as well as software and hardware analyses, are described. Problems encountered have ranged from nuisance situations associated with existing plants to the more basic consideration of licensing of new units.

Expansion of an existing N.Y. station

The effect of changes in air quality control philosophies between 1968 and 1976 is illustrated by the expansion of the Oswego Steam Station of the Niagara Mohawk Power Co. The station was constructed as a coal-fired power plant between 1939–1951. Four 100-MW units burning eastern coal with sulfur content of 2.5–4% discharged their stack gases through 365-ft stacks after treatment in mechanical dust collectors.

During the late 1960's, the economic and environmental acceptability of coal as a fuel for the station became problematic. Also, the system load of Niagara Mohawk was expanding significantly, and increased generating capacity was needed. A major construction program from 1971–1976 resulted in conversion of the existing coal-fired capacity to oil and the addition of two 875-MW oil-fired units (Units 5 and 6).

Prior to 1970, the State of New York had developed a comprehensive air pollution control program that included a statewide system of land-use zones, each requiring a specified level of air quality. The zones ranged from heavy industrial areas that allowed the lowest (level V) air quality through pristine areas that required the highest (level I) air quality. The Oswego Station is located in a small upstate city identified as a level III zone. The boundary of the level II zone is, however, less than a mile away, and level I areas are within the influence of the plant.

A second feature of the regulations affected emissions of air contaminants. The primary requirements dealt with the sulfur content of fuels and particulate matter. Particulate emissions for coal-fired boilers were regulated on a sliding scale, with low stack concentrations allowed for larger boilers; for oil-fired boilers, compliance with opacity standards was the governing criterion.

The National Ambient Air Quality Standards, promulgated in early 1971, affected planning for the first 875-MW unit. In 1972 the New York State implementation plan for compliance with federal requirements was filed. Considerable tightening of the New York new source emission regulations affected Unit 6. During 1972–1973, controversy over these standards caused uncertainty over relative emission levels for the various units. The use of oil instead of coal did, however, allow planning and construction to continue since adjustments in fuel specifications could be made with limited effect on plant equipment.

When burning coal, the four original units were unable to meet New York State emission requirements in 1969 without upgrading from the existing mechanical dust collectors to electrostatic precipitators. The mechanical collectors could be retained for oil burning, however.

Because of their size, the potential impact of emissions from Units 5 and 6 would be far greater than that from the four older units. Also, regulations emphasized emission controls for newer and larger units. Alternatives considered included control of both new and old units, of only the new units, and of manifolded common flue gas. The decision was to install high-efficiency electrostatic precipitators on the new units only. The precipitators were sized to maintain low particulate emissions, even with the poorest expected fuel, and to prevent any stack opacity problems.

Other emission controls, in accordance with good engineering practice, were selected to meet air quality standards. Although nitrogen oxide emission standards were not applicable at the time, Unit 5 was specified to emit no more than 250 ppm, which was quite close to the federal new source standard subsequently

61

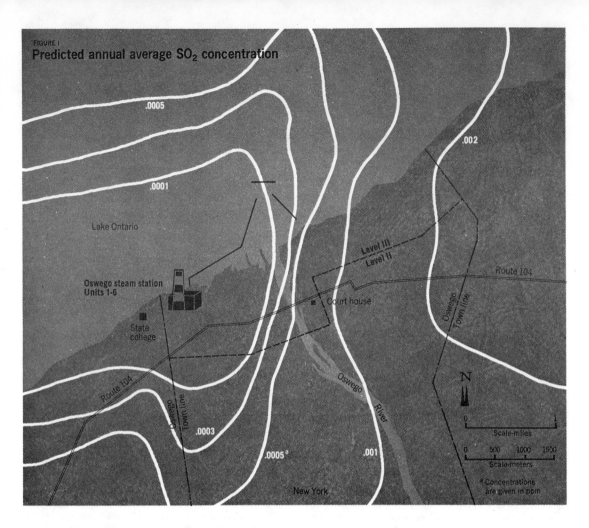

FIGURE 1

Predicted annual average SO₂ concentration

.0005

.0001

.002

Lake Ontario

Level III
Level II

Route 104

Oswego steam station
Units 1-6

Court house

Oswego Town line

State
college

Route 104

Oswego Town line

Oswego River

N

.0003

0   1
Scale-miles

.0005ª

.001

0   500   1000   1500
Scale-meters

New York

ª Concentrations
are given in ppm

required for Unit 6. Sulfur dioxide emissions for all units were to be controlled by the sulfur content of the fuel.

The magnitude of emissions from the expanded plant, together with the proximity of relatively clean "level II" areas, required comprehensive dispersion analyses.

Diffusion climatology in the Oswego area is generally quite good; however, occasionally adverse conditions in combination with the large size of the units and some local terrain features create the potential for higher short-term concentrations. Unstable atmospheric conditions capable of causing rapid downward mixing of the plume can occur fairly often. Also, in spite of a high-stack-discharge velocity, plume rise can be prevented by high-wind velocity under certain conditions.

Full compliance with air quality standards, when burning higher sulfur oil, would require a common 700-ft stack for the four existing units and Unit 5. Alternative strategies included lower sulfur fuel and flue-gas desulfurization. However, the reliability and ease of operation of the tall stack approach were overriding considerations.

Regulatory conditions had changed when the sixth unit was being planned. Emission limits for new sources now required use of 0.75% sulfur fuel, or roughly one-quarter the design value for Unit 5. The governing factors were quite different from those addressed just a few years before. However, analysis of the new unit with the required lower sulfur fuel and the same 700-ft stack height demonstrated that impacts would be acceptable.

The predicted impact of the entire plant is illustrated by the annual SO₂ isopleths shown in Figure 1. Operation of the plant

has confirmed the low contribution of air contaminants to the area, and no air quality standard violations are known to have occurred.

Addition to an existing Boston plant

Another example of an existing plant expansion is the Mystic Station of Boston Edison Company (BECO). In 1970, BECO authorized Stone & Webster (S&W) to design and build a seventh fossil-fuel steam-electric generating unit at its Mystic Station. Detailed design and engineering for the new unit were undertaken from 1970–1975. Construction activities began in 1972 and were completed by 1975. Environmental engineering and licensing activities began in 1970, and several post-operational and environmental studies are still under way.

The Mystic Station in Everett, Mass., is about 3 mi north of downtown Boston. The immediate area surrounding the plant is heavily industrialized. Prior to the addition of the new unit (Unit 7), the Mystic Station's existing six units had a total capacity of 618 MW. Initially designed for coal and oil firing during the period from 1943–1961, the units were completely converted to oil firing by 1966.

Applicable emission standards included SO₂ and particulate matter standards of the Boston Metropolitan Air Pollution Control District and new source performance standards for large fuel-burning installations. The ambient air quality standards addressed included state air quality standards and EPA primary and secondary air quality standards for SO₂, suspended particulate matter, and NOₓ.

FIGURE 2
Station sites for monitoring sootfall from the Brayton Point Station

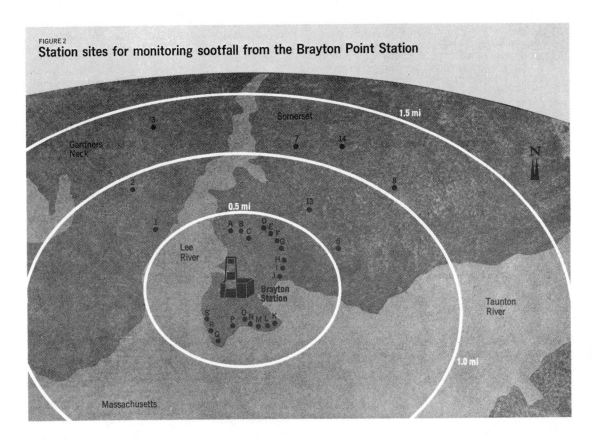

Although many federal and state air quality regulations were addressed during design and construction, a non-degradation policy adopted in 1970 by the Commonwealth of Massachusetts posed a major challenge to obtaining a construction permit. Under this policy, the air quality had to be restored and maintained at a "minimal high degree." The interpretation of this policy was that it would be applied to the maximum practicable extent for stack emissions, with strict adherence to the principle of decreased ambient impacts (reduction in ground-level concentrations of SO_2, NO_2 and suspended particulate matter). This interpretation, which emphasizes ambient impacts rather than emission tradeoffs, is somewhat different from the position currently being espoused by the EPA. The greater flexibility of the interpretation seems to offer considerable advantages.

For the expanded station, two air quality control plans were developed to maintain and/or enhance existing ambient air quality. The primary plan included reduced operation of Units 1, 2, and 3 from 3500 h/y to a projected 500 h/y, installation of flue-gas desulfurization systems on Units 4, 5, 6, and 7, and construction of a 500-ft stack for Unit 7. The alternate air quality control plan also included reduced operation of Units 1, 2, and 3 from 3500 h/y to 500 h/y, reduction in the fuel sulfur content from 0.5 to 0.3% for Units 1, 2, and 3, installation of an electrostatic precipitator on Unit 7 and construction of a 500-ft stack for Unit 7.

Both the primary and alternate air quality plans were approved by the Mass. Department of Public Health, and a construction permit was issued in 1971. Based on the results of a demonstration flue-gas desulfurization system installed on Unit 6, BECO has chosen to implement the alternate air quality control plan for the Mystic Station.

Sootfall from oil-fired plants

In contrast to the more comprehensive cases already described, S&W often handles problems of much more limited

scope. An air quality problem experienced by many northeastern utilities that use fuel oil is particulate matter fallout. Attempts to find satisfactory solutions have been infrequently successful. Through a comprehensive experimental program, New England Power Company, with the assistance of S&W, developed an operational procedure that provides for virtual elimination of sootfall occurrences.

After complaints on fallout were received in the area of New England Power's Brayton Point Station (Somerset, Mass.), extensive field sampling and meteorological data established that the power plant was the source of particulate matter fallout in certain well-defined areas during spring and fall months. Figure 2 shows the location of the plant in relation to the areas monitored. Since sootfall could not be correlated with plant operational procedures, a test program was designed to correlate stack emissions with variations in plant operations.

Because of wide variations in fuel oil physical characteristics, precombustion oil temperature was maintained at 250 °F. As a result, the fuel viscosity for proper atomization was not always obtained. Particulate matter effluent as a function of oil viscosity varied considerably from a low in the range of 0.005 lb/MM Btu, with carbon content ranging from 35–85%. The results, shown in Figure 3, indicated that precise control over viscosity is an important factor in optimizing combustion. New England Power now uses viscosity controllers, which are effective under all load conditions.

Transient conditions are frequently cited as possible causes of sootfall. Therefore, these conditions were simulated and tests were performed throughout the entire load range of the units during "light-off," soot blowing operations, and various burner arrangements and firing angles. Only small changes were observed in the rates of particulate matter emissions when compared with rates at corresponding loads under steady-state conditions.

Test results show a rather small but clear trend. As available

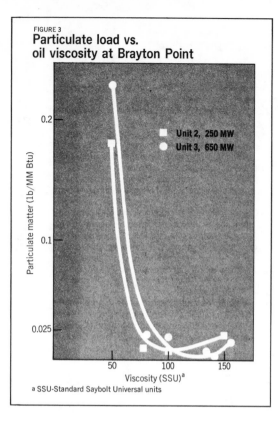

FIGURE 3
Particulate load vs. oil viscosity at Brayton Point

Particulate matter (lb/MM Btu)

0.2

■ Unit 2, 250 MW
□ Unit 3, 650 MW

0.1

0.025

50   100   150

Viscosity (SSU)[a]

a SSU-Standard Saybolt Universal units

values. Since the effect is most pronounced under conditions that are generally avoided, the usefulness of the additives appears to be greatest during periods of control failures or boiler upsets.

Implementation of the knowledge gained in this test program has provided for essentially trouble-free clear stack operation at the station, and has also been used to alleviate sootfall problems at a number of other utility plants.

Expansion in a sensitive area

Another example of an existing plant expansion concerned the Canal Plant located in Sandwich, Mass., on the south bank of the Cape Cod Canal, approximately 50 mi southeast of Boston. This expansion required resolving problems similar to those encountered at Oswego and Mystic but, from the standpoint of air quality, involved a more environmentally sensitive area.

In the early seventies, Canal Electric Co. and Montaup Electric Co. initiated plans to jointly construct an additional 560-MW oil-fired steam electric generating unit adjacent to the existing 560-MW oil-fired unit of the Canal Plant. The existing base-loaded unit had been in operation since early 1968. The second unit was to be constructed as a cyclic unit, with commercial operation originally scheduled for late 1975.

Monitoring initiated by the utility in the vicinity of the Canal Plant indicated that annual SO_2 concentrations were far below state and federal standards and 24-h concentrations never exceeded the Mass. standard or the federal secondary standard. However, the state hourly standard (0.28 ppm) in effect at that time had been exceeded on one 3-h occasion at a single monitoring location when the Canal Plant was burning a high-sulfur oil (2.2%). Revisions to the Mass. sulfur emission standards have greatly reduced the allowable sulfur emissions since the period (1970–1971) of the recorded violations.

The addition of a second 560-MW unit burning 1% sulfur fuel oil had the potential of approximately doubling atmospheric emissions from the Canal Station, and adversely affecting ambient air quality. Another design criterion that also required attention was the state non-degradation policy previously referred to with respect to the Mystic Station.

To address these constraints, an extensive stack dispersion analysis was performed. From this study it was concluded that the optimum alternative was to dismantle the existing 300-ft steel stack for Unit 1 and construct a dual-flue 500-ft concrete chimney that would serve both units. Separate flues were selected because of the operating differences of the two units.

oxygen is increased with reduced load, particulate matter emissions are reduced. Carbon monoxide and oxygen monitors were evaluated for their potential as operator tools and were found to be useful in maintaining optimum air/fuel ratios to assure minimum emissions of SO_3, NO_x, and particulate matter.

Four combustion catalyst additives were tested for their effect on particulate matter emissions. Particulate matter concentrations were reduced by 15–30%, with the larger reduction observed for tests run with fuel viscosity at other than optimum

Canal Plant. *Expansion in a "sensitive" area called for replacement of the original stack and construction of a dual-flue chimney*

The net effect of this stack configuration was to enhance the plume rise and dispersion of the effluents to a degree that would permit both the existing and new unit to burn the maximum allowable 1% sulfur fuel oil and still meet the state non-degradation policy.

The 500-ft stack similarly reduced the impact of other air contaminants such as NO_x and particulate matter emissions. However, it was deemed desirable to further control particulate matter emissions from oil firing because of isolated incidents of acidic soot fallout in the vicinity of the plant. These occurrences were usually associated with upset conditions or extreme load changes. Therefore, the new unit was also designed to operate with low excess air and minimum stack heat losses, made possible by separate insulated flues and stack liners. Also, a highly efficient electrostatic precipitator, one of the first ever specifically designed for oil firing, was included in the design to achieve a clear plume.

These features, in conjunction with sound operating and maintenance practices, are effective in curbing emission excursions. The actual start-up of Unit 2 occurred in late 1975, with commercial operation delayed to early 1976. Compliance with all air quality requirements has been achieved. The plant operates with an extremely clear plume.

In conclusion

The examples discussed some of the problems encountered by northeastern utilities over the last six years. These examples were selected to illustrate the broad range of problems and the manner in which the problems change. In the future, even greater attention may be required in initial siting and selection of generation type. Comprehensive environmental planning will be necessary at a very early stage in the conceptualization of a utility plant project.

Special attention will be required in the northeastern region to demonstrate that new plants or major modifications to existing plants are consistent with environmental goals. Continuous evaluations will become progressively more important to assess the cost and consequences of achieving those goals.

Joseph J. Cramer *is Project Engineer in S&W's Environmental Engineering Division. Dr. Cramer has contributed to the air quality and licensing aspects of several fossil-fueled power plants.*

Frank B. Kaylor *is Chief Environmental Engineer responsible for the administration of S&W's Environmental Engineering Division. Mr. Kaylor is Vice-Chairman of the ASME Air Pollution Control Division.*

Edward J. Schmidt *is currently on educational leave at the University of New Hampshire. He was formerly Chief Environmental Engineer at S&W.*

Ernest R. Zabolotny *is a consultant to S&W's Environmental Engineering Division. Dr. Zabolotny, among other things, is involved in the assessment of pollution control requirements for new and existing power plants.*

Coordinated by LRE

Reprinted from ENVIRON. SCI. TECHNOL., **10,** 532 (June 1976)

A southwest power plant saga

ES&T's Stan Miller finds that electric utilities
have installed certain environmental safeguards
because their officials also want clean air and clean water;
however, they find uniform federal guidelines and
regulations of little value to site-specific requirements

The current thrust of Federal energy policy is to increase the use of domestic fuel resources. Low-sulfur western coal is largely an untapped domestic resource and a new energy development of this type will severely impact that part of the U.S.

On one hand, utilities, in their opinion, find both justification and need for fossil-fuel development in the Southwest. The Western Energy Supply and Transmission Associates (WEST), an unincorporated organization of 22 public and investor-owned electric utilities, provide electricity in all or part of 11 western and south-western states. They have filed with the Federal Power Commission their planned expansions for the next 10 years, as required by law.

On the other, environmentalists are concerned with threatened air quality, strip-mining effects, and decrease in visibility—from 160 miles today to perhaps 40 miles—if and when expansions occur.

A big issue, largely unresolved at this time, is whether this coal will be stripped and shipped to eastern users or stripped and burned on the spot to generate elec-

tricity for transmission across interstate boundaries to users locally and as far away as Los Angeles and San Diego.

What, of course, is badly needed for power plant development decision-making are superregional air emissions data for the entire Southwest. It is important to know not only the contribution to air quality from individual power plants but to be able to assess the overall effect that new power plants might add to an area already burdened with existing power plants.

Recently, baseline monitoring for the Southwest has been renewed; it goes by the name, Western Area Environmental Monitoring Study. The EPA Environmental Monitoring and Support Laboratory (EMSL) in Las Vegas, Nev. (*ES&T,* December 1975, p 1109) is performing a 5-year monitoring study; it involves more than 50 airborne monitoring systems.

Earlier, the Department of the Interior conducted a Southwest Energy Study in 1970, which measured total air, water, and land impacts of existing and proposed plants in the Four Corners area. Also, a joint ambient air program with eight monitoring stations was started by the

utilities at the Four Corners and San Juan stations in 1972 and is still going on.

But it is likely that complete monitoring data will neither be available nor evaluated before decisions on expansions are finalized. It is also important to note that this monitoring activity will be the first time EPA has obtained its own baseline data for the superregional area of the Southwest, including many states of WEST involvement. However, monitoring activity on a more local basis has been performed for many years by local agencies and independent consultants.

This development is needed; there can be little doubt. Dr. Chauncey Starr, president of the Electric Power Research Institute (*ES&T,* January 1975, p 13), says, "Coal-derived fuels and nuclear sources must account for most of U.S. electric power production for the next quarter century." He continues, "Substantial dependence on foreign fuel sources is undesirable for reasons of national security, balance of payments, and unforeseeable costs. Therefore, electricity generation options must concentrate on indigenous fuel resources, which, between now and the year 2000, are principally coal and uranium."

Putting the clean energy potentially in perspective, Starr says, "Advanced technological concepts such as fusion and solar energy cannot play an important electricity production role before the year

2000. Although most scientists and engineers actually working on the development of these new concepts fully agree with this point, unfortunately, the lay public has been grossly misled by others to believe that such novel approaches are viable in the near term as alternatives to coal and nuclear power."

This position was reiterated recently by the Electric Edison Institute (EEI), a principal national association of investor-owned electric utility companies, and the National Coal Association (NCA), the national association of commercial coal producers and allied interests. In a joint statement EEI president W. Donham Crawford and NCA president Carl Bagge said that the demand for electricity will continue to increase, and that the utilities must meet this demand by expanded use of coal and faster growth of nuclear power.

By 1984, they said, 781 million tons of coal a year will be required for making electric power, almost twice as much as last year—and this assumes the predicted growth of nuclear power. If nuclear power is delayed, they said, the demand upon coal will be even greater. In its report, "Economic Growth in the Future," EEI describes several scenarios including high growth, moderate growth, and low growth. It is being published by McGraw-Hill, Inc.

Some of the major power plants in the Southwest include Mohave, Navajo, San Juan, Cholla, Kaiparowits (whose construction was cancelled this April) and, of course, Four Corners (*ES&T,* June 1974, p 516). Their experiences in protecting the environment are noteworthy.

Mohave and Four Corners

Operated by Southern California Edison (SCE), the Mohave plant has two units of 790 megawatts each for a total plant generating capacity of 1580 megawatts —equivalent to the electricity needs of about 1.5 million people. Bechtel Corp., the major architectural-engineering (a-e) firm, started construction on the plant, which includes C-E boilers, in 1967. Unit 1 started up in the Fall of 1970, Unit 2 six months later.

Mohave is unique; it is the only power plant in the U.S. that uses slurry coal as fossil-fuel input. The slurry is about 50% finely divided coal and 50% water. The coal comes from the Black Mesa area, which has been leased by the Peabody Coal Company from local Indian tribes. It is transported by pipeline for a distance of nearly 275 miles from Black Mesa, near Kayenta, Ariz., via an 18-inch pipeline that is owned by the Black Mesa Pipeline Company, a subsidiary of the Southern Pacific Railroad. The trip takes three days; it begins at an elevation of 6500 ft and ends at an elevation of 700 ft.

On arrival at the Mohave plant, the slurry is stored in four working tanks that provide a smooth, uninterrupted flow of fuel to the boilers. From these tanks the slurry is pumped into centrifuges that

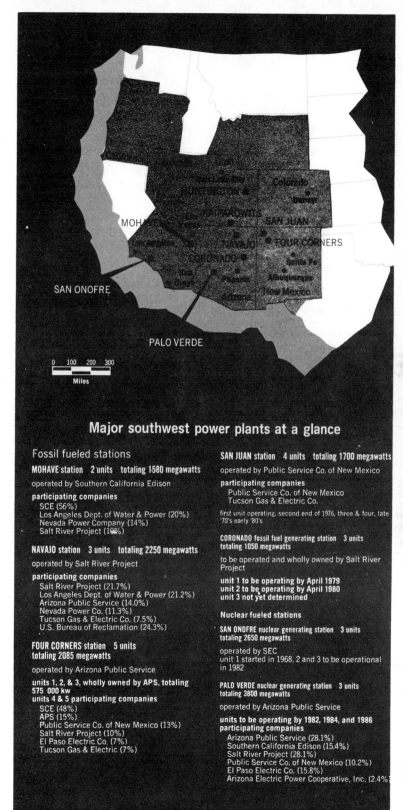

Major southwest power plants at a glance

Fossil fueled stations

MOHAVE station   2 units   totaling 1580 megawatts

operated by Southern California Edison

participating companies
SCE (56%)
Los Angeles Dept. of Water & Power (20%)
Nevada Power Company (14%)
Salt River Project (10%)

NAVAJO station   3 units   totaling 2250 megawatts

operated by Salt River Project

participating companies
Salt River Project (21.7%)
Los Angeles Dept. of Water & Power (21.2%)
Arizona Public Service (14.0%)
Nevada Power Co. (11.3%)
Tucson Gas & Electric Co. (7.5%)
U.S. Bureau of Reclamation (24.3%)

**FOUR CORNERS station   5 units
totaling 2085 megawatts**

operated by Arizona Public Service

**units 1, 2, & 3, wholly owned by APS, totaling
575 000 kw
units 4 & 5 participating companies**
SCE (48%)
APS (15%)
Public Service Co. of New Mexico (13%)
Salt River Project (10%)
El Paso Electric Co. (7%)
Tucson Gas & Electric (7%)

SAN JUAN station   4 units   totaling 1700 megawatts

operated by Public Service Co. of New Mexico

participating companies
Public Service Co. of New Mexico
Tucson Gas & Electric Co.

first unit operating, second end of 1976, three & four, late '70's early '80's

**CORONADO fossil fuel generating station   3 units
totaling 1050 megawatts**

to be operated and wholly owned by Salt River Project

**unit 1 to be operating by April 1979
unit 2 to be operating by April 1980
unit 3 not yet determined**

Nuclear fueled stations

**SAN ONOFRE nuclear generating station   3 units
totaling 2650 megawatts**

operated by SEC
unit 1 started in 1968, 2 and 3 to be operational in 1982

**PALO VERDE nuclear generating station   3 units
totaling 3800 megawatts**

operated by Arizona Public Service

**units to be operating by 1982, 1984, and 1986
participating companies**
Arizona Public Service (28.1%)
Southern California Edison (15.4%)
Salt River Project (28.1%)
Public Service Co. of New Mexico (10.2%)
El Paso Electric Co. (15.8%)
Arizona Electric Power Cooperative, Inc. (2.4%)

produce a force of 1000 G's to separate the coal from the water. Approximately 75% of the water is removed, enough to permit combustion of the coal.

Geographically, the Mohave plant is about 90 miles south of Las Vegas, the point being that it is still in the state of Nevada, but more importantly in Clark County. After construction was started, Clark County imposed an SO_2 emissions regulation of 0.15 lbs of SO_2/million Btu, eight times more stringent than the EPA new source performance standards (NSPS) (1.2 lbs of SO_2/million Btu) for new electric generating plants.

In January 1973, the Mohave Generating Station was granted a variance to provide for full compliance, by December 1976, with Clark County emissions regulations (subsequently amended to July 1977). This variance provided for the installation and testing of SO_2 emission scrubbers.

The State of Nevada legislators upset the regulatory apple cart by looking into the regulations and the extremely high cost to the public implied by the regulations. When the state legislature realized that the Clark County regulations were eight times more stringent than EPA's NSPS they put a 2-yr moratorium on meeting emissions standards to allow time for thorough review of the emission standards and their relation to ambient air quality.

Meanwhile, the Mohave experience reveals that SCE made a decision in 1970 to test eight different pilot scrubbers with four different chemicals. One of those was a 1 megawatt-size scrubber which, by now, has become known as the Weir horizontal scrubber, named after Alexander Weir, Jr., the inventor, who is principal scientist for air quality at SCE.

In 1972, SCE and Salt River Project (SRP) proposed a full-scale program to test two scrubbing modules, each larger than had ever previously been built anywhere in the world. Each was to scrub one-fifth of the total air flow from one of the two units at the Mohave plant. In power plant parlance this flow is referred to as a one-fifth slip stream. Simultaneous with the Weir scrubber evaluation was an evaluation of the UOP vertical turbulent contact absorber (TCA), again on a one-fifth slip stream, equivalent to about 170 megawatts of power generation. The testing decision became known as the Mohave-Navajo test program, the Navajo plant being operated by the Salt River Project with main business offices in Tempe, Arizona, although the plant is located at Page.

SRP director of communications and public affairs, Stanley E. Hancock, says that the Mohave-Navajo test scrubber evaluation was jointly paid for by the participating companies and agencies of the two plants and was based on a kilowatt distribution—the split being 60% Navajo and 40% Mohave. He says it was a voluntary program and cost about $30 million. Testing lasted 1.5 years, and was started on the Weir horizontal scrubber on January 16, 1974 and on the UOP vertical on November 1, 1974, delayed from the spring of 1974 because of a fire during startup in January 1974. Testing was completed on the horizontal on February 9, 1975 and on the vertical on July 1, 1975.

Mohave station superintendent Greg L. Fraser, a former tank commander with service in the Italian and African campaigns during World War II, says that the plant investment is $234 million. "To put scrubbers on the entire flow stream of the Mohave plant would necessitate an additional investment of many millions of dollars," Fraser says. The plant employs about 250 people, and 80 are at the plant at all times. He indicated that putting scrubbers on the two units for full rated emissions would not only increase the capital outlay but estimates that his present maintenance staff of 150 people would have to be increased by as much as 60%.

Mohave is a base-load plant and operates routinely between 60% and 70% of rated capacity, on an annual basis. Power from this plant is transmitted via station switchyard facilities to a transmission system from which it flows into the electrical system that supplies the southwestern U.S.

SCE research scientist Dr. Dale Jones says that the cost of constructing and testing the two scrubbers ($30 million) included $1.8 million fire damage repair to the UOP vertical. He indicates that the space requirement is 15 ft by 28 ft for the horizontal and 18 ft by 40 ft for the vertical. He also says that there is a lower tonnage of steel on the horizontal system, resulting in lower capital costs.

When ES&T visited the Mohave plant last September, the Weir scrubber had been sold at salvage value, disassembled, and reassembled and refurbished for further evaluation on Unit 5 of the Four Corners plant, near Farmington, N. Mex., which is operated by the Arizona Public Service Co. (APS).

The five units at Four Corners are equipped with various pollution control equipment. Originally, Units 1, 2, and 3

were equipped with mechanical collectors capable of removing 77% of ash. However, with the construction of Units 4 and 5, and the advancement of control technology, APS retrofitted Units 1, 2, and 3 with wet scrubbers at an estimated cost of about $5 million. A Chemico system was added to these units at a final cost of approximately $30 million (ES&T, June 1974, p 516).

The scrubbers meet particulate removal regulation (99.2%) and inadvertently remove approximately 30% of the SO_2 because of the inherent pH control and alkalinity of the ash. The result is a relatively clean stack with a vapor plume only visible during the cold winter months.

Units 4 and 5 are currently operating with electrostatic precipitators removing 97% of the particulate matter. Additional control of both particulates and SO_2 are required by state and federal agencies. The first stage of this program is the testing of the horizontal scrubber from Mohave.

APS vice-president T. G. Woods, Jr., says that testing began this February on unit 5 operating on coal with 22% ash and 0.6–0.75% sulfur. One objective, Woods says, is to operate as continuously as possible for a minimum of 6 months testing and operational evaluation. The purpose of the testing is the determination of the commitment that will be made for SO_2 removal, and also the technical and economical feasibility. Woods also mentioned that one difference between the two scrubbers is energy use. The horizontal scrubber requires less energy in its operation than the vertical design.

But SCE is not enamored with the use of scrubbers on its plants. Nor is APS for that matter. Dr. Raymond E. Kary, manager of the environmental management department, says Arizona Public Service, which has had costly experience with scrubbers at the Four Corners Power Plant in northwestern New Mexico, takes exception to the EPA claims. He described APS' pioneering efforts to develop low-sulfur coal beds of the Southwest.

The original cost estimate for the scrubbers for the plant's three original units was $5.5 million, Kary said. But the final cost skyrocketed to $30 million, because of last minute design changes, delayed deliveries, unproductive overtime, and unavailable equipment and materials, he added. "When we add the installation cost, the cost of the money over 30 years, and the O & M costs over the same period," Dr. Kary pointed out, "the scrubbers would cost APS customers up to $546 million."

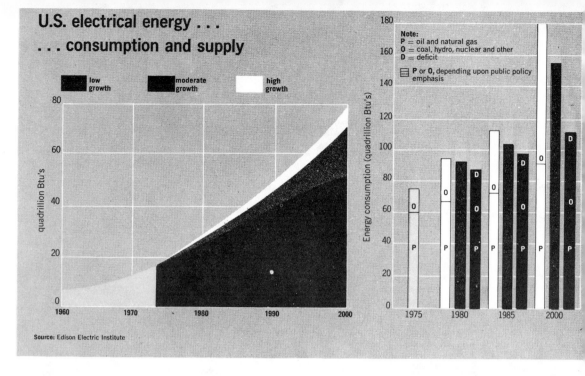

U.S. electrical energy . . .
. . . consumption and supply

low growth
moderate growth
high growth

quadrillion Btu's

1960 1970 1980 1990 2000

Source: Edison Electric Institute

Energy consumption (quadrillion Btu's)

Note:
P = oil and natural gas
0 = coal, hydro, nuclear and other
D = deficit

P or 0, depending upon public policy emphasis

1975 1980 1985 2000

The installed value for each kilowatt at Mohave is $147/kW, including the cost of electrostatic precipitators for fly ash removal. In general, western coal is less than 1% sulfur, averaging 0.5% sulfur. The WEST consensus questions the need for scrubbers for removing SO_2 from plants utilizing low sulfur western coals. It would take an additional $50–100/kilowatt to put on scrubbers that would push the cost of power generation to $400–500/kilowatt; there would also be higher operating costs, including more maintenance to service the units.

Pullman Kellogg, formerly the M.W. Kellogg Co., one of the tall stack builders (*ES&T*, June 1975, p 522) has an exclusive license to build the Weir scrubber in the U.S., West Germany, and Japan. At press time, only one order had been placed. At the same time, all participants in the Mohave-Navajo test program could obtain a royalty-free license on the scrubber, meaning that the operators and owners of these plants can contract for the equipment with any other a-e firms.

Navajo

Operated by the Salt River Project, the Navajo plant is newer than the Mohave plant; its environmental safeguards are also noteworthy. There are three units at the Navajo plant; unit 1 started operation on May 30, 1974, the second nearly a year later—specifically on April 3, 1975—and the third unit started this April. Again, the a-e firm is Bechtel. Construction began in 1970 and each unit has a 775 ft stack built by the tall stack builder Pullman Kellogg.

Navajo, the largest electric generating plant in the state of Arizona, where only 15% of the land is privately held, is classed as an existing source rather than a new source and hence not regulated by EPA new source performance standards.

Navajo station superintendent is Dean Johnson, but Navajo plant environmental supervisor, Mike Webb, a chemist from Arizona State University, was *ES&T's* guide. Like Mohave, the Navajo plant gets its coal from the Black Mesa, which is only 75 miles from the plant; the coal is transported by the BM&LP (Black Mesa & Lake Powell) railway. The trains are electrified; they travel 24 hours/day. The BM&LP railroad cost nearly $75 million, and was constructed by Morrison-Knudsen Co., (Boise, Idaho).

The coal that is delivered is about 0.5% sulfur and 8.0% ash. It is ground to the consistency of flour before burning in the C-E (Combustion Engineering) boilers. With the Navajo plant being 4360 ft above sea level and tall stacks dispersing its emissions at an elevation of 5135 ft, one might still ask what the emissions will be.

The EPA, based on the application of a NOAA model, promulgated regulations required 70% SO_2 removal. SRP challenged the regulations in Federal Court since SRP recognized that such models were subject to potential inaccuracies in modeling and input data. As a result of litigation and discussions with the federal agencies involved, an agreement was reached that the degree of SO_2 removal required at Navajo should be established by a scientific method of testing. Accordingly, SRP initiated an SO_2 monitoring program to provide actual field data and to develop analytical procedures to determine SO_2 ambient concentrations impacts.

In September 1974, Rockwell Science Center and their subcontractors, Meteorology Research Inc. and Systems Applications Inc., began monitoring SO_2 emission and ambient concentrations at Navajo and in the surrounding area. The field program, which was conducted from September 1974–February 1975, was comprehensive and included not only SO_2 measurement noted above, but plume tracking by aircraft and collection of detailed meteorological data.

The reductions and processing of all data were monitored by EPA-RTP as well as an independent consultant. Also, EPA was involved in the establishment of analytical procedures as were SRP and Bechtel Power program manager for Navajo. The cost for this study was $2 million. The principal conclusion of this monitoring program is that no SO_2 removal is needed to meet federal ambient SO_2 standards. Resolution of program results are now registered with the EPA regional administrator.

Hancock explains that NSPS requirements are not applicable to the Navajo plant. Ground breaking for construction of this plant was in April 1970, prior to promulgation of federal NSPS regulations of August 31, 1971. On the other hand, Hancock says that the state of Arizona differentiates between existing and new generating stations. The state also required annual testing of emissions control devices. The state of Arizona views the three units at Navajo as one stationary source and the state has a sliding scale for particulate emissions based on unit size.

According to the calculations of Richard F. Durning, a professional engineer and staff consultant of the Salt River Project, there are 7000×10^6 Btu/h/unit.

The heat input for the three units would be on the order of 21 000 million Btu/h and the allowable particulate emissions would be 0.60 lbs/h/million Btu, based on a 24-h arithmetic average. Durning also says the needed controls depend on the ash content of the coal. For coal with 8% ash the needed control would be 99.1% whereas for coal with 13% ash the removal would be 99.46%.

The three units at the Navajo plant represent an investment of $659 million of which more than $200 million has been budgeted for environmental protection and safeguards. The electrostatic precipitators alone cost more than $40 million. Conventional scrubbers on the Navajo plant units could come to an additional cost of $100 million, if installed. Southwest utility decision-makers are not unreasonable in asking for the rationale behind spending such large sums of money to provide protection to the ambient air for periods of concern that amount to only a few days per year. They make their point, but will it be heeded?

For cooling purposes, the plant may use up to 34 000 acre ft of water each year and takes that water from Lake Powell. The efficiency of a fossil-fuel power-generating plant is 38%, because of the limitations of Carnot cycle efficiency. The remaining 62% of the energy is not available for electric power generation, and is rejected to the ambient air via cooling towers and the stacks. Each unit has a bank of two cooling towers. Built by the Marley Co. (Mission, Kansas), these plant towers cost nearly $9 million.

Also in Page, Arizona, is the Glen Canyon hydroelectric plant. Its generating capacity is 980 megawatts. The Glen Canyon Dam blocks up the water of the Colorado River for as far as 186 miles upstream; in essence, it filled in the canyon. The Glen Canyon Dam was started in 1956 during the Eisenhower administration, and the last bucket of concrete was poured in 1963. It started producing electricity in 1964.

But the Navajo plant will be generating 2.5 times the electricity of its hydroelectric neighbor. One of the big users of the Navajo electricity is the federal government. Its Bureau of Reclamation gets 24% of the power and plans to use it for the Central Arizona Project. Some of this electricity will be used to pump water to Phoenix and Tucson, the latter being one of the largest cities in the U.S. that has no natural flow of water!

The Navajo plant is located on 1021 acres of land leased from the Navajo tribe. Peter MacDonald is chief of the tribal council, which meets at Window Rock,

Delivering coal. *The BM&LP railway runs 24 h/day bringing coal from the Black Mesa to the Navajo plant, a distance of 75 miles*

Arizona. Although the water for the plant is taken from Lake Powell, it is important to note that there is no discharge to the lake. The cooling water is softened before use and after use is placed in holding ponds that cover 125 acres. The evaporation rate in that part of the U.S. is on the average 8 ft/year. Also, there is an ash disposal area of 765 acres.

Starting last September, the ash collected by the electrostatic precipitators is hauled away from the plant by Cement Transporters, Inc. (Phoenix). Then, The Tanner Co. of Phoenix incorporates the material in their concrete formulations, similar in concept to the process of IU Conversions Systems, Inc. (*ES&T*, October 1972, p 874).

The plant uses approximately 24 000 tons of domestic coal each day when all three units are operating at full load. On a Btu basis, this is equivalent to 90 000 barrels of imported oil. Webb says that the Black Mesa coal costs about $3/ton. Assuming that oil costs $15/barrel, its cost would be $1.35 million/day, but the cost of the coal would only be $0.072 million each day. In other words, based on those costs, the Navajo fossil-fuel generating station could save $1.278 million each day on fuel costs alone, and also produce

clean electric power from a noninterruptable domestic source of fossil fuel.

San Juan & Cholla

Located in northwestern New Mexico, the San Juan project involves four coal-fired units totaling 1700 MW. Operated by the Public Service of New Mexico, in conjunction with Tucson Gas & Electric Co., Unit 1 is now in operation, and Unit 2 is under construction. The third and fourth units, each 500 MW, are scheduled for completion in 1980.

The units have Foster Wheeler boilers, provide 99.8% removal of fly ash by electrostatic precipitators, and will remove 90% of the SO_2 from stack gases. The coal is 18% ash, 0.8% sulfur, and 10% moisture. Unit 1 cost more than $107 million, and unit 2 cost $80 million. The SO_2 removal systems will add $40 million to the final cost of each unit. Arizona Public Service Co. is expanding its Cholla Power Plant in northern Arizona, with two 250 MW units to be completed in 1978 and 1979. A fourth unit (350 MW) is slated for operation in 1980.

Kaiparowits

Without any question, the most controversial development in the Southwest was the planned Kaiparowits power plant on federal land in southern Utah. With 4 units totaling a 3000 megawatt generating capacity, it would have been the largest coal-fired power plant in the world. This April, utility companies withdrew their backing for the plant. Originally, the first unit was planned for 1980, a second for 1981, and the full plant by 1982.

William R. Gould, executive vice president of Southern California Edison—the project director for Kaiparowits—said, "It didn't die slowly; it was beaten to death by the environmentalists." (The battle lasted 13 years.)

The plant would have been located in an otherwise inaccessible area of southern Utah. WEST utility men go to remote places to build their generating stations because the areas have underground coal supplies. It is more economical for the customers using electric power in the Southwest to receive that power from mine-mouth, coal-fired power plants.

Recoverable coal reserves in the Kaiparowits Plateau are 15.2 billion tons. About 1.8 billion tons of coal reserve lie within the lease holding of the project participants. The plant would take about 12 million tons annually of raw coal from four underground mines.

Utah governor Calvin L. Rampton called the project's demise, "A serious blow at the attempts of the U.S. to secure energy independence."

Environmentalists warned that the plant would be located within a 100–200 mile radius of the Golden Circle of National Parks and monuments. This impact area includes eight national parks, 26 national monuments, three national recreation areas, and two historic sites, in addition to numerous primitive areas, de facto wilderness areas, Indian sacred grounds, and state parks, according to one environmentalist's tabulation. This scenic area is under the jurisdiction of the Bureau of Land Management or the Navajo Tribal Council. Visibility is usually very high and noise levels are minimal.

What about nuclear?

One of the largest nuclear power plants in the U.S. will be the San Onofre plant (near San Clemente, Calif.) of Southern California Edison and San Diego Gas and Electric Co. Construction of the second and third units began last year; nuclear fuel is to be loaded in 1980, and the second of three units is scheduled to begin operation in 1982. The first 450 MW unit was initially operated in 1968. Each of the units is rated at 1100 megawatts for a total generating capacity of 2650 MW. Palo Verde is 3810 MW.

APS is also project manager and operating agent for the proposed 3810 megawatt Palo Verde Nuclear Generating Station to be built 50 miles west of Phoenix. Completion dates of the three 1270-MW nuclear units are 1982, 1984, and 1986. Other participants include: Salt River Project, Southern California Edison, Public Service Co. of New Mexico, El Paso Electric Co. and Arizona Electric Power Cooperative, Inc. Favorable action on APS' application for a construction permit, may see the project get underway this spring.

The only other non-nuclear planned project is that of the Coronado generating station. Wholly owned by SRP there will be three 350 megawatt units. Construction began in August 1975. It will be located on private land in northeastern Arizona.

Looking ahead

Fifteen years ago, a fossil-fuel power plant could be built for about $100–150/kilowatt of generating capacity. At that time, the argument against nuclear energy was that it was too expensive, on the order of $1000/kilowatt of generating capacity.

But what has happened since then is that the cost of a fossil-fuel power plant, what with all the environmental safeguards, has escalated to the comparable cost of a nuclear plant. The remaining difference is that a nuclear plant takes a longer lead time from ground breaking to operation than a fossil-fuel plant.

But recent development in the Southwest can be put into the perspective of economics. The Mohave station represents an investment of $147/kilowatt of generating capacity, The Navajo plant, constructed somewhat later, represents about $300/kilowatt of capacity. The Kaiparowits plant is in the $900–1000/ kilowatt range. The proposed Coronado plant will cost $1000/kilowatt.

Key utility officials fail to see the need to remove SO_2 from the burning of western low-sulfur coal, and why controls need be the same for this coal and high-sulfur eastern coal. What is the rationale for transporting low-sulfur western coal to eastern utilities? Transportation costs only add to the price of this coal and might jump the price to about $20–25 per ton. $20/ton of coal is equal to $6/bbl of oil on the same heating value basis.

On the record, utility officials also allege that eastern utilities justify the purchase of the low-sulfur western coal on the assumption that no SO_2 removal is needed. Again, where is the logic in requiring western utilties to scrub?

Wet scrubbing of fly ash and SO₂

Reprinted from ENVIRON. SCI. TECHNOL., **11**, 1054 (November 1977)

Combustion Engineering scrubbers have logged a

92% availability on units 1 and 2 of Northern States Power

at its Sherco plant in Becker, Minn.

The scrubber business is heating up—there are currently 15 vendors. Recently, TVA went outside on a contract proposal to put a scrubber on its coal-fired Widows Creek No. 7 power plant, a 550-MW unit. The proposals will be evaluated within the next four weeks; but an announcement was not made before press time.

At a recent power plant tour *ES&T* viewed the operation of one of today's largest scrubber operations. The scrubber on the Sherco plant of Northern States Power Company at Becker, Minn.—45

miles NW of Minneapolis-St. Paul—has been operating for more than one and a half years. The Sherco station of the NSP in Sherburne County has two 700-MW coal-fired units. Unit 1 started commercial operation May 1976 and Unit 2 in May 1977.

In an earlier PEDCo report of available scrubber technology the Sherco plant of NSP was not originally listed. It was argued that the firing of low sulfur western coal did not qualify the operation. Now, however, this plant is included in the

PEDCo reports, as it removes in excess of 50% of the SO₂ emitted.

In fact, the units fire about 400 tons of coal each hour and use about 3 tons of limestone per hour in the scrubbing operation. The coal, containing about 9% ash and 0.8% sulfur, comes by unit train from the Colstrip area of Montana, some 600 miles away. NSP placed one of the first bids for low sulfur western coal and subsequently purchased an initial 20-y supply for the Sherco plant.

This semi-base-load power plant supplys power to a 4-state area—Minn., N. Dak., S. Dak., and Wis. The plant's on-line capacity factor is about 78%.

In operation

On each unit there are 12 scrubber modules for a total of 24 modules on the two units currently producing power. Although each unit has 12 scrubber modules, only 11 are required for full load operation. The dimension of a single module is 64 feet high, 18 feet wide, and 27 feet deep.

Stack gas exit conditions on units 1 and 2 are 0.04 grains/dscf of particulate matter and 350 ppm SO₂. The scrubbers remove in excess of 50% of the SO₂ and 99% of the particulate matter to achieve this condition.

Units 1 and 2 have C-E boilers, GE turbines, and C-E scrubbers. These units represent a $368 million plant investment; the air pollution controls, including the C-E scrubbers, cost $55 million. Two additional units, 3 and 4, are planned for 1981 and 1983, respectively. These units will have B&W (Babcock and Wilcox) boilers, Westinghouse turbines, and C-E scrub-

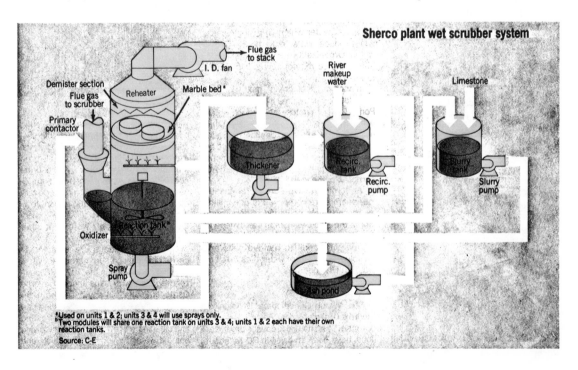

Sherco plant wet scrubber system

*Used on units 1 & 2; units 3 & 4 will use sprays only.
Two modules will share one reaction tank on units 3 & 4; units 1 & 2 each have their own reaction tanks.
Source: C-E

bers. The newer scrubbers will get the particulate loading down to an equivalent stack gas exit condition of 0.02 grains/dscf.

The present plant stack is common for units 1 and 2; it's 650 feet tall with a bottom diameter of 72 feet and tapering to 40 feet at the top. Opacity is measured continuously by an optical transmissometer located in the plant stack. With a particulate emission of approximately 0.04 g/dscf, opacity readings have typically been in the 40–45% range. Opacity values remain well above the 20% standard of the EPA (the units were controlled before issuance of the EPA standard) and the Minnesota Pollution Control Agency. Actions to resolve the difference with the regulatory agencies are underway.

The scrubbing action

Called the Air Quality Control System (AQCS) by the company, the C-E flue gas desulfurization system is a wet scrubbing, non-recovery system using lime or limestone as the chemical additive. James Martin, manager of AQCS performance design, and A. Jackson Snider, product manager for environmental control products, pointed out that this scrubber operation on units 1 and 2 is in fact 1971 technology. The work on units 3 and 4 will get at the newer improvements in scrubber design and operation. For example, each of the 800-MW planned units will have C-E scrubbers, but this time there will be eight scrubber modules, seven of which will be needed for full load operation.

How the scrubber works

The raw limestone is ground in wet ball mills and delivered as a slurry to each of the module's reaction tank. The slurry in the reaction tank contains the limestone, fly ash, calcium sulfate and dissolved ions (especially calcium, magnesium, and sulfate). The slurry is continually mixed in the reaction tanks by one vertical entry and two horizontal entry mixers. The reaction tank slurry is then pumped, by a single spray pump, to the first stage spray nozzles and the underbed spray nozzles, at flows of approximately 3540 and 1900 gallons per minute, respectively. The slurry collects the fly ash and the sulfur dioxide from the flue gas and returns to the reaction tanks.

The process of removing the SO_2 from the flue gas is achieved by using two additive sources; the calcium oxide in fly ash and tail-end addition of limestone.

The Sherco cooling system is a closed cycle. Water drawn into the plant for cooling purposes is reused and essentially *none* of the heated water effluent (see diagram) will flow into the Mississippi River. This winter, the heated water effluent from the plant will be used to heat greenhouses already installed on the plant site for growing plants such as tomatoes and the like.   SSM

Reprinted from ENVIRON. SCI. TECHNOL., **10**, 416 (May 1976)

Stack gas cleaning: A 1976 update

New Orleans' meeting goers learn that the technology of flue gas desulfurization is gaining acceptance

Flue gas desulfurization (FGD) technology has two major aspects: The first involves the "throw-away" approach, and the second aims at absorbent regeneration, and perhaps other materials recovery. This was one view expressed among the nearly 700 attendees at the 7th Flue Gas Desulfurization Symposium (New Orleans, La.). Sponsored by the EPA's Industrial Environmental Research Laboratory (IERL, Research Triangle Park, N.C.), the symposium heard a reminder from the Electric Power Research Institute (Palo Alto, Calif.) that FGD (scrubber) installations—particularly first-of-a-kind demonstrations—should be viewed as experimental until the technology fully evolves.

Nevertheless, the EPA told the symposium, consisting of people from the U.S., Canada, France, Japan, Sweden, and West Germany, that only FGD scrubbing can furnish SO_x cleanup within requirements of the Clean Air Act. Stephen Gage, EPA's deputy assistant administrator for energy, minerals, and industry, said that the agency certainly keeps abreast of alternatives to FGD technology, but generally foresees no appreciable impact of these alternatives, such as coal precleaning, until well into the 1980's. Speaking for administrator Russell Train, Gage also described the EPA's mission to see that the public is protected against SO_x emissions. A number of symposium attendees privately expressed opposing views to an *ES&T* editor.

In keeping with the EPA position, the IERL is going ahead with its own FGD studies. Richard Stern, chief of the IERL's Process Technology Branch, and symposium chairman, told the meeting about the laboratory's forthcoming demonstration of a 100-MW double-alkali scrubber, as well as that of a 100-MW regenerable alkali scrubber. He said that FGD will be the only available technique for general use in meeting SO_2 emission control requirements for the next 10 years. However, as this goes to press, sites and contractors for these demonstration scrubbers have not yet been selected.

As of January, FGD systems in the U.S. used mostly limestone, with lime as a close runner-up. By 1980, limestone is still expected to be the dominant absorbent for scrubbers, with lime in second place, according to a study completed this year for IERL by PEDCo Environmental Specialists, Inc. (Cincinnati, Ohio). Next come sodium carbonate and magnesium oxide (MgO) systems, followed by Wellman-Lord (sodium sulfite) and catalytic oxidation (Cat-Ox) techniques. SO_x removal efficiencies as high as 98% were reported, with particulate removal of 90+%, according to the PEDCo study. Latest PEDCo information gives a figure of 22 FGD units (3828 MW) operational, and 20 units (7096 MW) under construction.

On line

In upper Montgomery County, Md., Potomac Electric Power Co. (PEPCO) has a two-stage scrubber that uses MgO as SO_2 absorber at its Dickerson Generating Station, Unit No. 3 (95 MW). This MgO system was supplied by Chemico Air Pollution Control Co. (New York, N.Y.), now a division of Envirotech Corp. (Menlo Park, Calif.). The system's objects are to remove 90+% of SO_2; determine effects of particulates on SO_2 absorption; and make magnesium sulfite for reconversion to MgO, which would be reused as an SO_2 absorbent. Other objectives are to produce commercially marketable sulfuric

---

FGD systems in the U.S.

Here is where they stand now . . .

Status	Number of units	MW
Operational	21	3796
Under construction	20	7026
Planned		
Contract awarded	10	3761
Letter of intent	10	3911
Requesting or evaluating bids	7	3837
Considering FGD systems	40	19 797
Total	**109**	**42 128**

. . . and these are the processes they will use[a]

	MW 1976	MW 1980
Limestone	4077	11 221
Lime	3442	7716
Sodium carbonate	375	375
MgO	365	1091
Wellman-Lord	115	1830
Cat-Ox	110	110
Lime/limestone	30	1740
Dilute acid	23	23
Activated carbon	20	20
Dual alkali	20	20
Total	**8577**	**24 416**

[a] FGD megawattage for which a process has not been selected is not included.
Note: In 1976, about 65% of FGD megawattage will be retrofit, and about 35% will be on new plants. By 1980, 30% and 70% will be retrofit and new installations, respectively.
Source: PEDCo

FGD costs

A utility survey by PEDCo and the Edison Electric Institute (New York, N.Y.) came up with scrubber cost ranges of $33–197/kW, with an average of $94/kW; standard deviation, $40/kW. Where lime or limestone is the absorbent, costs reported range $34–116/kW, with an average of $78/kW; standard deviation, $27/kW. These figures are tentative, and would change with adjustment to 1975 dollars, removal of particulate control costs, and indirect cost items.

In an exhaustive economic analysis of FGD, the TVA (Muscle Shoals, Ala.) assumed a 25–30-yr scrubber life, a midwest location, a clay-lined sludge disposal pond, and certain other boundary conditions. Costs and resource needs can vary widely with scrubber megawattage, fuel, percentage of sulfur, SO$_2$ removal efficiency, and scrubbing absorbents used. Nevertheless, the TVA came up with capital costs of $32–118/kW (1975–1978 cost basis), and annual rate revenue requirements of 1.87–7.14 mills/kWh (1978 cost basis).

acid from the captured SO$_2$, and to obtain data and design parameters for scale-up to 600 MW or more.

The 295 000-acfm FGD system first started up in September 1973, and operated on and off until late 1975. During operations, SO$_2$ removal efficiency was above 90% whenever gas flow and inlet concentrations were more than 1000 ppm. PEPCO spent $493 000 for operation and maintenance (O&M) of this pilot scrubber in 1974. If MgO units (for 850 MW) start up in the spring of 1982, PEPCO foresees total capital costs of $44.81 million and revenue needs of 5.10 mills/kWh, 1975 dollars. An assumption is made that sulfuric acid sales will offset fixed charges and O&M costs of the associated acid plant; some other conditions, including operation at a 75% capacity factor (8.8 billion kWh/yr), are also assumed.

Last October, FMC Corp. (Glen Ellyn, Ill.) received a U.S. patent for its "concentrated double-alkali" process, which uses sodium sulfite and lime; the sodium sulfite is regenerated. According to FMC, large pH fluctuations, oxidation to sodium sulfate with attendant sodium loss, and scaling are inhibited, while SO$_2$ removal is efficient. System components are virtually identical to those of a lime/limestone system. Scrubber wastes resemble a dry filter cake and are easily handled and disposed of. FMC's biggest installation so far is at Caterpillar Tractor Co. (Mossville, Ill.). This installation collects SO$_2$ and fly ash concurrently from boilers of 460 000 lbs/h of capacity, and is the largest such installation in the U.S., at present. It went into operation last October and has had a very high availability.

Research-Cottrell, Inc. (R-C, Bound Brook, N.J.) told *ES&T* that its two-stage limestone scrubber is now in a technical position to operate at a pH low enough to allow complete *in situ* oxidation to calcium sulfate (gypsum). According to R-C, this gypsum can be disposed of either blended with, or separated from, fly ash. The company considers this capability important, since most eastern high-sulfur coals do not have enough ash to stabilize a sulfite sludge not fully oxidized to gypsum.

R-C engineers said that they believe that their most exemplary limestone scrubber installations will be on two 750-MW units at the Martin Lake Station, Tex. Using 2% S lignite containing 15–22% ash, the units are to start up this winter. The ash is blended with limestone scrubber wastes to form a transportable and landfillable product.

With regard to SO$_x$ scrubber sludge fixation (*ES&T*, July 1975, p 622), the "big three" at present are IU Conversion Systems (Philadelphia, Pa.), R-C, and Dravo Corp. (Pittsburgh, Pa.). IU is to stabilize and dispose of SO$_2$ scrubber sludge from two 400-MW units of Columbus and Southern Ohio Electric Co., at Conesville, Ohio. Dravo estimates that in terms of 1975 dollars, stabilization with its "Calcilox" system and disposal could cost $2–4/ton, if high-sulfur coal is burned.

Other approaches

One well-known system that recovers sodium sulfite absorbent and concentrated SO$_2$ uses the Wellman-Lord process, now available from Davy Powergas Inc. (Lakeland, Fla.). SO$_2$ reacts with aqueous sodium sulfite to form sodium bisulfite; the sulfite is thermally regenerated with recovery of SO$_2$ from a concentrated stream. This SO$_2$ can be liquefied, used for a sulfuric acid plant, or reduced to sulfur.

EPA's Stern
"It's FGD for next ten years"

As this goes to press, five Wellman-Lord installations, with 90+% SO$_2$ removal and 97+% on-stream time are operating in the U.S., and 12 more in Japan. Moreover, six additional Wellman-Lord systems are under construction in the U.S., and five in Japan; and two are under design or construction in West Germany. The symposium heard, however, that the largest such installation will be at the San Juan Station Units 1 and 2 of Public Service of New Mexico, at Fruitland, N. Mex. This installation is to go on stream in the summer of 1977. Similar installations at Units 3 and 4 are to start up in November 1978 and April 1980, respectively, with a total of 1770 MW by then.

Japan had about 100 FGD units, some for 250–500 MW, in operation by late last year. Jumpei Ando of Chou University (Tokyo) told the meeting that 97+% availability was the rule; however, costs were high, though less so than low-sulfur fuel. Among Japanese companies involved in FGD are Chiyoda (ferric ion oxidant/limestone), Dowa (aluminum sulfate-gypsum), Fuji Kasei, Kureha (sodium acetate-gypsum), Mitsubishi Heavy Industries (MHI), Mitsui Miike, and TSK. Many of these installations produce sulfuric acid and gypsum, for which a market exists in Japan at present.

What about flue gas denitrification (NO$_x$ control) in conjunction with desulfurization? Chiyoda, Fuji Kasei, and MHI are going this route. MHI uses a lime/limestone approach, and has successfully completed an SO$_x$-NO$_x$ pilot test (1200 scfm) last year.

Resource recovery

The TVA analyzed economic possibilities of using controlled SO$_x$ emissions for resources, and appeared optimistic. One factor that the TVA cited to the symposium was a possible Frasch sulfur shortage 10–20 years from now. Another conceivable gain would be in sulfuric acid, which the authority says that the market can absorb at present, even though the TVA assumed a zero incentive for calculation purposes.

These TVA predictions should be tempered with possible clean-fuel market developments, future sulfur prices, and further economic data-base refinement. The clean fuel factor, especially as it involves washing, solvent refinement, liquefaction, and gasification of coal, must be taken into account. However, the EPA does not foresee these precleaning technologies as being significant soon— although it does leave its options open in selected cases—and thus calls for constant control scrubbers. Resource recovery and by-product sale, perhaps, might prove to be a means by which some financial return may be derived from the scrubbers, which many utilities and industries would have to install to provide cleaner air and protect public health. JJ

Removing SO₂ from stack gases

Reprinted from ENVIRON. SCI. TECHNOL.,

7, 110 (February 1973)

The current crash program on sulfur dioxide control process development has produced a somewhat confused situation. A bewildering array of process types are under study; process concepts once widely accepted are being abandoned; claims and counterclaims for processes are widespread; the cost of sulfur dioxide control is being debated, some claiming what might be called shocking levels; and there is continuing disagreement as to whether a developed technology is available or not. This article is an effort to bring all these differences into focus and to develop the proper perspective for process evaluation.

Stack gas cleaning technology has become so complicated that it is difficult to set up any simple classification as a basis for discussion and evaluation. The usual classification is between wet and dry processes although other considerations are also important.

One of the major considerations is the status of development—how near the process is to being ready for use. To evaluate on this basis, only methods that have been tested in a full-scale unit can be considered. The question then becomes one of designating the size that can be classed as full scale. A reasonable critical size would seem to be about 100 MW; scrubber modules probably will be at least this large when the technology is fully developed. On this basis, 10 processes can qualify (or will qualify when projects are completed): lime scrubbing (boiler injection), lime scrubbing (scrubber introduction), limestone scrubbing, magnesia scrubbing, catalytic oxidation, sodium salt scrubbing (thermal regeneration), sodium salt scrubbing (throwaway), manganese oxide absorption, carbon oxide absorption, and carbon adsorption.

The next size category covers the prototype projects, at the intermediate size level of roughly 10–50 MW. There are some 20 processes or process variations being

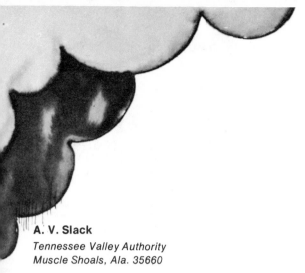

FEATURE

A. V. Slack

Tennessee Valley Authority
Muscle Shoals, Ala. 35660

tested in projects of this category around the world, for example, molten salt (Atomics International, Consolidated Edison), ammonia scrubbing–lime regeneration (Kuhlmann, Electricite de France), lime–limestone scrubbing (EPA-TVA-Bechtel), lime scrubbing (special scrubber—Bischoff; STEAG—Germany), and copper oxide absorption (Shell, Showa Yokkaichi Sekiyu—Japan). Some of the 20 or so test units are on slip streams, and others are on boilers small enough to fall in the prototype range. There are also several test units (or commercial in some cases) in industries such as smelting, paper pulp, and sulfuric acid; although there are important differences, the technology should be generally applicable to electric power production.

The next category is the pilot plant size, of which there are a large number. Many of these are concerned with the same process, for example, limestone scrubbing, but departures inevitably develop so that each can be said to cover a separate "process" or at least a process variation. Classification is also difficult because some are too small to be called pilot plants, and others are such short-term projects that including them is questionable. With a liberal classification, some 75 current or fairly recent pilot plant projects can be identified.

The category of processes on which only bench-scale work, or no work at all, has been done include many patented processes, those that surface only in discussions between research people, and many that have not been disclosed at all. No reasonable estimate can be made in this category.

A third major classification is between recovery and throwaway processes. In the U.S., the major emphasis, especially by utilities planning full-scale installations to meet environmental regulations, has been on the throwaway approach. Some 30 full-boiler installations (some involving full plants with several boilers) are known to be in various stages of planning, construction, or operation; of these, only four involve recovery. The uncertainties regarding market potential have made most utilities shy away from recovery until the regulation situation develops further.

Most throwaway processes involve use of lime or limestone since these are relatively inexpensive materials. One company in Nevada uses sodium carbonate. Lime is also used, in Japan, for sulfur dioxide recovery—the product is a marketable grade of gypsum.

Processes can also be classified on the basis of difficulty in applying to retrofit installation, a very important consideration if regulations require fitting existing plants with stack gas cleaning. Although all processes can be adapted to retrofit, some require the complication of gas

reheating if major alteration of the boiler is to be avoided. These include molten salt (Atomics International), catalytic oxidation (Monsanto), and some of the in situ methods for sulfur dioxide reduction in the gas stream.

Dry processes—metal oxides, lime–limestone

The dry processes have the major advantage that the gas is not cooled. However, most of them involve absorption in a solid followed by regeneration of the loaded absorbent to give either a final product or a stream highly concentrated in either sulfur dioxide or an intermediate compound such as hydrogen sulfide. The drawback to this is that moving the absorbent from the absorber to the regenerator and back again causes attrition and breakage. The absorbent can be left in place and a "swing type" operation used in which the gas stream is shifted from one vessel to another in a cycle with flow of regeneration gas. This reduces mechanical stress on the absorbent but requires a very expensive dampering system to prevent hazardous leaks from one gas stream to the other.

Solid absorbents also should be quite porous to get as much reaction per cycle as possible. Researchers have found it quite difficult to develop such porosity and activity without seriously weakening the granule structure. Moreover, activity is difficult to maintain because of the sintering effect of many regeneration cycles and because solid impurities in the gas stream tend to plug the pores.

A further drawback is that full oxidation usually occurs on solid sorbents whereas oxidation can be held to less than 10% in wet scrubbing. Thus, even to make sulfuric acid, some solid absorbent processes require a reducing agent to reduce the sulfate to sulfur dioxide. If sulfur is the product, the overall reducing agent requirement is higher than if sulfite were formed during absorption.

There are five leading dry processes involving use of metal oxides. The copper oxide processes—Shell (full-scale unit planned for a 30-MW oil-fired boiler in Japan), Esso-B&W (150-MW unit planned in the U.S. on a coal-fired boiler), and U.S. Bureau of Mines (small scale)—give a rich stream of sulfur dioxide that can either be reduced to sulfur or used as feed to a sulfuric acid plant. All have the problems mentioned earlier for granular sorbents; in addition, since copper oxide sorption must be carried out at 600–750°F. either stack gas heating (to reaction temperature) or major alterations would be required for a retrofit. All this indicates that wet processes may be superior even with the plume reheat requirement.

The alkalized alumina process ($NA_2O \cdot Al_2O_3$) was once considered a leading contender. Although developed originally by the U.S. Bureau of Mines, the Central Electricity Generating Board (CEGB) in England worked independently and planned a full-scale installation. Difficulty in maintaining granule strength, however, was found to be a serious problem. CEGB dropped the project, and today there appears to be little activity on the process.

The Mitsubishi Heavy Industries (Japan) manganese oxide process, which is being tested in a full-scale, oil-fired 110-MW unit, makes ammonium sulfate, which is not a good product in the U.S. except for limited production. The finely divided absorbent used in an entrainment reactor avoids the problems of granular absorbents but introduces some problems in removal from the gas stream. The process is somewhat complicated and requires relatively complex equipment.

The major emphasis in attempts to develop a dry sulfur dioxide removal process has been on use of lime or limestone, in most cases injected directly into the boiler. The main efforts have been in Czechoslovakia, Germany, Japan, and the U.S.; the most extensive work has been done in the federal EPA-sponsored project at the Tennessee Valley Authority (TVA) Shawnee Steam Plant, completed last year.

Results have varied widely. In some of the earlier tests in relatively small units, 50% or higher removal of sulfur dioxide was reported. In the EPA-TVA test work (limestone injection) on a 175-MW boiler, however, only 10–15% removal per stoichiometric amount of limestone could be obtained. Moreover, boiler fouling occurred, and the limestone reduced electrostatic precipitator efficiency. The method does not appear usable where high sulfur dioxide removal is required, but may be applicable in special cases, especially when partial removal is acceptable.

Another major project is the test of the Foster Wheeler–Corson lime injection process under way on an 80-MW coal-fired boiler of the Dairyland Power Cooperative in Wisconsin. The lime is hydrated before injection. Results are said to be encouraging, but no data have been released.

Dry processes—carbon, molten salt

Adsorption on activated carbon has several potential advantages, including low-temperature operation and a simple method for direct oxidation to give sulfuric acid. Although the carbon processes have some important intrinsic advantages, there are also some troublesome problems. For example, the Reinluft prototypes (with a concentrated H_2SO_4 product) were subject to uncontrollable carbon combustion; Chemiebau, the present owner of the process, has reported improvement in this respect, but there have been no large-scale tests. A basic disadvantage for all carbon methods is the low gas velocity allowable and consequently the high carbon volume required; the velocity in most installations has been in the range of 0.5–1.5 ft/sec, as compared with as much as 10 ft/sec in wet-scrubbing processes.

An improved version of the Lurgi process will be tested in a small pilot unit on a coal-fired plant in Czechoslovakia. Although several commercial Lurgi units are in operation in nonpower industries, application to coal-fired power plants brings in some difficulties. The process also has the basic disadvantage that the water-wash method of getting the product acid out of the carbon pores yields weak sulfuric acid (10–20%).

Tests of the Bergbau-Forschung method (producing concentrated H_2SO_4) are planned at the STEAG Lunen plant in Germany next year. Like the Reinluft, the loaded carbon is heated to reduce sulfuric acid to sulfur dioxide by reaction with the carbon. Enlargement of the pores by the consumption of carbon eventually reduces activity and strength. Foster Wheeler has licensed the process in the U.S.

The Westvaco method is unique in that it utilizes fluidized beds and employs hydrogen sulfide developed in the regeneration train to reduce sulfuric acid to sulfur. The process cannot be adequately evaluated until experimental work is done on a larger scale.

The Hitachi (water wash) and Sumitomo (reduction to SO_2) methods are in full-scale operation in Japan. The weak Hitachi acid is neutralized with limestone to give a salable grade of gypsum. In the Sumitomo process, also

Magnesia slurry. *This demonstration SO_2 absorption system, partially funded by EPA, is at a Boston Edison plant*

similar to the Reinluft, hot combustion gas causes reduction of sulfuric acid and desorption of sulfur dioxide, which is then converted to a commercial grade of sulfuric acid. The carbon is recycled between the absorber and the desorber; attrition has been excessive in this process.

The carbon processes afford a means for oxidizing sulfur dioxide to sulfur trioxide in the gas stream at low temperature. Getting the resulting sulfuric acid out of the carbon pores, however, complicates the process so—and the absorbent volume required is so large—that the economics are not encouraging.

The molten salt process developed by Atomics International (North American Rockwell) involves scrubbing with a eutectic melt of Na_2CO_3, K_2CO_3, and Li_2CO_3 at about 800°F. For coal-fired boilers, the ash must be first removed in a hot precipitator. The loaded melt is reduced with petroleum coke to give hydrogen sulfide, which is then converted to sulfur.

The method is being tested on a 10-MW slipstream at the Arthur Kill plant of Consolidated Edison. It is a unique process, with considerable promise; solutions to the peculiar problems involved will be sought in the prototype tests.

Atomics International has also carried out pilot plant work (at the Mohave station of Southern California Edison) with sodium carbonate in a spray-dryer type of operation that might be classed as a dry system in that it cools the gas only partially. Throwaway operation is contemplated.

Wet processes

Wet scrubbing methods are of both the clear solution and slurry types. Although solid material in the scrubber liquor introduces problems, it is necessary in some processes because of low sorbent or product solubility. In others, operation with a slurry of highly soluble crystals in a saturated solution may be desirable as a means of reducing recirculation requirement through the regeneration section.

In contrast to dry sorption, metal oxides have received little attention in wet scrubbing. Lack of advantage over the more reactive alkali absorbents is the main reason. Zinc oxide has been considered for special situations in the smelter industry, but the several small-scale projects on applying zinc oxide scrubbing to power plant gases have not progressed very far. One approach, studied at TVA, is to oxidize $ZnSO_3$ to $ZnSO_4$ and decompose the sulfate thermally to SO_3 and ZnO, thereby eliminating need for a sulfuric acid plant.

The main absorbents are the alkalis (sodium and ammonium) and the alkaline earths (calcium and magnesium). In addition, weak sulfuric acid is used in one process and an unidentified organic material in another. Most of the absorbents are used both in throwaway and recovery process types.

Another consideration is whether the boiler is fired with coal or oil, which affects not only the dust content of the gas but also the presence and nature of minor impurities. The content of soluble impurities will differ, which affects the purge situation in the scrubber system, and oxidation inhibitors and promoters will be present in varying amounts. Data obtained on one gas type are not necessarily applicable to the other.

The main competition in the alkali absorbent area is between sodium and ammonia. Both are excellent absorbents with a high affinity for sulfur dioxide and, with both, the sulfite-bisulfite equilibrium can be used to optimize the absorption-regeneration cycle. Moreover, the regeneration can be carried out in a liquid system at relatively low temperature, whereas the alkaline earths require high temperature for sulfite decomposition and absorbent re-

TABLE I

Sodium-based methods for stack gas cleaning

Developer	Status	Type and size of test unit	Location	Product
Wellman-Power Gas	Commercial operation, oil-fired, 60 MW	125 MW, coal-fired planned	Japan and U.S.	H_2SO_4 or S
Stone & Webster-Ionics	Piloted	70 MW planned	U.S.	H_2SO_4
UOP	Prototype in operation	. . .	U.S.	S
U.S. Bureau of Mines	Small scale	Piloting on coal-fired steam boiler planned	U.S.	S
Kureha Chemical	Commercial	Oil-fired; 75 MW	Japan	Crystal Na_2SO_3
Showa Denko	Commercial	Oil-fired; 45 MW	Japan	Na_2SO_3 solution
Oji Paper	Commercial	Oil-fired, several, up to 55 MW	Japan	Na_2SO_3 solution

The dust content of the gas is a very important consideration in wet scrubbing processes. For a new boiler, it is desirable to remove both dust and sulfur dioxide in the scrubber system to avoid the cost of an electrostatic precipitator. The question then is whether to use only one scrubber or two—one designed for dust and the other for sulfur dioxide. If the dust has relatively small particle size, the two-scrubber arrangement may be unavoidable. If not, it may still be desirable to avoid getting the dust mixed in with the sulfur dioxide recovery product. There is a major advantage, however, in use of only one scrubber and filtering the scrubber effluent liquor to separate the dust; the hot gas evaporates water from the scrubber solution and thus reduces heat requirement in the regeneration section. This applies only to recovery processes, but there is an advantage in throwaway operation also—more makeup water can be used in the scrubber (for example, for washing the mist eliminator) without incurring "open loop" operation. Single-scrubber operation in recovery processes of the slurry-scrubbing type probably would not be feasible, since it would be difficult to separate the collected dust from the solid absorbent.

For existing plants, the same considerations hold generally, except that most boilers will be already equipped with precipitators when the sulfur dioxide unit is installed. Some existing precipitators, however, will not meet regulatory standards for dust removal, and the scrubbers thus will have to remove the residual dust.

generation. Another advantage is that the clear solutions used avoid the problems in alkaline earth slurry scrubbing systems.

Sodium has the advantage over ammonia that the cation is not volatile and therefore a simpler scrubbing system can be used. Also, no fume problem occurs as in ammonia scrubbing. Ammonia, on the other hand, solves the sulfate problem. Since some oxidation inevitably takes place in the scrubber with any absorbent, the resulting sulfate must be either regenerated, sold, or purged as a waste material. Purging is generally considered unacceptable because of potential water pollution; for regeneration or sale, particularly the latter, ammonium sulfate is much more desirable than sodium sulfate.

There are seven leading sodium-based methods for stack gas cleaning (Table I). The Wellman-Power Gas method (in commercial operation on an oil-fired, 60-MW power plant in Japan) involves thermal regeneration and is the most advanced. It is used on Claus and sulfuric acid plants as well as in the power industry; the product is sulfuric acid or sulfur. Most of the process problems appear to have been worked out with the exception of sulfate disposal; in Japan, for example, the sulfate solution is discarded to the ocean. Some progress has been made toward reducing oxidation, particularly in Japan where Sumitomo Chemical has developed an inhibitor that reportedly cuts oxidation in half. Nevertheless, even with inhibition a large power plant is likely to produce re-

TABLE II
Processes using ammonia to absorb sulfur dioxide

Developer	Status	Type and size of test unit	Product	Location
NIIOGAZ	Commercial; abandoned	Coal-fired; 75 MW	Liquid SO_2, $(NH_4)_2SO_4$	USSR
Cominco	Commercial; operation since 1930's	Smelter	H_2SO_4, $(NH_4)_2SO_4$	Canada
Cominco	Commercial	H_2SO_4 plants	H_2SO_4, $(NH_4)_2SO_4$	U.S.
JECCO	Commercial; abandoned	H_2SO_4 plants	$(NH_4)_2SO_4$	Japan
IFP	Prototype planned	. .	S	France
EPA–TVA	Pilot in operation	Coal-fired; 1 MW	S	U.S.
RIIC	Pilot operated	Coal-fired; 4 MW	H_2SO_4, $(NH_4)_2SO_4$	Czechoslovakia
RIIC	Commercial	H_2SO_4 plant	H_2SO_4, ammonium nitrate	Czechoslovakia
IPRAN	Commercial	H_2SO_4 plants	H_2SO_4, ammonium phosphate	Romania
Monsanto	Under development	. . .	H_2SO_4 or S	U.S.

several tons of sodium sulfate per day. Various regeneration methods are available, but all appear quite expensive. An important phase in development of the process is a project cosponsored by EPA at Northern Indiana Public Service for a 125-MW, coal-fired test.

The Stone and Webster-Ionics method involves use of electrolytic cells to convert Na_2SO_4 to $NaHSO_4$, which is then added to the scrubber effluent solution to decompose sulfite and bisulfite—evolving sulfur dioxide and giving sodium sulfate (Na_2SO_4) for recycling. Sulfite oxidized to sulfate in the scrubber is converted in other cells to weak sulfuric acid which can be used in the sulfuric acid plant if acid is the main product. Otherwise, the waste acid probably would have to be neutralized with limestone.

The method will be piloted at a Wisconsin Electric Power plant in a joint project with EPA. If successful, a full-scale unit may be built. The main problem, other than possible high cost for power, appears to be sulfate disposal if the sulfur dioxide is reduced to sulfur.

The UOP process has not been disclosed in any detail; a prototype unit is in the startup phases. Sulfate disposal may be a problem here also.

The Bureau of Mines method is unique in that sodium citrate is the absorbent rather than the usual NaOH or sodium sulfite (Na_2SO_3). Use of the citrate allows regeneration of the scrubber effluent liquor by treatment with hydrogen sulfide, giving sulfur as the product without need for first evolving sulfur dioxide from the solution. Larger pilot plants, one at a smelter and another on a coal-fired steam boiler, are planned.

Sale of Na_2SO_3 as such to paper pulp companies, as in Japan, eliminates the cost of regeneration. The market for sulfite in the U.S., however, appears to be limited. Three companies in Japan operate commercial units that yield Na_2SO_3 in solution or crystal form.

Use of ammonia to absorb sulfur dioxide is a very old process, used in the fertilizer industry long before the modern emphasis on sulfur dioxide emission control developed. Much of the use of ammonia, as shown in the following project tabulation, has been in fields other than power production (Table II).

For power plant application, the NIIOGAZ project in Russia seems to have been fairly successful, but the unit was abandoned when the power plant with which it was associated was converted to natural gas several years ago. So far, Russian government authorities have not seen fit to build another. The same situation holds in Czechoslovakia; the test unit is no longer operated, and there is no prospect for building a larger one. The emphasis on sulfur dioxide control does not appear so intense as in the U.S.

The IFP (Institut Francais du Petrole) method involves thermal decomposition of scrubber effluent liquor to give a gaseous mixture of ammonia, sulfur dioxide, and water, which is passed through a "wet Claus" reactor at elevated temperature to reduce the sulfur dioxide to sulfur by reaction with hydrogen sulfide. The ammonia is recycled. Sulfate produced in the scrubber can also be reduced. The method seems promising, although two thirds of the sulfur must be reduced to hydrogen sulfide and recycled. A full development program is planned.

The EPA-TVA pilot plant project involves treatment of scrubber solution with ammonium bisulfate (NH_4HSO_4) to evolve sulfur dioxide and give $(NH_4)_2SO_4$, which is crystallized and thermally decomposed to NH_4HSO_4 for recycling. The process has several attractive features, including low energy requirement and relatively simple separation and disposal of sulfate (by diverting part of the sulfate crystals to fertilizer use). Fume development in the scrubber, however, is a problem in all ammonia-scrubbing methods.

The only potassium-based recovery method known to be under development is the "formate" process of Consolidation Coal. The sulfur dioxide is absorbed in potassium formate solution, and the scrubber effluent is regenerated by a somewhat complicated procedure to give sulfur as the product. A pilot plant is in operation at the Phoenixville, Pa., station of Philadelphia Electric; data for evaluation are not yet available.

Lime-limestone. *Three 30,000 acfm scrubbers operate at test facility at TVA's Shawnee (Ky.) coal-fired plant*

The Monsanto NOSOX process is based on an undisclosed organic absorbent rather than an alkali. The scrubber solution is decomposed thermally to give sulfur dioxide; a separate, undisclosed step removes sulfates. Only bench-scale studies have been made.

Alkali absorbents (throwaway)

The major throwaway method, lime-limestone slurry scrubbing, is beset by many problems. Because of this, a considerable amount of development work has been done on what are generally called "double alkali" processes. A clear solution of sodium or ammonium salts is used to absorb the sulfur dioxide, and the resulting solution is treated with lime or limestone to precipitate calcium sulfite–sulfate for discard. Scrubber operation is improved considerably although an extra processing step is added.

Activity has been so intense in this area that it is diffi-
cult to present an accurate assessment of the status. In Table III, some of the processes listed are not strictly double alkali because the absorbent is something other than sodium or ammonium salts; all, however, have the calcium sulfite–sulfate precipitation step.

The pros and cons of sodium vs. ammonia in the double alkali field are about the same as in recovery. For sodium salt scrubbing, the sulfate can be regenerated if lime is used as the regenerant, but only a dilute solution of NaOH can be produced for recycling. Moreover, calcium dissolved in the regeneration step can cause scaling in the scrubber, a problem that can be avoided by treating the return solution with Na_2CO_3 to precipitate calcium as $CaCO_3$—but at the expense of a purge requirement to rid the system of the added sodium.

The General Motors (GM) work on the sodium system, with imput from A. D. Little, is the most significant in the U.S. GM used lime as the regenerant, as do several others whose work has not been disclosed sufficiently for inclusion in Table III, but limestone has also been used. In this case, only the $NaHSO_3$ is regenerated; the Na_2SO_3 is recycled to the scrubber, and the Na_2SO_4 builds up in the circuit. The Japanese companies have a special system for regenerating the sulfate, involving treatment of a side stream with $CaSO_3$ crystals and H_2SO_4; $CaSO_4$ precipitates, and $NaHSO_3$ is produced for recycling.

The sulfate problem can be handled much more easily if ammonia is the absorbent. In the EDF-Kuhlmann process, for example, the lime reacts well with $(NH_4)_2SO_4$ to volatilize ammonia and precipitate calcium sulfate. The evolution and removal of ammonia drives the reaction to completion. The EDF test unit is being moved from Paris to Champagneoise for further testing. The cost of the process can be reduced by using limestone as the regenerant. Again, only the bisulfite reacts but the ammonium sulfate can be separated and sold as fertilizer. TVA has done small-scale work to evaluate this approach and plans to carry out pilot plant studies as soon as possible.

Chiyoda uses dilute sulfuric acid rather than alkali as the absorbent. A catalyst in the liquor promotes oxidation of sulfurous acid to sulfuric, part of the sulfuric is reacted with limestone to precipitate calcium sulfate, and the residual acid is recycled. The main drawback is the very

TABLE III
Alkali absorbents (throwaway)

Developer	Status	Type and size of test unit	Absorbent/ regenerant	Location
General Motors	Piloted	Coal-fired, 40 MW planned	NaOH/CaO	U.S.
Chemico	Piloted	Various	$Na_2SO_3/CaCO_3$	U.S.
TVA	Small-scale	Pilot planned; coal-fired, 1 MW	$(NH_4)_2SO_3/CaCO_3$	U.S.
EDF–Kuhlmann	Prototype	Oil-fired, 30 MW	NH_4OH/CaO	France
Showa Denko	Pilot in operation	Oil-fired, 3 MW 175 MW planned	$Na_2SO_3/CaCO_3$	Japan
Kureha Chemical	Pilot in operation	Oil-fired, 1.5 MW	$Na_2SO_3/CaCO_3$	Japan
Chiyoda	Piloted	Commercial planned; oil-fired, 25 MW	Water/$CaCO_3$	Japan
Monsanto	Small pilot	Pilot planned; coal fired, 1 MW	Organic/CaO	U.S.
Eimco	Piloted	Coal-fired	NaOH/CaO	U.S.

TABLE IV
Lime–limestone SO$_x$ scrubbing large-scale projects

Company	Developer	Status	Unit type and size	Absorbent
FULL SCALE				
Arizona Public Service	Research-Cottrell	Under construction	Coal-fired, 115 MW	$CaCO_3$
Commonwealth Edison	B & W	Operating	Coal-fired, 175 MW	$CaCO_3$
Detroit Edison	. . .	Planned	Coal-fired, 160 MW	$CaCO_3$
Duquesne Light	Chemico	Under construction	Coal-fired, 150 MW	CaO
Kansas City Power and Light	Combustion Engineering	Operating	Coal-fired, 2 × 120 MW	CaO
Kansas City Power and Light	B & W	Operating	Coal-fired, 117 MW	$CaCO_3$
Kansas City Power and Light	Combustion Engineering	Operating	Coal-fired, 125, 430 MW	CaO
Key West	Zurn	Operating	Oil-fired, 37 MW	$CaCO_3$
Louisville Gas and Electric	Combustion Engineering	Under construction	Coal-fired, 70 MW	$Ca(OH)_2$
Northern States	Combustion Engineering	Under construction	Coal-fired, 2 × 700 MW	$CaCO_3$
Ohio Edison	. . .	Planned	Coal-fired, 2 × 880 MW	CaO
TVA	TVA	Under construction	Coal-fired, 550 MW	$CaCO_3$
Union Electric	Combustion Engineering	Abandoned	Coal-fired, 120 MW	CaO
USSR (state-operated)	NIIOGAZ	Operating	Smelter, 900 MW equiv	$CaCO_3$
Nippon Kokan KK (Japan)	Mitsubishi	Operating	H_2SO_4 plant, about 20 MW equiv	CaO
Kansai Electric (Japan)	Mitsubishi	Operating	Oil-fired, 30 MW	CaO
Tomakomai Chemical (Japan)	Mitsubishi	Planned	Smelter, 15 MW equiv	CaO
Mitsui Aluminum (Japan)	Chemico	Operating	Coal-fired, 165 MW	CaO
Sodersjukhuset (Sweden)	Bahco	Operating	Oil-fired, 3 × 6 MW	CaO
Egypt (state-operated)	Chemiebau	Abandoned	H_2SO_4 plant, small	CaO
Public Service of Indiana	Combustion Engineering	Planned	Coal-fired, 650 MW	$CaCO_3$
Southern California Electric	SCE	Planned	Coal-fired, 2x150 MW	$CaO/CaCO_3$
PROTOTYPE				
EPA–TVA	Bechtel, EPA, TVA	Operating	Coal-fired, 3 × 10 MW	$CaCO_3$, CaO
STEAG (Germany)	Bischoff	Operating	Coal-fired, 35 MW	CaO
Ontario Hydro (Canada)	Ontario Hydro	Planned	Coal-fired, 30 MW	$CaCO_3$

large volume of liquor that must be recirculated through the scrubber. Moreover, the process has not been tested on gas from a coal-fired boiler; such gas is known to contain oxidation catalyst poisons.

The Monsanto absorbent is an undisclosed organic compound. The resulting sulfate is said to react easily with lime. A 2000-scfm pilot plant is being installed at an Indianapolis Power and Light plant.

Alkali scrubbing has also been used in installations where the alkali sulfite–sulfate is discarded. This is the process being installed at the Reid Gardner plant (125 MW) of Nevada Power. The plant is located in a very dry area where a soluble salt can be discarded with less question about water pollution than in other areas. Discard of sodium sulfate (to the ocean) is also planned by Kawasaki Heavy Industries in Japan. Investment is relatively low in such an approach, but raw material cost is high. The spray-dryer process tested by Atomics International can also be operated as a wet scrubbing system.

Magnesia (recovery)

Magnesia has received special emphasis as an absorbent for sulfur dioxide recovery. Lime and limestone have also been considered but have some disadvantages as compared with magnesia; the main use of lime in recovery is in Japan where calcium sulfite is oxidized to the sulfate and sold to the cement and wallboard industries.

All MgO processes involve scrubbing with $Mg(OH)_2$ slurry, separation of the resulting $MgSO_3$, and calcination of the sulfite to evolve a stream of sulfur dioxide and regenerate MgO. The calciner gases dilute the sulfur dioxide to a range that is acceptable for sulfuric acid production but which would increase cost if the sulfur dioxide were reduced to sulfur.

Venturi scrubbers are used in the Chemico-Basic process which is a full-scale operation. Demonstration projects, partially funded by EPA, are under way at Boston Edison (Mass.)—oil-fired—and Potomac Electric (Md.)—coal-fired. The separated and dried $MgSO_3$ is shipped from these plants to a sulfuric acid producer where the material is calcined and acid produced; the MgO is returned to the power plant. This is the central processing concept under which several power boilers

supply loaded absorbent to a central regeneration plant large enough to be economical. The feasibility depends on whether a very large central plant, for instance, 2000–3000 tons of acid per day, could sell all the acid at reasonable net profit.

The Grillo process employs manganese as well as magnesium in the absorbent, purportedly to get faster absorption and promote reduction of sulfate. A spray-type absorber is used. The method has been tested on a fairly small scale (about 7 MW) on an oil-fired boiler in Germany. In Japan, Mitsui Shipbuilding has taken a license and operates a small pilot plant; a 60-MW commercial unit (oil-fired) is planned.

The United Engineers process has not been publicized. A 120-MW unit will be constructed at Philadelphia Electric's Eddystone (Pa.) station.

The magnesia processes have some advantages over the alkali type, including absence of a sulfate problem; addition of coke or other reducing agent in the calcining step reduces any sulfate present. On the negative side, heat requirement is high, and the gas dilution would make sulfur production expensive.

Lime-limestone (throwaway)

The bulk of the research and development effort is being applied to throwaway processes using lime or limestone as the absorbent. There are so many projects, of various magnitudes and in various stages of progress, that it is difficult to summarize them. Tables IV and V list the more significant projects under way; no claim is made for completeness or for a high degree of accuracy.

The main problem by far in scrubbing with either lime or limestone is deposition of solids on surfaces in the scrubber and associated equipment from crystallization of calcium sulfate or sulfite (scaling) and from mechanical sticking of slurry solids to surfaces. The latter is especially troublesome at the two wet-dry interfaces—as the gas contacts the slurry and as it leaves the scrubber.

Scaling is a much more severe problem when lime is used rather than limestone. Because scaling is such a complex phenomenon, the reasons are not yet fully understood. Presumably they are associated with the generally higher pH in the system when lime is used and with the tendency of freshly hydrated lime to stick to surfaces.

For either lime or limestone, there are certain operating parameters that have an important effect on scaling. These effects were worked out by Imperial Chemical Industries (ICI) in England some 40 years ago, and none of the recent work controverts the ICI findings. The following factors help to reduce scaling:

• high liquor rate. If enough liquor is circulated (per unit of sulfur dioxide absorbed) to avoid exceeding a certain level of supersaturation wth sulfite and sulfate, and if this supersaturation is dissipated before the liquor returns to the scrubber, then no scaling can occur

• adequate delay time before the liquor is returned to allow dissipation of the supersaturation

• high content of sultite and sulfate crystals in the circulating slurry to provide surfaces on which dissolved sulfite and sulfate can crystallize

• high slurry velocity to scour off any deposits that tend to form

• dilution with fresh water to reduce the degree of supersaturation

• use of open-type scrubbers with no obstructing surfaces that might cause liquor stagnation or silting of solids

• use of scrubbers or scrubber stages in series, with separate slurry circulation to each, thereby reducing the

TABLE V
Lime–limestone pilot studies

Company	Absorbent	Station (or location)
Babcock and Wilcox	$CaCO_3$	Ohio
Combustion Engineering	CaO, $CaCO_3$	Connecticut; also others
Chemico	CaO, $CaCO_3$	Various
Ontario Hydro	$CaCO_3$	Toronto
American Metals Climax	$CaCO_3$	Pennsylvania
Smelter Control Research Association	CaO, $CaCO_3$	Nevada
American Air Filter	CaO, $CaCO_3$	Kentucky
Bechtel	CaO, $CaCO_3$	Nevada
UOP	CaO	Connecticut
Southern California Edison	CaO, $CaCO_3$	Nevada
Pennsylvania Power and Light	$CaCO_3$	Pennsylvania
Detroit Edison	$CaCO_3$	Michigan
Krebs Engineering	CaO	Minnesota
Marblehead Lime	CaO	Illinois
National Lime Association	CaO	Nevada
Research-Cottrell	$CaCO_3$	Ohio
Joy Manufacturing	$CaCO_3$	California, Pa.
EPA	$CaCO_3$	North Carolina
Du Pont	CaO, $CaCO_3$	Tennessee
TVA	$CaCO_3$	Colbert Station, Ala.
Environeering	$CaCO_3$	Illinois
Zurn	$CaCO_3$	Key West
Bahco	CaO, $CaCO_3$	Sweden
Mitsubishi	CaO	Japan
CHEPOS	$CaCO_3$	Czechoslovakia
NIIOGAZ	$CaCO_3$	USSR

amount of sulfur dioxide absorbed per volume of liquor per pass.

Because of the higher reactivity of lime, there has been a major effort to juggle these factors in such a way to avoid scaling and still get the advantages resulting from the superior reactivity, including smaller scrubbers, lower liquor recirculation rate, and better absorbent utilization. The results have been mixed. In most cases, severe scaling has occurred. In a few cases, notably in the plants in Japan, scale-free operation has been obtained. Pilot plant tests with a spray scrubber on gas from low-sulfur western coal have also been successful. It appears, however, that no tests cover completely the situation faced by many utilities in the eastern part of the United States:

• high sulfur content (3.5–4.5%). The higher the sulfur content, the higher must be the circulation rate to avoid excessive supersaturation as the liquor passes through the scrubber. Thus lime scrubbing is much more applicable to low-sulfur western coals.

• closed liquor loop operation (except for liquor discarded with waste solids). Some of the successful tests have been with partial open-loop operation (discard of scrubber liquor and replacement with water, which probably cannot be tolerated in future practice because of the resulting water pollution

• lime kiln operation at the power plant. Freshly hydrated lime may stick to scrubber surfaces more than the "aged" hydrated lime used in most of the tests.

One problem is that two of the advantages expected from lime—low pumping rate and low solids content in the slurry (which reduces entrainment problems)—cannot be fully realized because of the high liquor rate and high solids content required for controlling scaling. However, smaller scrubber size (for a given degree of sulfur dioxide removal) and better utilization, which are important considerations, can be attained.

If lime is to be used, it appears that a self-scouring type of scrubber such as a venturi or spray-type (with high liquor velocity) is preferable and that stage operation would be helpful. Most of the more successful lime tests have been with equipment of this type. Producing the lime in a separate kiln also seems preferable to injection into the boiler, since in some of the tests the latter has caused boiler fouling.

Because of the question regarding scaling propensity of lime, there has been a trend toward use of finely ground limestone. Although this reduces the scaling problem, it is still there—and the general effect of variables is the same as for lime. The more successful projects have involved high liquor rate (on the order of 50 gal/mcf), high solids content (7–15%), 5–10 min delay time, and open scrubbers of the spray type or equipped with a very open type of packing (or mobile packing).

The "mud" deposition problem can be controlled by soot blowing or by washing with water or slurry. (Use of recycled waste pond water has generally resulted in severe scaling.) This problem is especially serious in the mist eliminator, where entrained solids can cause rapid plugging. The safest procedure, although expensive, is to place the mist eliminator in a horizontal duct section at the top of the scrubber and wash it with a separate flow of water. This opens the loop slightly but perhaps not to an unacceptable degree. Another approach is to place a valve tray or sieve tray in the top of the scrubber and irrigate it with water, which is thus kept separate from the scrubber liquor. The tray prevents solids from reaching the mist eliminator above.

Cat-Ox. *Monsanto's catalytic oxidation system is under test in an EPA-sponsored project at Illinois Power Co.*

By holding gas velocity and slurry solids content at relatively low levels, the mist eliminator (in the standard position) has been kept clean in some tests by washing with only that amount of makeup water available in closed-loop operation. Whether or not this is feasible in large-scale, long-term operation remains to be seen.

A major unresolved problem for both lime and limestone is sludge disposal. The solids do not compact well, and large tonnages are produced. This is probably the main drawback to the throwaway type of operation.

At the moment the industry is faced with the choice between relatively safe procedures on the one hand (use of limestone, special mist eliminator arrangements, high slurry solids content) and on the other, procedures that may save money but which increase the chances of unreliable performance. Reliability can also be improved by open-loop operation but at the expense of water pollution.

Technology is far from fully developed; however, the EPA-sponsored project at TVA's Shawnee station should provide many of the answers needed. The problems of both lime and limestone scrubbing are to be thoroughly explored.

Catalytic oxidation

Catalytic oxidation does not fit well into either of the preceding categories. The method is an old one but the only significant work is that done in the past few years by Monsanto. The gas is thoroughly cleaned of dust in a hot precipitator and passed through sulfur dioxide–oxidation catalyst at high temperature. The resulting sulfur trioxide condenses with moisture to form about 80% H_2SO_4 when the gas becomes cooled to the dew point by the normal cooling in the boiler. Acid mist is removed in a Brink mist eliminator.

The process is being tested in an EPA-sponsored project (100 MW) at the Wood River station of Illinois Power. There are several advantages, including simple operation, low labor requirement, and no cooling of the gas. The main drawback appears to be the grade of the acid produced—the concentration is lower and less pure than the usual commercial acid.

Costs

Although numerous cost estimates have been made for cleaning stack gases, the cost situation has become so complex that most of the published information is not definitive. The stage of development, for example, is a factor, because the projected cost of any process rises steadily as more is learned about it.

Throwaway investment has been generally assumed to be lower than for recovery because no regeneration facilities are needed. As more has been learned about the problems of scaling, mist eliminator plugging, and sludge disposal in lime/limestone scrubbing, however, this assumption has become less firm. It may be that the simpler recovery methods will actually be less expensive to install, but the product disposal problem still remains. If the utility and smelter industries should install recovery systems widely, the market structure would be upset no matter what the product. An extended period of market adjustment will be necessary, during which time throwaway installations will be essential on part of the capacity converted to stack gas cleaning.

As between lime and limestone for throwaway operation, the main factor is the scaling problem. For those situations in which lime is acceptable, the choice will be an economic one. If the delivered raw limestone cost is high, then calcining to reduce scrubber cost and improve utilization may reduce overall cost; if limestone cost is low, however, as it is in much of the eastern U.S., use without calcining would be indicated.

It is difficult to evaluate the recovery processes since so much depends on problems still to be identified during development. This situation should change soon, however, since there are now 10 processes (including both recovery and throwaway) on which design and operating data for units of 100 MW or more either are available now or will be when planned projects are completed. This should provide an adequate basis for a meaningful comparison both between throwaway and recovery and between the various processes in each group. It seems likely, however, that the situation will be more important than the process; in other words, although a particular process may be superior under one set of conditions, it will not be under some other. Moreover, even if the best process is chosen for the particular combination of conditions, the cost will vary widely depending on factors such as plant size and location, load factor, cost of raw material, difficulty of product disposal, and whether the installation is new or retrofit. The range cannot be projected with any assurance of accuracy until more definitive cost estimates are available. However, the present indication is that for power plants investment can easily be as high as $75/kW and operating cost as high as 2.5 mills/kWh (about $6.50/ton of coal or 27¢/10 million Btu). Cost both lower and higher than this can be expected, depending on the situation.

A. V. Slack, *chief chemical engineer, Division of Chemical Development, Tennessee Valley Authority, directs efforts to develop and evaluate technology related to control of sulfur oxides in stack gases. He has been with TVA since receiving his MS degree from the University of Tennessee in 1941, working mainly in the field of fertilizer research and development and now in sulfur oxide control. Mr. Slack's paper originally appeared in* Electrical World.

Reprinted from ENVIRON. SCI. TECHNOL., **9**, 627 (July 1975)

Sodium scrubbing wastes

Insolubilization processes improve disposal options

Jacques M. Dulin
Industrial Resources, Inc.
Chicago, Ill. 60603

Edward C. Rosar
Industrial Resources, Inc.
Denver, Colo. 80215

Sodium alkalis for scrubbing of flue gases to control SO_x emissions are better sorbents than calcium alkalis—lime or limestone. Yet, sodium alkalis, the compounds of choice in recycle systems, which are slightly less desirable in "throwaway" systems, have one major disadvantage: the "high" solubility of the end-product sodium sulfates and sulfites.

Industrial Resources, Inc. (IRI), a Chicago company, has developed methods to "insolubilize" sodium sulfate, both through an aqueous and a sintering process. The resultant compounds offer new methods of disposing of sulfates.

The impetus

IRI's interest in developing new sodium disposal techniques stems from the fact that the company holds federal leases for the mining of nahcolite (sodium) ore in western Colorado. This nahcolite ore assays 60–85% $NaHCO_3$, and can be used as an SO_x sorbent in scrubbing processes.

In discussions with utilities on the use of nahcolite ore as an SO_x sorbent, it appeared that the problem of sodium sulfate disposal could pose a worse problem than the SO_x emissions. Without a low-cost regeneration process, sodium as an SO_x sorbent was not attractive to utilities unless the solubility of resulting sodium sulfur oxide wastes could be reduced. Utilities did not want to substitute a water pollution problem for an air pollution problem.

To have an available sodium scrubbing process that competes as a throwaway process with lime or limestone wet scrubbing requires:
- a plentiful source of crude sodium at a reasonable cost
- a potentially approvable method of disposal of sodium sulfate wastes.

Raw nahcolite ore answers the first requirement. And IRI has 8,358 acres in western Colorado under federal lease on which there is an estimated resource of 100 million recoverable tons of nahcolite ore in the uppermost four beds of the deposits, and substantially more nahcolite is available if the remaining bedded and disseminated deposits are included. Because the raw ore is useful in a crushed or powdered form without need to further beneficiate, the cost does not include a premium for upgrading. A high quality product, such as glassmakers-grade soda ash, is really unnecessary for flue gas SO_x scrubbing. Raw ore in lump and powdered forms are shown in Figure 1.

To achieve the second requirement necessitated a 100-fold reduction in the solubility of sodium sulfates and sulfites from the 100–500 g/liter solubility level of these salts to the 2.3 g/liter level of calcium sulfate—the accepted industry standard. This reduction in solubility has been accomplished through research sponsored by IRI and conducted at Battelle Memorial Institute's Columbus (Ohio) Laboratories under the direction of Drs. Joseph M. Genco, Harvey S. Rosenberg, and Russell B. Bennett.

Aqueous insolubilization

One method for reducing the solubility of sodium sulfates/sulfites, primarily the idea of Edward C. Rosar and called the FERSONA process, involves the formation of relatively insoluble, granular crystalline precipitates; this is accomplished by reacting waste sodium sulfate with acidic ferric ion solutions to form double salts of sodium ferric hydroxy disulfates; as found in nature these species are known as sideronatrite and natrojarosite (Figure 2). The reaction conditions are relatively mild—atmospheric pressure and temperatures of 120–150°F. Unlike calcium sulfate/sulfite wastes, the resulting precipitates do not form a thixotropic, water-retaining sludge, but may be easily dewatered to dryness. Solubilities of sideronatrite and natrojarosite (see table) are 10 to 100 times less than calcium sulfate; this is equivalent to a 10^3–10^4 reduction in the starting solubility of sodium sulfates/sulfites, and is well below the calcium sulfate standard.

Continued research into this FERSONA process has extended its range of applicability. The process chemistry has successfully insolubilized sodium sulfate wastes from a variety of scrubbing processes, using both sodium carbonate and bicarbonate as the sorbent. Indeed, it appears that neither the source of the sodium, nor the means or manner in which the scrubbing process produces the sodium sulfate/sulfite is particularly critical.

Perhaps the most startling application is with nuclear power plant wastes. Nuclear plants may typically use cooling towers for heat dissipation by evaporation of water from natural sources. Such sources may contain about 15–250 ppm Na^+ and $SO_4^=$ ions, which are concentrated during the recirculation in the tower to about 10,000–50,000 ppm. This waste "magma" poses a serious disposal problem since it amounts to an estimated 15 tons per year/MW. IRI's FERSONA process technology has been applied to a waste typical of cooling tower magma to produce an insoluble precipitate. The technology can be applied to similar wastes from any cooling tower, whether associated with a nuclear or fossil fuel-fired plant. The FERSONA process is also a solution to the problem of disposal of sodium sulfate-containing wastes from demineralizers (for boiler feed water), an indispensible part of any power or industrial plant. These applications of the FERSONA process—disposal of cooling tower and demineralizer wastes—are the subject of U.S. patent 3,876,537 granted to IRI in April 1975.

FIGURE 1

IRI disposal processes for SO$_x$ emissions

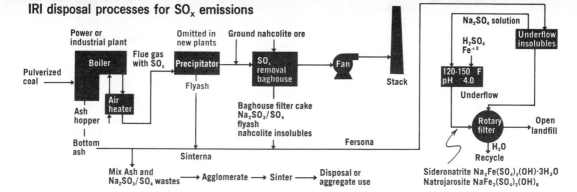

Power or industrial plant — Boiler — Flue gas with SO$_x$ — Precipitator (Omitted in new plants) — SO$_x$ removal baghouse (Ground nahcolite ore) — Fan — Stack

Pulverized coal — Boiler

Air heater

Ash hopper

Bottom ash

Flyash

Baghouse filter cake Na$_2$SO$_3$/SO$_4$ flyash nahcolite insolubles

Sinterna

Fersona

Mix Ash and Na$_2$SO$_3$/SO$_4$ wastes → Agglomerate → Sinter → Disposal or aggregate use

Na$_2$SO$_4$ solution

H$_2$SO$_4$
Fe^{+3}

Underflow insolubles

120-150 °F
pH 4.0

Underflow

Rotary filter → Open landfill

↓ H$_2$O
Recycle

Sideronatrite Na$_2$Fe(SO$_4$)$_2$(OH)·3H$_2$O
Natrojarosite NaFe$_3$(SO$_4$)$_2$(OH)$_6$

A hidden bonus of the FERSONA process is its applicability as a method of disposal of other wastes, such as waste crude sulfuric acid, steel pickling liquors, acid mine waters, and pyrites dump leach waters. These wastes typically contain sulfuric acid and/or sources of ferrous or ferric ion useful as FERSONA process technology.

Further, in some situations such as mine mouth power plants where there is a pyrites dump leachate or acid mine water problem, the sodium sulfate/sulfite wastes from sodium scrubbing of flue gases could be combined with the waste leach or mine water to dispose of both simultaneously.

The FERSONIUM process is the sister of the FERSONA process. As the name suggests, it deals with disposal of am-mon*ium* sulfates and sulfites by aqueous reaction with acidic ferric solutions. The FERSONIUM process produces the analogous ammonium ferric hydroxydisulfates as granular precipitates, with a similar order of magnitude reduction in the solubility

SINTERNA process

Another approach pioneered by IRI for the disposal of sodium sulfate wastes from SO$_x$ emissions control processes that use sodium alkali sorbents is its SINTERNA process. This process involves mixing the sodium sulfate wastes with powdered fly ash and/or bottom ash, pelletizing or briquetting the mixture, and sintering the briquette. A sintered briquette is shown in Figure 3.

The developmental work, also done at Battelle Columbus Laboratories principally under the direction of Dr. Russell Bennett, was concerned with simultaneously disposing of the sodium sulfate values while reducing the fly ash dusting problem. In accelerated leaching tests, pellets have shown sodium solubility to be reduced to about equal to or less than that of calcium sulfate. Again, this is a reduction of about two orders of magnitude in the solubility of the sodium salt.

The SINTERNA process is a method of disposal of wastes in a dry form, and the sintered briquettes show indications of having potential value as aggregate materials. Further, during the sintering, there is as high as 99% retention, that is, less than 1% of the sodium sulfate is lost as evolved SO$_2$ gas.

Calcium disadvantages

The implications of the IRI developments must be viewed from the background of penalties paid for the use of calcium alkali flue gas scrubbing. These penalties have been severe. Lime or limestone must be used in a wet system because of the relative inactivity in a dry system, except at unrealistically high temperatures. Wet systems require reheat of the flue gas after scrubbing. This reheat costs 2 to 5% of the total power plant energy output (not counting scrubber energy requirements). Scrubber internals tend to set up aerodynamic vibrations within the scrubber-stack systems. Chlorides in the water or coal tend to cause stress corrosion of stainless steel tubes.

In addition, there is the problem of scale formation inside lime or limestone scrubbers caused, in part, by the inverse solubility of calcium sulfate. Since the calcium source, lime or limestone, is a low solubility particulate, its presence in scrubber water adds to abrasion problems. The alternative is to increase the quantity of water, particularly where high quantities of fly ash are to be removed from the flue gas. This means greater water usage, which in the southwest may not be possible. It also means greater pumping costs and additional holding tanks and recycle lines to improve the utilization efficiency of the calcium within the scrubber. Single pass scrubbing becomes too inefficient at high dilution factors.

Finally, the calcium sulfate/sulfite sludge problem may be the penalty "straw that breaks the camel's back." Of course, the double alkali process is an attempt to use soluble sodium alkalis *inside* the scrubber to reduce the capital costs of the scrubber chamber itself and to improve reactivity while lowering the water requirements. However, the double alkali process merely moves the problem downstream since the external reaction with lime (and optionally, also, limestone) in turn produces the same calcium sulfate/sulfite sludge.

There are recent data to believe that calcium sulfate/sulfite sludge poses other disposal problems. Since the sludge is thixotropic, it cannot be piled above the ground and must be ponded. Typically, the ponds are constructed in clay formations. However, gypsum has been used for years to improve

FIGURE 2

FIGURE 3

one inch

Relative Solubilities

Component	Solubility and conditions		
Anhydrous Na_2SO_4	488	g/liter @	40°C
Monoclinic columnar			
Rhombic (thenardite)	427	g/liter @	100°C
	47.6	g/liter @	0°C
$Na_2SO_4 \cdot 10H_2O$ (Glaubers salt)	110	g/liter @	0°C
$Na_2SO_4 \cdot 7H_2O$	195	g/liter @	0°C
$Na_2SO_3 \cdot 7H_2O$	328	g/liter @	0°C
Na_2SO_3 anhydrous	125.4	g/liter @	0°C
$CaSO_4 \cdot 2H_2O$ (Industry standard)	2.3	g/liter @	R.T.[a]
$CaSO_3 \cdot 2H_2O$	0.043	g/liter @	R.T.
$MgSO_4 \cdot 7H_2O$	710	g/liter	
Sideronatrite	0.55	g/liter @	R.T.
Natrojarosite	0.37	g/liter @	R.T.
Ammoniojarosite	0.37	g/liter @	R.T.

[a] Room temperature

the permeability of clay. Thus, there is a potential that calcium sulfate sludges may act to increase the permeability of the clay formations underlying the ponds, permitting leaching of calcium sulfate/sulfite, magnesium sulfate, and heavy metals.

Sodium disposal methods

In contrast, sodium sulfate/sulfite wastes are not sludges and, thus, do not pose sludge-related handling problems. However, they do have a solubility problem. Ocean dumping is one method of disposal of sodium sulfate wastes; other methods include deep well disposal, isolation clay cell landfill and playa lake dumping in the western U.S. Playa dumping has been approved in California, and has been practiced by the Reid Gardner Station of the Nevada Power Co. in Clark County, Nev., for over one year with no adverse environmental effects. Reid Gardner is using once-through sodium scrubbing on two 125-MV units, and a scrubber for a third 125-MW unit is under construction. Since playa lakes contain very substantial quantities of sodium salts, including sulfates, sulfites, chlorides, and borates, return of those waste materials to the playas is considered essentially a conservation of the sodium salts, and does not significantly change the overall percentage of sodium sulfites or sulfates in the dry lake deposits.

To current disposal methods should be added IRI's FERSONA and SINTERNA insolubilization processes. However, these latter processes may be more attractive since they are not site-specific, as are dumping, well disposal or landfilling.

The implications of dumping, landfilling, and well disposal, permanent disposal methods, are important. The nation in effect may have delayed the adoption of proven methods of sodium scrubbing by requiring the utilities to solve all the problems of disposal before doing any scrubbing. A better approach may be a more classical, two-generation development.

Throwaway scrubbing

In the first generation, utilities can concentrate on scrubbing, while practicing interim disposal of the scrubbing wastes. This gives pollution control or chemical industries time to develop second generation processes for recycle, regeneration or utilization of wastes. Typically, sodium regeneration costs more than scrubbing, and requiring the utilities to pay these costs now seems to be misplaced. Rather, the development of regeneration and the capital costs thereof are more appropriately the function of chemical companies. Chemical companies are better suited to the business of regional regeneration of SO_x sorbents for utilities or industrial plants, or processing of wastes to produce useful products.

Utilities could reduce their emissions control capital costs by using wet or dry sodium scrubbing followed by throwaway disposal or storage. The disposal of the wastes could be by ocean dumping, deep well disposal, playa lake dumping, isolation clay cell landfill, FERSONA or SINTERNA processing.

These disposal methods represent first generation technology that IRI believes is available.

By way of example, a flow diagram shows a dry sodium scrubbing with nahcolite ore followed by IRI's FERSONA and/or SINTERNA disposal processes. The baghouse serves as a particulate collection device as well as a site for the nahcolite-SO_x reaction, while the nahcolite functions as a particulate filter aid and an SO_x sorbent. The dry reaction eliminates flue gas reheat, reduces process energy (water pumping) costs, and permits handling of dry wastes rather than sludges. Success of baghouse collection of fly ash under power plant conditions has been demonstrated at the Sunbury Station of Pennsylvania Power & Light (PP&L). These baghouses have been operating continuously at better than design specifications for two years. Typically, PP&L burns a mixture of petroleum coke (15–35%) and anthracite coal. PP&L's anthracite coal ash is 25%, the pressure drop through the bags is between 3–4 in. W.G. (design was 5 in. W.G.), and ash removal exceeds 99.5%. Apparently the bags have not been damaged by superheater tube leakage or temperature drop below dewpoint. After startup, maintenance has been less than design expectations. While these baghouses do not use nahcolite ore, the nahcolite ore baghouse injection process chemistry has been proven in full-sized bags.

Second generation technology would be the development of sorbent regeneration, or extractive processing of the sodium and/or sulfur values from the waste materials. This, it would appear, is the function of chemical, mineral supply or pollution control companies on whom would fall the burden of second generation capital investment. Their recycled product would then be returned to the utilities for reuse in the scrubber. For example, a single processing plant centrally located in Utah or Arizona could service the present or proposed 16 power plants of the southwest region.

Current plans

IRI is currently working toward development of an estimated 100 million tons of its reserves of low-sulfur coal in the Book Cliffs coal field, along with related water rights on the Colorado River near Loma, Colorado. This is in addition to its work with nahcolite ore from the Piceance Creek Basin for SO_x emissions control and as a feedstock for soda ash production. IRI is a partner in a joint venture for copper and silver exploration in Oklahoma. IRI also holds claims on substantial reserves of chemical grade limestone, about 98% $CaCO_3$, in the Glenwood Canyon, Colorado area.

Within this context, IRI is carrying on its pollution control research in the utilization of nahcolite ore as an SO_x emissions control sorbent. IRI is reviewing potential commercial applications of the FERSONA, FERSONIUM and SINTERNA processes to sodium sulfite/sulfate wastes disposal. However, since IRI is not a pollution control hardware manufacturer or systems supplier, it also seeks further development of these processes through joint ventures, licensing, or sale of the technology.

Jacques M. Dulin is president of Industrial Resources, Inc. He has a background in chemical research and development work, and has been in private practice as a patent attorney. Mr. Dulin is a co-author on a number of articles in the pollution control and legal fields.

Edward C. Rosar is vice president-exploration and mining operations of Industrial Resources, Inc. Mr. Rosar is a patentee and co-author of a number of papers in the mining and pollution control field. He formerly served in the U.S. Army Corps of Engineers.

Coordinated by LRE

Dry scrubbing of utility emissions

Foster Wheeler process of flue gas desulfurization is now underway at a Florida utility; demonstration started this May

Reprinted from ENVIRON. SCI. TECHNOL., **9**, 712 (August 1975)

Second generation flue gas desulfurization (FGD) systems may still be in their infancy, but at least seven systems are being investigated (*ES&T,* April 1974, p 306). Although the dry adsorption process was listed in the developmental category then, considerable new results are paving the way for the use of this process on utility boilers. In that same listing, four regenerable processes were listed as commercially available.

The dry adsorption process is applicable to all fossil fuel-fired utilities. Primary application is found in the area of large coal-fired boilers because of the established capability of the system to handle all three pollutants associated with such boilers. The process could also be used in large size refinery units as well as chemical process plants and metals smelting operations.

Mr. G. O. Layman, manager of power production for Gulf Power Co., an affiliate of the Southern Company, says that construction on the Foster Wheeler process started on February 15, 1974; the construction was at the Scholz Steam Plant, a 40 MW plant (Chattachoochee, Fla.). The testing program began this May.

Two years earlier, in January 1973, the Southern Services Company awarded a contract to Foster Wheeler Corp. to build a 20 MW prototype dry adsorption system. The actual contract called for the design, engineering, and construction and testing of a system that would accept 50% of the flue gas produced by a 40 MW boiler firing coal with 3% S, 14% ash, and a heating value of 12,400 Btu/lb.

The system at the Scholz plant consists of a 20 MW adsorber section and a 47.5 MW regeneration and RESOX (reduction of SO_2 to elemental sulfur) section. RESOX is a trademark of the

Foster Wheeler Energy Corp. The 20 MW adsorber is designed to accept 50% (half of the boiler flue gas flow) when the coal fired boiler is operating at a nominal 40 MW load.

This Scholz FGD unit is designed to meet the Florida Code for SO_2 emissions, which is 1.2 lbs/million Btu. The unit can handle coals with sulfur content as high as 5% because of the oversized design requirements of the regeneration and RESOX sections, and the inherent flexibility of the system.

Gulf Power's Layman
"regeneration—using coal to reduce SO_2 to elemental sulfur—is a real breakthrough"

Layman says that Southern Services, a private investor utility that relies on coal for more than 90% of its power generation, chose the system because it had already accomplished a very successful and lengthy pilot plant operation.

"One of the most important features of the system that has great long-range potential is the RESOX section. In this reduction step, coal directly reduces the SO_2, adsorbed by the front end of the system, to elemental sulfur." Layman adds, "It's a real breakthrough."

He explains that "all other systems . . . for reduction of SO_2 to elemental sulfur require hydrogen in some form, usually natural gas, as well as catalysts. Foster Wheeler's unique approach to using coal to act as the reductant avoids the use of a costly and scarce natural resource."

In addition to FGD, the process has the potential for NOx and fly ash removal. Also, it produces a commercial grade elemental sulfur as a by-product from the regeneration section (see box).

The process

Foster Wheeler, a designer and manufacturer of steam–generating equipment and a process plants contractor, has been licensed for the dry adsorption SO_2 removal process. The removal process was developed by Bergbau-Forschung (B-F), the central research institute for the German coal mining industry. Bench scale work at B-F began in 1966. Based on this pilot plant work, SOx removal of up to 95% has been achieved. But Foster Wheeler independently developed the backend system, RESOX, for the reduction of SO_2 to elemental sulfur.

A second dry adsorption unit has been installed by Bergbau-Forschung, the licensor, at the Kellerman Power Plant of STEAG in Lünen, W. Ger. The unit will treat 10% of the flue gas from a 350 MW unit that will be burning coal of about 2% sulfur content. Here, a modified Claus unit will be used to process the SO_2-rich off gas. RESOX was developed by Foster Wheeler after B-F had gone ahead with the decision to install the modified Claus unit.

How it works

The adsorption section is based on the well-known suitability of activated char as an adsorption and filter media.

Dry scrubbing removes SO_2, fly ash, and NOx[a]

The steps:

| Adsorption of dilute SO_2 on char | Regeneration of the char with the resultant production of SO_2-rich off gas | Reduction of SO_2 (RESOX) to elemental sulfur using coal |

The flow sheet:

Flue gas from dust collector (after electrostatic precipitator)

Cleaned flue gas to stack

Crushed coal

Ash   Sulfur

The equations:

$$SO_2 + \tfrac{1}{2}O_2 + H_2O \rightarrow H_2SO_4$$

$$2H_2SO_4 + C \xrightarrow{HEAT} 2H_2O + 2SO_2 + CO_2$$
$$2NO_x + C \xrightarrow{HEAT} CO_2 + N_2$$

$$SO_2 + C \rightarrow CO_2 + S$$

[a] Removal efficiency: 95%—SO_2, 90-95%—fly ash, 40-60%—NOx

The adsorber consists of vertical columns of parallel louvre beds that support and contain the char. The char moves slowly downward in mass flow while the pollutant-laden gases pass through the adsorber char bed in cross flow at 250–300°F. Sulfur dioxide, oxygen and water vapor contained in the flue gas are adsorbed into the char pores. Adsorbed SO_2 then reacts with the O_2 and H_2O to form H_2SO_4, which is firmly retained in the interior pore system of the char pellets. Oxides of nitrogen are also adsorbed by the char pellets. Clean gases, which are unchanged in temperature, are exhausted to the stack via an induced draft fan.

As the char progresses slowly down the adsorber, it becomes saturated and must be regenerated. It is heated in the regenerator vessel to 1200°F in an inert atmosphere. Sand at 1500°F is mixed with char to bring the char to 1200°F. All the reactions that have occurred in the adsorber are reversed at this elevated temperature (see box). Sulfuric acid is reduced to SO_2, oxides of nitrogen dissociate to oxygen and nitrogen, and CO_2 is produced as a result of chemically combining carbon in the char with the oxygen liberated from the reduction reactions. The regeneration section yields adsorbed SO_2 in concentrated form, approximately 20% by volume. This SO_2-rich gas stream is sent to the RESOX section for reduction to elemental sulfur.

In the RESOX section, a Foster Wheeler proprietary process, SO_2-rich gas is passed through a vessel containing crushed coal. Here, SO_2 is reduced to gaseous elemental sulfur and the liberated oxygen combines with a portion of the coal to form carbon dioxide. Reduction in the RESOX reactor is done at 1200–1500°F. The gases leaving the reactor enter a condenser where gaseous sulfur is condensed and the tail gas recycled back to the adsorber to remove residual sulfur values, thus providing a closed-loop system.

This dry scrubbing process offers a number of advantages over conventional wet scrubbing. These include:

• significant NOx and fly ash removal

• no slurry handling or pH controls required

• no stack reheat is required; the flue gas enters and leaves at the same temperature

• reduced power requirements stemming from the special size and shape of the char pellets, and the adsorber louvres design gives low pressure drop on the gas side

• the char adsorbing material used in the process has a high ignition temperature of about 700°F that ensures safe operation in the 250–300°F designed temperature range

• a regenerable system with a saleable by-product (The char can be used for many cycles, the average life of a pellet being six months. Commercial grade sulfur is produced by the RESOX system)

• less space (The system uses considerably less space than a wet scrubbing system by virtue of its vertical integration concept and absence of sludge conditioning requirement.)

How much does it cost?

Installed capital costs are somewhat sensitive to the sulfur content of the fuel and the SO_2 percent removal to meet local codes. However, they range from $25–35/kW for low sulfur level that requires a low removal efficiency to $70–75/kW for high sulfur fuel that requires a high removal efficiency. Costs are competitive with wet scrubbing systems and may be lower depending on the price of recovered sulfur and the final cost of sludge disposal.

Fluidized-bed combustion— full steam!

Several pilot plants are evaluating this cooler, cleaner way of using a variety of fuels

Reprinted from ENVIRON. SCI. TECHNOL., **10,** 120 (February 1976)

Go to technical conferences concerning the use of coal, and you will hear that it is economical/uneconomical to use SO_2 scrubbers, and that it is feasible/infeasible to clean coal prior to use. But what about coal cleaning *during* the combustion process? This concept was discussed at great length and from varying aspects at the Fourth International Conference on Fluidized-Bed Combustion, sponsored by the U.S. Energy Research and Development Administration (ERDA), and coordinated by The MITRE Corp. (McLean, Va.). More than 200 people from the U.S., Canada, England, Germany, Japan, and other countries attended this conference, held at McLean last December.

Conventionally, a combustible material is burned in the boiler in a fixed bed, or by suspension firing. By contrast, in fluidized-bed combustion (FBC), air is passed upward from underneath the fuel mixed with a bed of limestone/dolomite material with sufficient velocity to cause the bed to become highly turbulent. The surface ceases to be well-defined; rather, it appears diffuse, with bubbles of "gas" similar to those seen in a boiling liquid, rising through the bed. Such a "boiling" bed actually behaves like a liquid, seeking its own level and having a hydrostatic head. Rapid mixing of particles occurs during FBC. While this turbulence is desired, care should be taken to prevent gas velocity from becoming so high that the gas stream entrains particles and carries them off.

Burning cool

The optimum FBC temperature seems to be 750–950 °C (1382–1742 °F), depending on application and fuel characteristics. By comparison, pulverized coal-burning temperatures encountered at conventional utility boilers, such as those of Union Electric Co. (Meramec, Mo.), for example, could be as high as 1400–1500 °C (2552–2732 °F). Lower combustion temperatures in FBC reduce NO_x formation. They also inhibit slag and clinker formation, and sharply decrease the volatilization of

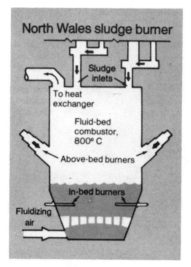

North Wales sludge burner

Sludge inlets

To heat exchanger

Fluid-bed combustor, 800° C

Above-bed burners

In-bed burners

Fluidizing air

highly corrosive alkali salts and trace metals. This process produces easily handled soft ash, rather than vitrified ash or slag; the coarse fraction of this ash can serve as part of the inert mineral matter needed for the fluidized bed.

In order to control SO_x emissions, the FBC process uses a sorbent bed material—limestone or dolomite. Rapid mixing occurring in the fluidized bed allows for effective capture of sulfur by the limestone. Consequently, FBC systems could eliminate the need for SO_x scrubbers.

According to Combustion Systems Ltd. (CSL, London, England), FBC offers some other advantages. Among these can be up to 75% reduction in boiler volume; lower capital and maintenance costs; shorter construction periods because of factory assembly; a variety of power generation cycles; and, very significantly, multifuel operation. By multifuel operation it is meant that combustible solid wastes, residual oil, organic materials from anthracite mine waste, and other combustibles, as well as coal, can be burned, as long as, if solid, they are crushed or shredded to proper-sized particles—¼ in. top size.

Sulfur removal

The removal of sulfur from coal or high-sulfur oil during combustion is the engineering crux of FBC. Temperatures must be kept at such levels that the "stones", be they limestone or dolomite, calcine, at least partially; that is, change to a porous form of calcium oxide, which is the effective SO_x absorber. At the same time, the temperature must not be so high that equilibrium for calcium sulfate formation is no longer favorable.

At present, in order to bring these systems on-line at an early date, once-through use and acceptable disposal of spent sorbent "stones" is the initial route taken. However, work on optimizing sorbent regeneration is being continued. Regeneration offers not only reuse of "stones", but also useful by-products, such as sulfur, through the Claus or carbon reduction processes.

An idea of how sorption works was given by John Montagna of the Argonne National Laboratory (ANL, Argonne, Ill.). The ANL data showed that about one lb of "stones" are sulfated for 5 lb of 3% S coal burned. Montagna suggested use of spent sorbent as a soil additive or building material, but he expressed a strong preference for "stone" regeneration and recalcining. Even with regeneration, however, 100% reuse does not appear to be feasible, and some makeup sorbent would have to be added and calcined to keep the bed active.

More work, less planning needed

Ernest Daman, vice president of Foster Wheeler Energy Corp. (FW, Livingston, N.J.), complained to the conference that in FBC, there has been "much planning and no work, and what work there is is often duplication." He called for hardware development with large units put in the field in order to define problems, "and you can't know what problems you have until you get equipment into the field."

One place where equipment is being "gotten into the field" is at Rivesville, W. Va., site of a Rivesville Monongahe-

Fluid-bed system. *PER president Pope stands in front of the system at the Rivesville power station*

la Power Co. power station. This equipment is a 30-MWe, 300 000 lb/h multicell FBC boiler system, a cell for an 800-MW FBC steam generator designed by FW (*ES&T,* November 1974, pp 968–970). The system is being installed by Pope, Evans and Robbins (PER, New York, N.Y.) who have pioneered fluidized-bed coal combustion in the U.S.; its steam conditions will be 1350 psig, 925 °F with no reheat, and it can work with existing turbine generating equipment. PER president Michael Pope foresees system startup this coming July.

The FBC combustor bed itself will be kept at 1550 °F to control SO_2 and cut agglomeration of ash. No estimate of noise level could be given yet, but FW's Robert Gamble expects no noise problem from the bed itself. The Rivesville FBC system will operate at atmospheric pressure. PER president Pope said that tests showed no SO_3 detected from the system, but if inorganic chlorides are present, about 40 ppm of hydrogen chloride could be emitted. He also said that carbon monoxide (CO) has been as high as 0.3% when 18% excess air was used in a small (5000 lb/h) FBC pilot plant where rapid combustion quenching was present. Scale-up projection of CO and hydrocarbon emission at Rivesville has not yet been accomplished.

Putting on pressure

FBC development is proceeding rapidly in England. Raymond Hoy, director of research at BCURA Ltd. (Leatherhead, Surrey), and conference dinner speaker, said that the FBC state-of-the-art in the UK is based on 20 000 hours of operating experience on various pilot plant and combustion rigs. Major technological contributions were made by BCURA, the National Coal Board, and British Petroleum Ltd. CSL and Babcock & Wilcox (B&W) have formed a joint venture to design and construct, and are now operating the largest present working FBC steam boiler in the world, a 44 000 lb/h stoker boiler conversion.

This project is a 10 ft by 10 ft atmo-

FW vice president Daman
"Too much planning, too little work"

spheric FBC boiler at Renfrew, Scotland. The Renfrew plant works at 1550–1650 °F. Oil heats the plant to 800 °F, at which point coal is introduced and ignited. At 1100 °F, the oil is cut off. There will be ash deposit, but Hoy said that this is "easily removed." Hoy also wants to develop gas cleaning methods at varying temperatures. The Renfrew plant at full load supplies steam to the B&W works at 400 psig and 44 000 lb/h. Another project in northern Wales will be for an FBC system to burn sewage sludge.

In the UK the greater share of expenditure for FBC research is going to pressurized FBC (PFBC), with pressures of up to 16 atm. By 1974, $7 million had been expended for this purpose.

In PFBC, dolomites seem very superior to limestones as SO_2 sorbents. PFBC would provide hot, pressurized flue gases to be cleaned and expanded to run a turbine for power generation. PFBC also could allow for substantial boiler size reduction. Hoy believes that PFBC is now well demonstrated on the bench-scale level.

Here in the U.S.

Combustion Power Co., Inc. (CPC, Menlo Park, Calif.) is concerned with cleanup of PFBC flue gases. CPC has

BCURA's Hoy
"20 000 hours of experience in the UK"

developed under an EPA contract, a PFBC installation, and is testing granular bed filter technology, under an ERDA contract, on its CPU-400 pilot plant (*ES&T,* August 1970, pp 631–633). EPA (Research Triangle Park, N.C.) itself has recently had completed by Exxon Research (Linden, N.J.), a 100-h run on a PFBC coal miniplant (12.5-in. inner diameter combustor) at 1670 °F, up to 10 atm pressure. The object was to develop a complete environmental characterization of the FBC process, and to define any required environmental control technology.

Also, the Curtiss-Wright Co. (Wood-Ridge, N.J.) will build and run a PFBC pilot plant to burn high-sulfur coal, and generate 13 MWe, under ERDA auspices. Estimated cost is $25 million, with startup by 1980.

The U.S. is fortunate to possess huge coal resources that could provide one early route to greater energy self-sufficiency. Much of that coal, unfortunately, is high-sulfur, and its use could have serious air pollution effects. Hopefully, various FBC process will offer a means to use even high-sulfur coal, clean it during combustion, safeguard air quality with less capital and operating expenditures, and perhaps produce useful, marketable by-products. JJ

Minimizing emissions from vinyl chloride plants

Available and future pollution control systems,
though expensive, are required by law to protect
the worker and his environment

Carl G. Bertram
Badger America, Inc.
Tampa, Fla. 33622

Reprinted from ENVIRON. SCI. TECHNOL., **11,** 864 (September 1977)

Vinyl chloride (chloroethene) is a synthetic organic chemical made from ethylene or acetylene and chlorine by using different processing schemes. The most common route is from ethylene; ethylene dichloride (EDC), an intermediate, is then thermally cracked to the monomer (VCM) and by-product hydrogen chloride (HCl). Hydrogen chloride is further reacted with ethylene (oxyhydrochlorination) to form more EDC.

In 1974, in the U.S., OSHA promulgated a stringent VCM standard. This standard requires that worker exposures be no more than 1 ppm on a time-weighted average and no more than 5 ppm for any 15-min period.

Table 1 shows the maximum worker exposure concentration permitted by various countries throughout the world. As can be seen from this table, the concern over angiosarcoma, a rare form of liver cancer, that has been linked to exposure to VCM, has become worldwide.

EDC, long considered a toxic substance, has also been classified as a suspected carcinogen that also causes kidney damage in animal tests. The U.S. government is presently considering a proposed EDC regulation that would set average exposure at 5 ppm and peak exposure at 15 ppm. It will not be long before other chlorinated hydrocarbons are similarly restricted.

Existing plants have had to make many modifications to reduce exposure to VCM, and are now taking steps to reduce exposure to EDC as well. New plants under construction face strict EDC/VCM exposure levels, restrictions and generally must meet stringent overall environmental pollution requirements. The greatest causes of air pollution from a VCM plant are continuous or intermittent process vent streams, which number about 10 per plant.

Abatement methods

Various methods have been proposed and actually used to treat process vent streams containing chlorinated hydrocarbons. From environmental and economic standpoints, one of the best solutions is the use of an incinerator-waste heat boiler followed by scrubbing of the flue gas, as shown in Figure 1. When the

TABLE 1
Worldwide maximum workplace VC concentrations [a]

Country	Year	VCM concentration
Belgium	1975	max. 200 ppm TWA [b]
Canada	1975	10 ppm TWA (8 h) to 25 ppm TWA (15 min)
Finland	1975	5 ppm TWA (8 h) to 10 ppm TWA (10 min)
France	1975	no limits
Germany	1975	technical standard value: max. 5 ppm [c]
Great Britain	1975	25 ppm TWA (8 h) to 50 ppm (15 min)
Italy	1975	50 ppm TWA (8 h) expected: 25 ppm (8 h) to 15 ppm TWA (15 min)
Japan		expected: 25 mg/m^3 (10 ppm)
The Netherlands	1975	10 ppm TWA (8 h)
Norway	1975	no value expected: 1 ppm TWA (8 h) to 5 ppm TWA (15 min)
Sweden	1975	5 ppm TWA (8 h) to 20 ppm TWA (15 min)
	1976	1 ppm TWA (8 h) to 5 ppm TWA (15 min)
Switzerland	1975	100 ppm max. expected: 10 ppm TWA (8 h)
U.S.A.	1976	1 ppm TWA (8 h) to 5 ppm TWA (15 min)
U.S.S.R.		30 mg/m^3 (12 ppm)

[a] Source: *Chem. Ing.-Tech.*, **47**, 1975.
[b] TWA = Time Weight Average
[c] Over the period of 1 hour the average concentration may not exceed three times the technical standard value.

steam generated can be taken as a credit, this solution minimizes the overall cost of pollution control.

Burning chlorinated hydrocarbons unfortunately results in HCl, along with CO_2 and water. The release of HCl to atmosphere, if permitted at all, is *not* desirable. Scrubbing the incinerator off-gas to form a dilute acid liquor or a salt solution is necessary; however, disposal of the scrubbing liquor can be a problem.

VCM plants located near the sea can dispose of a salt solution without major problems, but the cost of neutralizing significant amounts of HCl can become prohibitive. For this reason vent streams sent to the incinerator should contain a minimum of chlorinated hydrocarbons. Economic evaluation will determine the degree of treatment (clean up) that vent streams require prior to incineration.

Incorporating an incinerator-waste heat boiler scrubber (vent gas incinerator) into a VCM plant is not a simple matter because of the safety and corrosion aspects that need to be considered. Substantial instrumentation, including interlocks and automatic shutdowns, is required, plus a number of separate header systems to collect the process vent streams.

The process vent streams are not the only waste gas streams handled by the vent gas incinerator; a number of intermittent and miscellaneous vent streams can also be directed to the incinerator for disposal. The vent gas incinerator becomes the heart of the pollution-abatement system of the VCM plant.

Loading operations

The second largest source of emissions in a VCM plant is the loading operations; for example, loading of EDC or VCM into tank cars, tank trucks, barges or ships. The major problem areas associated with loading are shipping container venting and the cleaning of loading arms and lines after loading; several methods have been developed to minimize these emissions.

Loading EDC. Preferably, the atmosphere in the shipping container—in this example, a ship—should be inert before loading the EDC (Figure 2). When this has been established, the dual-pipe loading arm, containing a loading line and an equilibrium (balancing) line, can be connected. During loading the at-

mosphere in the ship is displaced to the EDC storage tank without net venting being required. After loading is completed the loading arm must be decontaminated so that it can be disconnected without venting EDC. The liquid loading arm is first drained into a jetty slop tank for eventual recycle back into the process. The loading arm is purged into the ship with nitrogen. The block valves on the ship's manifold are then closed and the equilibrium line purged with nitrogen back to the storage tank. The loading arm can then be safely disconnected and parked in a safe condition. Excess inert material in the EDC storage resulting from purging operations is disposed of in the vent gas incinerator.

Loading VCM. Since VCM is a condensed gas, it is slightly more difficult to load than EDC (Figure 3). The ship to be loaded *must* contain VCM vapors or inert gases, with VCM vapors preferred. Theoretically, if no inert material and only VCM vapors are present, VCM can be loaded without venting. As in the case of EDC, the atmosphere in the transport container is displaced in the storage tank, with no venting of VCM realized. After loading, the loading arm must be purged with nitrogen prior to disconnecting.

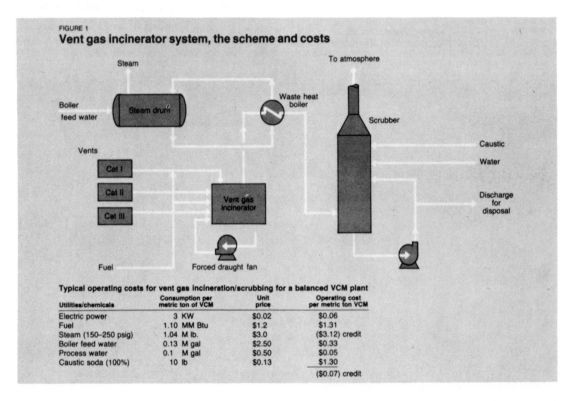

FIGURE 1
Vent gas incinerator system, the scheme and costs

Typical operating costs for vent gas incineration/scrubbing for a balanced VCM plant

Utilities/chemicals	Consumption per metric ton of VCM	Unit price	Operating cost per metric ton VCM
Electric power	3 KW	$0.02	$0.06
Fuel	1.10 MM Btu	$1.2	$1.31
Steam (150–250 psig)	1.04 M lb.	$3.0	($3.12) credit
Boiler feed water	0.13 M gal	$2.50	$0.33
Process water	0.1 M gal	$0.50	$0.05
Caustic soda (100%)	10 lb	$0.13	$1.30
			($0.07) credit

Inert material will accumulate in the VCM storage sphere and cause excessive pressure in the sphere as it is refilled. A refrigerated vent condenser recovers the greater amount of VCM in the vent stream from the storage sphere. The chilled vent stream is then directed to the vent gas incinerator for final disposal thereby minimizing the emission of VCM.

Storage vents, pump seals, sample collections

Fairly large volumes of EDC, stored in cone-roof or dome-roof tanks, are required for smooth operation of a VCM plant. The atmospheric venting of these tanks can result in a substantial source of chlorinated hydrocarbon emissions. At storage temperatures of 30–40 °C, the vents from EDC tanks can contain 13–20% EDC by volume. Incineration of these out streams (Figure 2) is one way to eliminate this source of air pollution. As cone-roof or dome-roof tanks are limited to an internal pressure rating of 150–200 mm of water, an ejector system is normally used to feed the vent gases into the incinerator. For safety, the storage tanks are blanketed with nitrogen, with in-and-out breathing regulated by a split-range pressure controller.

Pump seals (single or double mechanical seals) are another source of emissions if EDC or VCM is the seal liquid. Where possible, the use of a compatible seal oil that contains *no* chlorinated hydrocarbons has solved this problem. In services where seal oil cannot be used, glandless pumps (canned or magnetic clutch type) are now being used with fairly good success.

Because of the crude sampling techniques of the past, workers collecting samples in a plant were exposed to fairly high levels of EDC and VCM. To avoid exposure, all samples containing EDC and/or VCM should be collected in bombs in a closed-loop system.

The basic concept is to trap a representative sample in the bomb and then purge the piping with nitrogen, which must be disconnected in order to remove the sample bomb safely. In the case of a VCM sample, the piping is purged to the flare, while in the case of an EDC sample (less volatile) the piping is drained and purged to a convenient low-pressure source within the process or to the closed-process sewer.

Equipment decommissioning

All equipment will sooner or later have to be prepared for inspection or maintenance. To decommission equipment without excessive EDC or VCM emissions, a number of support systems, representing a considerable investment, are required. To cover all equipment containing EDC and/or VCM, the following support systems are required:

• Process sewer—this is a completely closed system, located below grade, which drains into a sump tank; the vapors from this tank are directed via a knock-out drum and booster ejector to the vent gas incinerator.

• Contaminated water sewer—this system, located below grade, drains to an open sump; this system collects aqueous drains contaminated with EDC; water is separated from the organic layer in the sump and treated in the wastewater unit; the organic phase is sent to storage or recycled back to the process.

• Contaminated water header—this is a closed system to collect highly contaminated wastewater and direct water to a holding tank that is vented to the incinerator via a booster ejector; the wastewater is then treated in the wastewater unit.

• Steam-out collection system—this system collects all steam-out vapors containing chlorinated hydrocarbons; steam and condensable impurities are condensed and the noncondensable vapors are directed by an ejector to the vent gas incinerator.

• High-pressure nitrogen—this system is required for purging operations.

• Slop header system—this system collects substantial quantities of EDC (which may or may not contain small amounts of VCM) when equipment is decommissioned; it is a closed-piping system that empties into a slop tank with the vent gases directed to the incinerator or flare and the liquid sent to storage or recycled to the processes.

• HCl neutralization system—all dry HCl reliefs and vents are collected in this system and either scrubbed with water or neutralized with caustic soda to minimize release of gaseous HCl.

Utilizing these systems, a number of procedures can be followed to decommission equipment. For example, the decommissioning of pumps will occur at frequent intervals for maintenance, normally done during plant operation (Figure 4). For pumps in EDC service with suction and discharge valve closed, drain EDC from the pump and associated piping to the process sewer. Hook up nitrogen to the suction line and blow out the

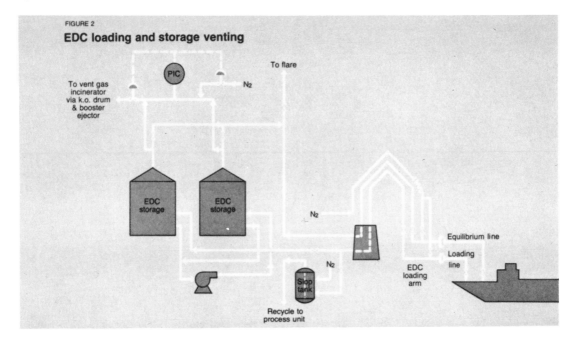

FIGURE 2

EDC loading and storage venting

remaining EDC to the process sewer. With the pump drain closed, blow through the pumps discharge line to the process sewer. Repeat operation via the pump drain valve. The pressure gauge drain connection can also serve as a purge point.

Disconnect the nitrogen hose and hook up water to the suction line of the pump. Flush to the contaminated water sewer, first through the discharge line, then via the pump drain. Shut off the water flow, open the connection at the pressure gauge and drain the pump. The pump can then be safely dismantled.

For pumps in VCM service, with suction and discharge valve closed, drain VCM from the pump and piping to the blowdown pot. Liquid will collect in the drum with the vapor venting into the flare.

Purge the lines with nitrogen connected to the suction line of the pump, blowing VCM and nitrogen into the blowdown pot. To permit a controlled release of VCM vapor to the flare, electric heating panels that permit close temperature control are used on the blowdown pot. The use of steam as the heating medium is not recommended.

After purging with nitrogen, crack open small valve (on the pressure-gauge drain on pump discharge) and test VCM level with a photoionization-type analyzer. If the level is unacceptable, continue the nitrogen purge until a safe level of VCM is obtained. The operator doing this test should wear a fresh-air gas mask. When safe level is reached, stop nitrogen flow. The pump can then be vented to the atmosphere and dismantled.

Reboilers should be decommissioned by first isolating the reboiler from the column. For all reboilers containing essentially VCM, depressure tube side to the flare; for all reboilers containing essentially EDC, drain to the slop header. Using nitrogen hooked up to the reboiler, blow out EDC or VCM to the appropriate header.

Then crack open the steam to the shell side of the reboiler. Hook up steam and steam out the process side from bottom to top into the steam-out header. Stop steam-out and hook up service water to the reboiler, flushing downward to the contaminated water sewer.

The reboiler is now ready for removing the heads. Despite the precautions taken, there can still be a momentary release of residual fumes as the reboiler head is opened. Workers performing this operation should, therefore, wear fresh-air gas masks.

To decommission filters or strainers, "block-in" the filter and depressurize to the flare. Blow contents of filter to the process sewer or flare (depending on volatility of filter contents) by using nitrogen. Continue nitrogen purge until safe levels of chlorinated hydrocarbons have been achieved. When testing for safe levels the wearing of a fresh-air gas mask is recommended. The filter can then be dismantled for clean-out.

Complete tower system and other sources

Various detailed procedures can be followed to clear a tower system for maintenance work; however, the basic procedural concept will undoubtedly be in accord with that outlined below for a mixed EDC/VCM-containing system.

First, reduce inventory of VCM in the system to a minimum. Then vent the residual VCM that accumulated in the overhead system to the flare. Next add EDC to the tower system and "boil up", withdrawing condensed EDC from the overhead system while measuring VCM content in the EDC.

The EDC contents should then be delivered to "off-spec" storage, and residual hold-up in system should be discharged to the process sewer (for small volumes) or to the slop header system (for larger quantities).

Add water to the tower system and "boil up" this system at total reflux, while venting the reflux drum to the steam-out header. Drain contents of the tower to the contaminated water header and repeat water boil-up operation until EDC is reduced to a safe level.

The tower system can then be vented to the atmosphere and water added to lower the temperature of the column. Residual water is then drained to the contaminated water sewer and, after positive isolation and adequate ventilation, the tower system is ready for inspection.

There are a number of other sources which, if not properly controlled, can add to EDC and VCM emissions. The following are just a few of these sources, and solutions that may be adopted.

Blowdown of level gauges. Depending on volatility, materials are directed to the flare or to closed-process sewer.

Control valves. Control valves that require occasional maintenance are provided with vent and drain connections to purge the valve so that the valve can be removed without the risk of exposure.

Relief valves. Relief valves in critical service are provided with installed spares, as well as rupture disks on the inlet side of the

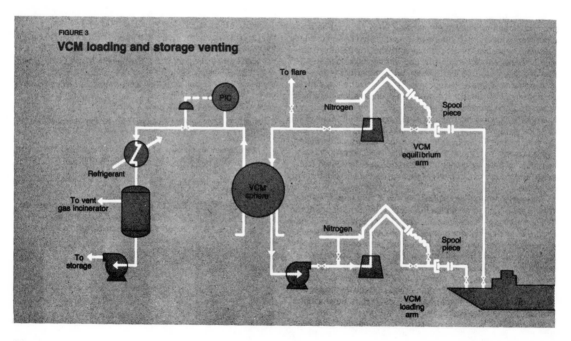

FIGURE 3
VCM loading and storage venting

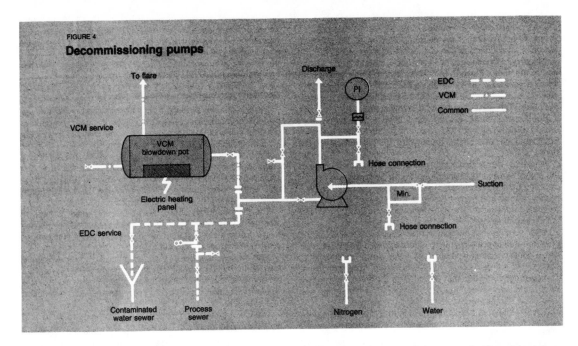

FIGURE 4

Decommissioning pumps

To flare

VCM service

VCM blowdown pot

Electric heating panel

EDC service

Contaminated water sewer

Process sewer

Discharge

PI

EDC
VCM
Common

Hose connection

Suction

Min.

Hose connection

Nitrogen

Water

relief valves. This reduces corrosion and leaks through the valves.

Monitoring and back-up systems

To ensure a safe environment for workers and demonstrate compliance with government regulations, frequent and thorough monitoring is required. A safe working environment throughout the process facility can be achieved, with an adequate monitoring system, in essentially two ways:

• The system provides information on exposure of plant personnel and records of individual workers can be kept.

• The system helps ensure that emissions are kept to a minimum by providing an early warning of leaks and pinpointing the source of emission.

Three types of monitors are in common use today. A *continuous-monitoring system* is automatic and consists of a gas chromatograph to analyze air samples drawn in rotation from a number of locations in the plant. A computer is often used to calculate exposure levels, compile and print out results to provide a permanent record. Alarms and warning lights can also be tied into this system to alert personnel in the affect area.

A *portable analyzer* based on a flame ionization system is used to pinpoint the exact source of emission, usually after a high alarm from the continuous monitor.

Personnel monitoring measures the amount of EDC or VCM to which an individual worker is exposed. The key to this system is an instrument worn by the worker that absorbs chlorinated hydrocarbons from the air. On a regular basis, after exposure, samples from the personnel monitor are analyzed in the laboratory and the amount of each worker's exposure level carefully recorded.

The design features mentioned in this article plus analytical surveillance help to minimize EDC and VCM emissions. However, accidental emissions cannot be completely avoided. Failures, mechanical or otherwise, of equipment or systems at a critical time are inevitable. To safeguard personnel should such failure occur—or for operations that include some risk of over-exposure—back-up systems are usually provided in the plant, such as a fresh air system or self-contained air masks. Since most gas masks do not provide adequate protection against a high-VCM concentration, gas masks have limited usefulness in a VCM plant.

The fresh air system consists of an "air header" that runs throughout the plant with numerous, easily accessible stations for connecting air-fed respirators. Lightweight, reasonably comfortable air respirators can be used with this system, thus eliminating the need for bulky air cylinders.

Summation

Because of the health hazards associated with exposure to VCM and EDC, and the resulting government regulations, today's VCM plants must have extensive pollution-abatement systems. Minimizing emissions is expensive. To ensure that the plant operates safely, the workers are also required to follow arduous procedures. One thing is clear: there are no easy solutions to emission control.

Additional reading

Reich, P., "Air or Oxygen for VCM?", *Hydrocarbon Process.,* **55,** No. 3 (March 1976).

Bell, Z. G., Lafleur, J. C., Lynch, R. P., Work, G. A., "Control Methods for Vinyl Chloride", *Chem. Eng. Prog.,* **71,** No. 9 (Sept. 1975).

Wheeler, R. N., Sutherland, M. E., "Control of In-Transit VCM", *Chem Eng. Prog.,* **71,** No. 9 (Sept. 1975).

Vervalin, C. H., "Curtail Vinyl Chloride Exposure", *Hydrocarbon Process.,* **55,** No. 2 (Feb. 1976).

Federal Register, **42,** 28154, June 2, 1977.

Huber, H., "Das Vinylchlorid-Problems," *Chem. Ing.-Tech.,* **47,** No. 19 (1975).

Acknowledgment should be given to Robert Vanheyst, a former Badger employee, who contributed significantly to the writing of this article.

Carl G. Bertram *is a senior process engineer with Badger America. He has extensive experience in chemical plant design.*

Coordinated by LRE

Progress in smelter emission control

Reprinted from ENVIRON. SCI. TECHNOL., **10**, 740 (August 1976)

As you approach the Canadian city of Sudbury, Ontario, from the west, along the Lake Superior Route of the Trans-Canada Highway, you pass two unique landmarks. On the left is a numismatic park at which large replicas of coins are on display; on the right is the copper smelter of International Nickel Co., Ltd. (INCO), which has the tallest stack in the world—1250 ft 9 in. The stack's builder was Canadian Kellogg Co., Ltd., a subsidiary of Pullman-Kellogg Co. (Houston, Tex.). In the U.S., the tallest smelter stack is that of Kennecott Copper at Magna, Utah, also built by Pullman-Kellogg. What these tall stacks have in common is high-level dispersion, and significant ground-level reduction of SO_x and other chimney emissions. Opinions as to how well tall stacks do this job at smelters, or power plants, vary widely, to say the least! The EPA, for example, generally endorses neither the use of such stacks nor intermittent controls.

The control picture becomes more involved with the concept of "best available control technology" (BACT) and "significant deterioration" rules as proposed in the Clean Air Act Amendments bill. If this bill becomes law, there could be very tight requirements concerning copper smelter siting and capacity, as well as separation between smelters. Moreover, these requirements, including 3- and 24-hour increments, could vary with air-quality class. For example, under House Class II increments, copper smelters must be at least 12–18 mi apart, whereas a 9–11-mi separation would be allowable under House Class III increments, by which air could deteriorate to National Air Quality Standards. The bill's Senate version would not allow such deterioration, since it omits Class III.

A U.S. Department of Commerce (DOC) staff study warns that House Class I provisions could preclude non-ferrous smelter sitting in that air-quality zone, even if the smelter has BACT. The study says that in that zone, a smelter might have to be so small as to be impractical, or so far from other smelters as to be uneconomical, particularly if copper smelters, for instance, must be located 15½–20½ mi apart. Also, for example, a 1500 tpd copper smelter would have to be 13–28 mi from a Class I boundary, if it meets source performance standards. Presently, 12 to 13 copper smelters—there are about 20–30 non-ferrous smelters in the U.S.—would not meet this requirement, according to the DOC study.

Associations

To work toward solutions to air pollution control (apc) problems, the copper smelting industry, in 1971, formed the Smelter Control Research Association, Inc. (SCRA, Columbus, Ohio), and this year, the Smelter Environmental Research Association (SERA, New York, N.Y.). Ivor Campbell of Clyde Williams and Co. (Columbus) is executive director of the SCRA, and Ralph Smith of the University of Michigan is technical director of SERA.

Much has been said about wet-limestone scrubbing of SO_2-laden flue gases. The SCRA added its view to what has already been said very often—essentially, "there has to be a better way," as far as smelter, and particularly copper reverberatory (reverb) furnace gases are concerned. Among complaints listed by the SCRA as of 1973 for wet-limestone scrubbing for this application, developed from pilot studies, are that

- The process is not reliable.
- Downtimes are excessive.
- There are no reliable criteria for scale-up to commercial models.
- SO_2 recoveries are not satisfactory.
- There is a real risk that limestone scrubbing may become obsolete before its problems can be resolved.

SCRA plans to devote its efforts to other processes that would avoid problems that it sees as inherent in limestone systems, and that would offer promise of greater reliability and SO_2 recovery.

Nevertheless, one must start somewhere, and SCRA certainly did not ignore the limestone option. Indeed, limestone was among the SO_2 reactants first tried for reacting SO_2 from reverb furnace tail gas in pilot studies at McGill, Nev.; other reactants studied were hydrated lime, and sodium sulfite with lime regeneration. In 1974, the SCRA and the U.S. Bureau of Mines (BuMines) engaged the C-E Lummus Co. to conduct an engineering evaluation of various soluble scrubbing processes.

More recently, SCRA sponsored a 4000-scfm pilot plant (start-up April 1975), using the ammonia-double-alkali

Ammonia double-alkali process

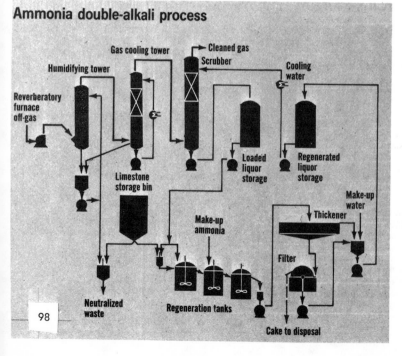

The industry is upgrading old practices, and introducing new techniques, in order to clear the air

scrubbing process. A feature is limestone regeneration of scrubbing liquor and avoidance of soluble salt production. This plant, at San Manuel, Ariz., removes more than 90% of the SO_2.

Reverb problems

A principal problem with reverb copper smelters is that their gas streams contain as much as 1–2% SO_2. Apparently too much for the wet-limestone scrubbers coming into use at electric power plants, this concentration is also unsuitable for acid plant feedstock. However, a solution must be found if air quality rules are not to sound the death knell for the reverb furnace.

According to a recent report by Carl Rampacek of the University of Alabama, and James Dunham of BuMines, the Bureau's Citrate process is being pilot-tested at a lead smelter. SO_2 is absorbed in a buffered citrate solution and reacted with hydrogen sulfide (H_2S) to produce sulfur. The citrate buffer is regenerated, and two-thirds of the sulfur is reconverted to H_2S for recycling. Remaining sulfur can be stored or sold. Costs are estimated to be 3–4¢/lb of copper, if this system was used in copper smelting, and no credit was taken for the sulfur. More than 2000 h of experience indicate SO_2 removal exceeding 95%.

SO_2 is not the only problem facing smelters. Particulate emissions can contain arsenic, which is toxic. For example, officials of Anaconda Co. (Anaconda, Mont.) themselves have said that the Anaconda smelter emits about 12 tpd of arsenic. Prior to 1971, the company trapped and collected arsenic, also a copper ore by-product, and offered it for sale. ASARCO (New York, N.Y.) stopped buying arsenic that year. However, Anaconda expects that new pollution control equipment, to go on stream later this year, will lower the arsenic emissions to a few lbs/h. This equipment uses a water-quenched gas stream to condense volatiles, and a baghouse to remove particulates. The company will keep the arsenic-laden material collected by the equipment, and see if any commercial use can be found for it.

Getting away from reverb

To comply with present and anticipated air-quality rules, and, perhaps, to save energy, copper companies are looking at other smelting techniques, including electric furnace smelting, flash furnace smelting, and continuous smelting. Additional approaches involve the Queneau-Schuhmann (Q-S) oxygen process, the AMAX pyrometallurgical process for chalcopyrites, and hydrometallurgy.

The Q-S process uses oxygen in flash melting to smelt copper and nickel sulfide ores in a single reactor from which gases cannot escape. This process produces much of its own heat, and is being used at INCO (Sudbury). Off-gas can have as much as 80% SO_2—enough not only for an acid plant, but for actual SO_2 liquefaction. INCO produces 1 million t/yr of sulfuric acid and 200 000 t/yr of liquid SO_2 from these tail gases, and has an arrangement for handling these by-products through Canada Industries Ltd., in Sudbury. Reputedly, INCO is upgrading its process to a proprietary technique for increasing the SO_2 concentration on an even more economical basis. So is AS-ARCO (Tacoma, Wash.) and Phelps Dodge (Ajo, Ariz.).

The AMAX (Carteret, N.J.) process now under development would use fluid-bed roasting at 850–900 °C to remove about 97% of the sulfur. The SO_2 concentration in the tail gas would be 12–14%—enough to justify an acid plant. Fugitive emissions would be eliminated. Experience is being gained from a 3.67 tpd pilot plant.

Wetting-down party

Air pollution considerations, as well as a need for processing methods that can be economical on a small scale, have encouraged R&D on hydrometallurgical processes. One such process, the Anaconda Arbiter process, extracts copper from sulfide concentrates by oxygen-ammonia leaching below 100 °C, with residue beneficiated to recover further copper and silver values. Leachate copper is extracted by solvents, and recovered through electrowinning. A prototype plant at Anaconda, Mont., was designed for 36 000 t/yr of electrolytic-grade copper produced by this method.

A commercial plant at the Hecla-El Paso Natural Gas Lakeshore Mine near Casa Grande, Ariz., in "shakedown" status earlier this year, will handle fluid-bed roasting of 400 tpd of sulfide flotation concentrate to obtain a sulfate calcine leachable with dilute sulfuric acid. Acid for vat-leaching of copper oxide ores will come from roaster off-gas; the copper itself is to be recovered by cementation.

Cyprus Mines is working on the Cymet ferric chloride process in which chalcopyrite (sulfide ore) is leached; iron and copper values are leached out, and elemental sulfur is produced. Ferric chloride is regenerated during copper electrowinning and electrolytic iron production. A pilot plant near Tucson, Ariz., with a 25-tpd feed produces 7 tpd of copper powder.

Also near Tucson, Duval Corp., a subsidiary of Pennzoil Co., has tested its Clear process for copper and iron leach and ferric chloride regeneration, and is building a 32 500-tpd facility at its Esperanza Mine. The Clear process also makes sulfur. Ferrous and cuprous chloride solutions are electrolyzed to produce copper. Ferrous chloride is further oxidized to regenerate ferric chloride for leach and precipitate excess iron in hydroxide form.

Fast change

In the last half-dozen years, copper ore processing practice in the U.S. has been changing fast. Among reasons for this change are the search for economical ways to recover copper from lower-grade ores, and protection of the public from adverse environmental effects. Perhaps a have-your-cake-and-eat-it situation will evolve, in which supplies of copper—so vital to the world economy since time immemorial—will be assured, while the nation's air and water are kept clean. JJ

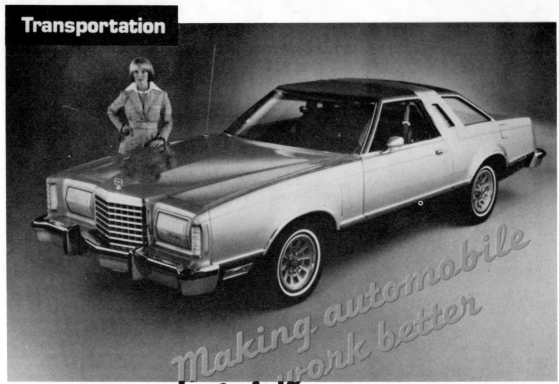

Making automobile rules work better

Reprinted from ENVIRON. SCI.

TECHNOL., **11**, 32 (January 1977)

Lee A. Iacocca
Ford Motor Company
Dearborn, Mich. 48121

How can motor vehicle emission and safety regulations be made more cost-effective? A leading industry spokesman offers his recommendations

Believe it or not, Ford Motor Company supports federal regulation of some aspects of the automobile industry. This acknowledgment will probably surprise many people who believe that the company opposes every law, rule, standard, or other directive coming out of Washington.

In fact, federal regulations did help to make it possible to build today's cleaner, safer cars. This could not have been done under private auspices alone, because of limited market appeal of emissions controls and safety features. Had Ford alone spent the hundreds of millions of dollars required to provide these features, and then priced all of its cars to recover at least the bulk of that investment, it would have been at a serious disadvantage. Competitive cars without those features would have been priced hundreds of dollars below those of Ford.

Added costs of standards

To be sure, the automobile has been regulated to some extent since the earliest years of its existence, mostly by state and local governments. Taxes were levied on vehicles and fuel. Drivers were required to purchase personal licenses. Over the years, such features as safety glass and sealed-beam headlamps were mandated.

The intensive phase of regulation began in the mid-1960's, with passage of the Federal Motor Vehicle Safety Act and, later, the Clean Air Act. The government had, of course, been involved with highway safety for a long time, but with these laws, federal agencies also became involved in car design. This involvement stemmed from federal requirements that all automobiles must meet certain emission and safety standards. Compliance with these standards resulted in additional costs, a good portion of which car manufacturers had to pass on to their customers.

Indeed, certain government-mandated standards have cost car buyers a great deal of money. Some of them have, perhaps, been worth the price, but others have not. However, before the question of costs and benefits of these standards is discussed, a look at the background of federal involvement in the automobile industry would be in order.

The laws passed during the mid-1960's, mentioned earlier, were what started the process of bringing the federal government into the industry with both feet, so to speak. Since that time, working closely with federal and state agencies, Ford Motor Company believes that the industry, as a whole, has done its part to help clean America's air. For instance, exhaust hydrocarbon (HC) emissions from 1977 cars are 90% lower than those from uncontrolled vehicles. Carbon monoxide (CO) is down 83% and nitrous oxides (NO_x) 67%.

The industry also significantly improved safety-production features of automobiles and reduced the vulnerability of cars to damage. Again, working with government, manufacturers have incorporated safety belts for front and rear seats, padded interiors, collapsible steering columns, warning flashers, stronger bumper systems, and many other features.

However, regulated improvements have not come cheaply. Through the 1977-model year. for example, Ford Motor Company would have had to have increased the retail price of a car by an average of about $580 in order to recover fully the costs of equipment needed to meet federally mandated safety, damageability, and emissions standards.

Moreover, as time went on, the industry reached a point of diminishing returns with respect to compliance with some government regulations. For instance, it was relatively inexpensive to remove the first 50% of tailpipe emissions. But as the 75% reduction level was reached, it was found that although costs were increasing, benefits were actually decreasing.

Air quality perspective

Before any discussion of emission standards, what Ford believes to be the role of the automobile in America's air quality should be put in perspective.

Now, some environmentalists concern themselves with only one aspect of the total air-pollution problem—the weight of pollutants from each source. Also, government researchers have calculated that in 1970, the automobile accounted for approximately 45% of all man-made pollution. However, it is generally recognized that a more realistic measure of air pollution from any given source should be based on its relative effects on human and plant life. For this reason, scientists, using information on present atmospheric levels and human tolerances, have developed systems that permit the rating of various pollutants in terms of these effects. They estimate that on a health-concern basis the automobile accounts for a maximum of 12% of all man-made air pollution.

In this light, it is unfortunate that the automobile is singled out too often as the source of most of the air pollution in this and other countries. Equally unfortunate, it seems, is the fact that the federal government's body of experts in air quality—the Environmental Protection Agency (EPA)—does not have the authority to set ultimate emission standards for motor vehicles. Instead, Congress has set those standards, and persists in retaining them, even though the EPA has said since 1973 that at least one of the statutory levels mandated by Congress—for oxides of nitrogen (NO_x)—may be more stringent than necessary.

This point also was made by Prof. Arthur Stern, recent president of the Air Pollution Control Association and chairman of a National Academy of Sciences panel that studied air quality and automotive emissions. Stern wrote that "... the panel felt that the statutory automotive emissions standards were too severe by a factor of about three to accomplish their intended purpose" (*ES&T*, June 1975, p 510).

In spite of such beliefs by respected authorities, Congress continues to keep the presently mandated standards. Automobile makers hope that Congress will seriously consider the EPA views when it deliberates on standards for 1978 and thereafter.

This consideration will obviously involve compromises—but reasoned compromises are needed in this matter. In other words, in cars, as in homes or anything else, it must be decided how clean is clean. For instance, it would be possible to keep a kitchen as clean as a hospital operating room, and there might be benefits in eating from sterilized dishes with sterilized utensils in a room whose air contents are carefully controlled. But the work and cost required to provide such an atmosphere at home would outweigh the benefits. Likewise with cars, the cost, time, and manpower required to eliminate all remaining traces of emissions—if they could be eliminated—would far outweigh the minute benefits to be derived.

Oil embargo fallout

A few years ago, the Arab oil embargo further complicated the emission picture by making fuel economy the auto industry's foremost priority. And last December, Congress passed the Energy Policy and Conservation Act, which included future requirements for vehicle fuel economy. The law requires a series of mandatory fuel-economy standards starting in 1978. By 1985 the cars each manufacturer builds and sells in the U.S. must achieve 27.5 mi/gal, averaged over its total output, on a production-weighted basis.

This requirement represents mileage improvement of 100% over 1974 and would be a greater change than has occurred since the birth of the industry. The average 1985 vehicle would have to achieve the fuel economy of today's small cars.

This means that the industry will have to commit large sums of money, although it does not have certain vital information. It is working hard on new technology that promises large fuel-

Some myths about automobiles

In many discussions of the automobile's past, present, or future, a number of myths often becloud rational thinking. For example, one such myth concerns the American people's overwhelming choice of the automobile as their principal means of transportation for so many years. Is it, as some contend, because of the industry's ability to force people to buy whatever it chooses to build, rather than what they really want?

Hardly. Americans like their cars because of the mobility they provide—the freedom to live and work where they choose, and to come and go as they please. In providing personal mobility, the car is far more flexible and convenient than the train, trolley, bus, or other conveyance.

Another important feature of the close relationship Americans have with their cars is the fact that they buy the kinds of cars they want, as recent wide swings between sales of big and small cars should demonstrate forcefully. People do not buy what they do not want. Many cars and car features have gone by the board because buyers rejected them. Indeed, if something better than the car comes along, the so-called love affair with the car will probably end abruptly.

Another myth is that the internal combustion engine is still on the scene because of some conspiracy. This engine has been dominant for 75 years—not because of a conspiracy, but because no other automotive power plant has been able to match it for dependability, efficiency, cost of manufacture, and economy of operation.

Yet another myth concerns the "gas guzzler." Although the big luxury car seems to be the cartoonists' model for all cars, it actually comprises only 4% of car sales. Most of the larger cars are the six-passenger sedans that people buy because they need them. One reason is that one-third of U.S. households consist of five or more people.

Such needs also explain the difference between American and European cars. European cars are built for short trips, mountainous terrain, narrow and crowded roads, moderate personal incomes, and gasoline prices that have long exceeded $1/gal.

By contrast, in the U.S., gasoline has been largely plentiful and reasonable in price. Indeed, between the end of World War II and the Arab oil embargo of 1973, "gas" prices declined in constant-dollar terms. During the same period, new car prices lagged behind the Consumer Price Index. Moreover, long travel distances, excellent highways, and rising personal incomes influenced the domestic automobile market.

economy improvements. But what is uncertain is how well each new approach will work. It is known, however, that major across-the-board fuel-economy gains will involve large investments and consumer costs. How large these costs will be is an open question. Nevertheless, consumers will have to pay for better fuel economy in one way or another. But what tradeoffs they will prefer to make among fuel economy, car size, performance, convenience features, and price cannot be forecast. The industry expects that the fuel economy demanded by law will be substantially greater in 1985 than the fuel economy that would be demanded in a free market. But how big the gap will be, or how many consumers will stay out of the market because they will not be able to buy the types of cars they want or because they cost too much, is not possible to predict at this time.

As for vehicle safety, seat belts—one of the first devices mandated, and a feature voluntarily offered by Ford as far back as 1956—have proved to be as valuable as all of the other government-required devices put together. The first dollars spent on safety improvement brought clear benefits at relatively reasonable cost; however, succeeding dollar outlays have brought benefits that were more and more marginal.

Cost-effective safety equipment

Although the goal is to continue building safe vehicles, one cannot lose track of what the customer can pay, or is willing to pay, for a new car. For instance, when a buyer spends a dollar for a safety feature, he should receive a dollar's worth of protection. But here are some examples which show that this is not always the way things work in real life:

• Each year, car buyers pay more than $100 million for head restraints, but studies estimate that the restraints eliminate only 3000 minor injuries. That represents an average expenditure of more than $33 000 to eliminate each minor injury.

• Car buyers also spend about $100 million/y for side-impact protection; estimates indicate that only 12 400 minor injuries are prevented.

• Car buyers pay more than $50 million each year for roof-crush resistance that provided little if *any* discernible benefit.

The National Highway Traffic Safety Administration (NHTSA) has considered requiring another device in cars—a so-called passive-restraint system most commonly represented by the air bag. The industry believes air bags should *not* be required by the government. Present safety belts provide protection comparable to bags plus lap belts; the lap belts would be needed *along with* air bags to meet a proposed government standard. Air bags would cost a buyer about $235 per car, in addition to the present cost of safety belts.

The government probably would not consider requiring air bags if more than half the drivers and passengers today did not ignore the safety belts they now have. Indeed, the lap-and-shoulder-belt combination is the best piece of safety equipment available, and if all people used it regularly, highway deaths and injuries would drop dramatically.

Government survey results which show that more and more people are "buckling up" are encouraging. But because some people do refuse to use the safety belt protection already available in nine of ten cars now on U.S. roads, those who *do* use their belts could be required to spend hundreds of dollars for air bags. This is unfair.

A more cost-effective approach—one that would show immediate results—would be for the 50 states to pass legislation requiring all occupants to wear safety belts. This approach has been successful in a dozen countries, including Australia, Czechoslovakia, France, New Zealand, Sweden, and Canada's province of Ontario. Since the law went into effect in Australia, for example, the accident fatality rate has dropped a dramatic 25%. And in Ontario, highway fatalities dropped 33% during the law's first three months. Perhaps establishment of carefully focused, Federally-supported educational campaigns would pave the way for mandatory belt-use laws.

Actually, a central question on highway safety is: How much attention should be focused on the car alone? A few years ago Ford spent millions of dollars building an experimental safety vehicle under a $1 contract with the federal government. The

Engine test. *For cleaner, more fuel-efficient operation*

car was designed with the goal of protecting "occupant" test mannequins in a barrier crash at 50 mph. Such a car could not be sold—its so-called "protection" features make it far too heavy. Instead, Americans' needs today are for lighter, more fuel-efficient cars, not cars that would resist impacts of military tanks, and costs that would put them out of reach for most people.

Other safety suggestions

It is time for Washington to turn its attention to factors other than the vehicle in highway safety. For example, imposition of the 55-mph speed limit has shown that significant fatality reductions are achievable by means other than setting vehicle standards. Highway deaths in 1975 were down 18% from those of 1973, the last year before the 55-mph law. Roads should be made safer, with more effective traffic signs and signals, and fewer unnecessary roadside hazards. Traffic law enforcement should be improved, and periodic inspection of cars and trucks should be required. Driver licensing requirements should be strengthened—mainly to get drunk and otherwise irresponsible drivers off the roads. Incidentally, with regard to licensing, there are 75- and 80-year old Americans whose driving ability hasn't been tested since they got their first driver's licenses at the age of 14.

Understandably, the government's preoccupation with vehicle changes causes problems. These problems do not stem solely from Washington's telling the industry, in effect, how to design cars. Rather, problems result from Washington's *philosophy* in establishing vehicle safety standards—a philosophy which is apparently vastly different from that envisioned by the Safety Act, which empowered the NHTSA to set reasonable standards for motor vehicles. Perhaps the NHTSA sometimes loses sight of that word "reasonable."

A case in point is Air-brake Standard 121 for large trucks and buses. The auto companies received about *20* rulemaking proposals and changes on this standard from NHTSA, none of which explained why such excessively difficult requirements were necessary. Justification was lacking—especially when buyers of complying trucks would have to pay an additional $1500 to $2000 for the mandated features. It was not until the standard had been in effect for several months that NHTSA relaxed the requirements for stopping distances to a more realistic minimum. Months later, it relaxed the requirements still more. Be that as it may, if NHTSA had initially established the requirements eventually promulgated, there would not have been all the "backing and filling" that caused significant confusion and problems, to say nothing of the engineering effort wasted on a number of designs required to meet the earlier standard. Truck makers and truck buyers would not have had to pay millions of unnecessary dollars to meet original requirements, which the industry and the truckers argued, vainly, were not needed to begin with.

Nevertheless, although portions of the air-brake standard are points of argument, it is re-emphasized that car makers have *not* disagreed with *all* safety requirements from Washington. Actually, they generally have agreed with the overall purpose of safety regulations. They take issue only with regulations or regulations that are believed not practical or cost-effective from the customer's standpoint.

The automobile in the future

The public—and, hopefully, the government—often profit from the industry's exchange of ideas with Washington. Standards achieved through such exchanges usually advance the cause of vehicle safety, while respecting the genuine needs of the auto makers, vehicle drivers and their passengers, and the American economy.

Except for the change to military production in World War II, the automobile industry is now in the midst of what is probably the most extensive product revamping in its history. New families of smaller, lighter, and more efficient cars are near at hand. Bringing them to market will require an all-out effort involving

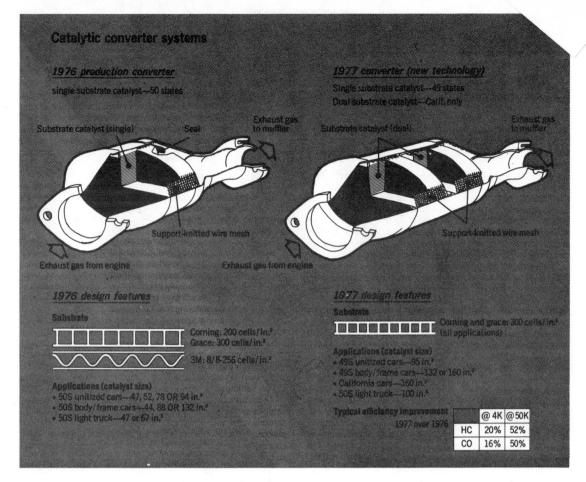

Catalytic converter systems

1976 production converter
single substrate catalyst—50 states

Substrate catalyst (single) — Seal — Exhaust gas to muffler

Support-knitted wire mesh

Exhaust gas from engine

1977 converter (new technology)
Single substrate catalyst—49 states
Dual substrate catalyst—Calif. only

Substrate catalyst (dual) — Exhaust gas to muffler

Support-knitted wire mesh

Exhaust gas from engine

1976 design features

Substrate

Corning: 200 cells/in.^{3}
Grace: 300 cells/in.^{2}

3M: 8/8-256 cells/in.^{2}

Applications (catalyst size)
• 50S unitized cars—47, 52, 78 OR 94 in.^{3}
• 50S body/frame cars—44, 88 OR 132 in.^{3}
• 50S light truck—47 or 67 in.^{3}

1977 design features

Substrate

Corning and grace: 300 cells/in.^{3}
(all applications)

Applications (catalyst size)
• 49S unitized cars—95 in.^{3}
• 49S body/frame cars—132 or 160 in.^{3}
• California—160 in.^{3}
• 50S light truck—100 in.^{3}

Typical efficiency improvement
1977 over 1976

	@ 4K	@ 50K
HC	20%	52%
CO	16%	50%

new capital investment, new design and engineering, installation of new tools, and the construction or renovation of numerous manufacturing and assembly plants.

For instance, for 1977 Ford has introduced a trimmer, sleeker Thunderbird. Later in the '77 model year, a new "mini," the Ford Fiesta, which will be smaller than the subcompact Pinto, will be coming out. Moreover, by 1980, standard-size Ford cars will be about two feet shorter and 700 to 800 pounds lighter than their present-day counterparts.

Development of alternatives to the internal combustion engine are also in progress. Back in May 1973, a U.S. Senate subcommittee was given a pledge that Ford Motor Company would build a new type of engine that would give superior fuel economy and good emission control if the federal government would change the statutory NO_x standard from 0.4 g/mi to 2.0 g/mi. After all, no prudent business people can invest hundreds of millions of dollars in a product that might not be marketable because it does not comply with a government regulation. Once rational long-term emission standards have been set, the company will know which of its development programs offers the best opportunity to meet those standards with a clean, high-mileage engine that represents a good buy for its customers.

As conditions change in automobile markets, the products will change, as they have all along. Ford entered the European market 70 years ago, and today is the leading seller of new cars in Europe. It holds this position because it has long had the engineering and management skills to respond to changing customer priorities. The same is true in North America and elsewhere around the world.

In the last analysis, people place much importance on the automobile and what it can do for them. They spend much of their earnings on motor vehicles. To be sure, expenditures for compliance with government regulations and for paying increasing labor and material costs have resulted in higher car prices, but the car still represents an excellent buy. Twenty years ago, it took 6.6 months of the median U.S. family income to buy the average Ford car. Ten years ago it took 5.1 months of income. Now it takes only 4.7 months. Also, that average domestic new car of today has more standard equipment than those of 10–20 years ago. Further, it offers more optional equipment, and is quieter, cleaner, safer, more comfortable, and substantially more maintenance-free.

To acquire objective judgments on the future of the automobile, Ford recently asked an outside research group to study the prospects. The study came up with a key conclusion: that the market for automobiles has not reached the saturation point; in fact, the future growth rate for the car population will exceed the growth rate for the "people" population, with annual sales of new cars doubling by the year 2000. Ford Motor Company believes that the future looks good for the automobile—especially if government regulators eventually fix on cost-effective requirements that give the new car buyer a dollar's worth of improvements for a dollar spent.

Lee A. Iacocca *is president of Ford Motor Company. He first joined Ford in 1946, and was elected to his present position in December 1970.*

Coordinated by JJ

FEATURE

Emission control and fuel economy

With the catalytic converter, 1975 and later model autos have cleaner exhausts, better driveability and improved fuel economy; however, problems remain to be solved

Ernest S. Starkman

General Motors Corp.
Warren, Mich. 48090

Reprinted from ENVIRON. SCI. TECHNOL., **9**, 820 (September 1975)

With the increasing intensity of attention to shortages of fuel and energy, every step taken for environmental reasons has become more subject to questioning by those concerned with the energy supply. Thus, President Ford urges a 40% improvement in fuel economy. But the automobile companies are increasingly concerned about the conflicts that accompany requirements for fuel economy and for cleaner exhausts.

Before requisite technology was available and before the emission limits became severe enough to necessitate use of the catalytic converters, emission control measures generally were accompanied by deterioration of both fuel economy and driveability (see Figure 1). With the 1975 introduction of catalytic emission control systems this trend was reversed. And, good fortune it was that a measure directed toward reducing exhaust emissions also helped fuel economy. After this fortuitous instance, however, a return can be expected to a condition of giving something up in exchange for that which was gained. If exhaust emission standards became more stringent than the 1975–76 standards, and no new practical system is developed for controlling emissions, a downward trend in fuel economy can be anticipated.

In the early days of emission control, modification of air-fuel ratio and retarding the ignition timing were two basic measures utilized: intake air was heated and the compression ratio was decreased. These steps, however, carried the penalties of higher fuel consumption (Figure 2) and poorer driveability. When substantial NO_x control became necessary, it was accomplished largely by spark retard and exhaust gas recirculation. By 1974, then, average or sales-weighted fuel economy for a General Motors car was about 16% less than for its 1970 counterpart. Sales-weighted fuel economy for 1974, however, improved over that of the preceding year, because the proportion of smaller cars sold had increased, even though the car-for-car decline continued.

Catalytic system development

Early in vehicular emission control development, it was generally believed in the industry that a catalytic or thermal aftertreatment system would eventually be needed, particularly when exhaust standards became too severe for other control methods. Consequently, development programs for such systems were instituted prior to the date for designing a specific system for a given set of standards. When William D. Ruckelshaus, then the EPA Administrator, imposed the 1975 standards, the catalytic system was the only technology available to meet compliance without making further major compromises in vehicle driveability and fuel economy. Catalytic systems, therefore, were almost universally applied to 1975 model year cars.

Platinum-palladium was selected as the catalyst combination. Other major changes were also made to the powerplant, including modification in carburetion and application of an electronic ignition system. Spark timing and fuel air mixture for 1975 cars were restored substantially to pre-control settings. Both fuel economy and driveability were improved as a consequence.

The oxidizing catalytic system introduced on most 1975 cars built in this country (see Figures 3 and 4) represents a major departure in emission control steps. It is the most discussed, the most praised, and the most criticized. It has also been the most effective.

The most important achievement of the catalytic system is that it has enabled automobile manufacturers to meet the 1975 interim exhaust hydrocarbon (HC) and carbon monoxide (CO) standards—1.5 g/mi for HC and 15 g/mi for CO. It reduced the 1974 exhaust emissions by 50% (Table 1).

Field trials

Both before and continuing on after the interim standards were established, field trials were conducted to evaluate the progress being made, and to learn what changes were needed to produce an effective, commercially acceptable catalyst emission control system.

An account of two field trials may help in understanding some of the painstaking and extensive programs that lead from design to tooling. The product cannot go from drawing board, or even from laboratory, directly to production.

Accumulation of major field experience on vehicles with systems much like that for 1975 began in November 1971, when GM's AC Spark Plug Division started a fleet of 18 Buicks equipped with underfloor catalytic converters in a field durability test intended to cover 50,000 mi per vehicle. They

FIGURE 1

Fuel economy trends

changes (gain/loss) in GM city-suburban schedule economy

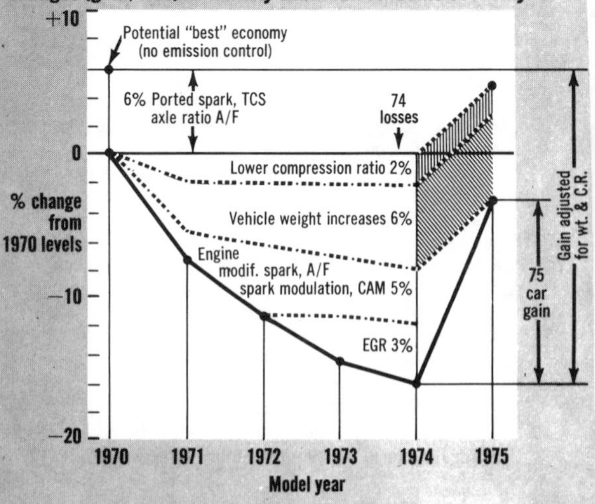

Fuel economy of GM cars has deteriorated in recent years as a result of weight increases and emission control. Use of the catalytic converter for exhaust emission control reversed that trend for 1975.

Analysis of fuel economy changes:

1975 **GAIN** results from engine optimization (EGR, spark and A/F ratio with catalytic converter)

1975 "Adjusted" to 1970 weight and compression ratio approaches 1970 best economy

FIGURE 2

Factors affecting emissions and fuel consumption

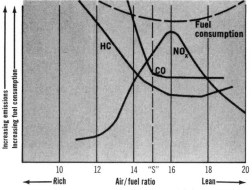

Air-fuel ratio affects both exhaust emissions and fuel consumption

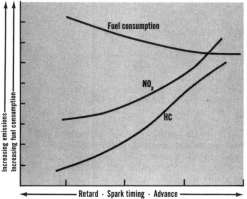

Ignition timing retard adversely affects fuel consumption

Table 1. Exhaust standards
(g/mi using 1975 Federal Test Procedure)

		HC	CO	NO$_x$
1974	Federal	3.0	28	3.1
	California	2.9	28	2.1
1975–6	Interim			
	Federal	1.5	15	3.1
	California	0.9	9.0	2.0
Statutory				
	All states	0.41	3.4	0.4

were started with base metal or non-noble metal catalysts. In six of the cars, a decision was made to replace the original base metal catalyst with platinum and palladium after only 12,000 mi of operation. The HC and CO emission levels had already far exceeded the statutory standards, which at that time were scheduled for 1975. This whole 18-car program was terminated after 602,000 mi of mixed results. It was one of the main information sources that subsequently lead to the adoption of the noble metal catalyst.

A larger and more comprehensive field durability program included almost 250 cars equipped with catalytic systems built under actual assembly line conditions. It was initiated with 13 Oldsmobiles built in October 1972, and put into full operation with 222 Chevrolets and 12 more Oldsmobiles as-

Table 2. Fleet mileage results

System	Standards	No. of cars	Mileage total
Group 1[a] Chev CCS Underfloor	'75–'76 Federal 1.5/15.0/3.1	75	3,772,000
Group 2 Chev CCS Manifold	'75–'76 Federal 1.5/15.0/3.1	71	3,189,000
Group 3[b] Chev AIR Underfloor	'75–'76 Calif. .9/9.0/2.0	36	1,265,000
Group 4 Chev CCS Manifold	'75–'76 Calif. .9/9.0/2.0	22	788,000
Group 5 Chev T-MECS[c]	'78 Federal .41/3.4/.4	18	171,000
Group 6 Olds CCS Underfloor	'77 Federal .41/3.4/2.0	13	408,000
Group 7 Olds AIR Underfloor	'77 Federal .41/3.4/2.0	12	259,000
	TOTAL	247	9,852,000

[a] Controlled Combustion System. [b] Air Injection Reactor. [c] Triple Mode Emission Control System—Oxidizing and reducing catalysts in single manifold-mounted unit.

FIGURE 3

260 cu. in. catalytic converter

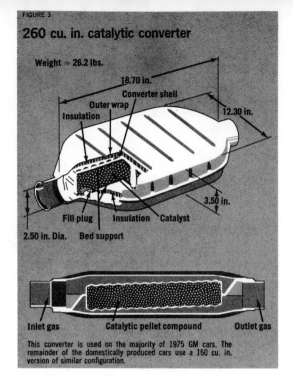

Weight = 26.2 lbs.

18.70 in.

Converter shell

Outer wrap

Insulation

12.30 in.

3.50 in.

Fill plug   Insulation   Catalyst

2.50 in. Dia.   Bed support

Inlet gas   Catalytic pellet compound   Outlet gas

This converter is used on the majority of 1975 GM cars. The remainder of the domestically produced cars use a 160 cu. in. version of similar configuration.

Table 3. Fuel economy
Oldsmobile converter fleets vs. certification data vehicles (5,000 lb inertia weight)

	Emission test fuel economy	Fuel economy penalty	No. of cars
1975 GM vehicles (Federal)	12.1	—	14
1975 GM vehicles (California)	10.9	10%	12
1977 Experimental Oldsmobile	9.6	21%	25

---

sembled in May 1973. This program was begun when the statutory 1975 HC and CO standards, 0.41 g/mi HC and 3.4 g/mi CO, were still on the books, and the decision to go from base metal to platinum group catalysts had been made; noble metal catalysts were used in all vehicles.

While this field trial was designed to provide some of the evidence needed to demonstrate that the 1975 catalytic emission control system could be made ready for release to the public, engineering for production, even before such trials were conducted, had to be in process in order to be ready for the 1975 model year. The fleet completed 10 million miles of operation early in October 1974, at about the time the 1975 production cars began to appear on the streets with the catalytic system (see Table 2).

In the course of the field test, the original 1975 standards were suspended. As a consequence, some of the vehicles, which were divided into seven groups, were recalibrated for different sets of standards. Two groups of cars were calibrated for the new 1975 interim standards for California—0.9 g/mi HC, 9.0 g/mi CO, 2.0 g/mi NO_x, and two for the 49-state interim standards of 1.5, 15.0, and 3.1 g/mi HC, CO, and NO_x, respectively. Two other groups were set up for the more severe standards then scheduled for 1977, and the seventh was continued at the statutory standards—0.41, 3.4, and 0.4 g/mi.

In the groups that were recalibrated, settings were adjusted for the improved fuel economy, driveability, and performance that could then be realized with the catalytic system because the emission constraints were less stringent. Important as this step turned out to be, it was still secondary to the trial's primary purpose: evaluation of durability and performance of catalysts and systems in a diverse range of service conditions while meeting the applicable exhaust emission standards. The fleet test also provided a means to evaluate maintenance requirements and any other problems resulting from operation in areas with special geographic characteristics. To assist in acquiring road experience quickly, the cars were placed in taxi, police, and governmental fleets throughout the country.

In general, the converter systems were successful in controlling emissions at their prescribed levels. The four groups calibrated for the 1975–76 emission requirements, both Federal and California, finished their test with average emissions meeting the limits for all pollutants (Figure 5). The two groups of Oldsmobiles set for exhaust emission standards of 0.41, 3.4, and 2.0 g/mi were running satisfactorily in mid-June 1975, with mileage accumulation of 26,000 to 37,000, with some individual cars exceeding the emission limits but the average meeting the standards. Keep in mind that to receive

---

Table 4. President's fuel economy objectives
(mi/gal)

Model year	GM	Ford	Chrysler	All other	Weighted average
1974	12.2	14.4	13.8	20.1	14.0
1975	15.7	13.3	15.8	21.2	15.9
1980 (goal)	18.7	18.7	18.7	24.7	19.6
percent improvement over 1974	53%	30%	36%	23%	40%

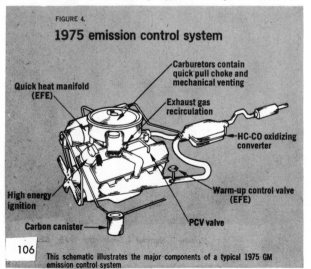

FIGURE 4.

1975 emission control system

Carburetors contain quick pull choke and mechanical venting

Quick heat manifold (EFE)

Exhaust gas recirculation

HC-CO oxidizing converter

High energy ignition

Warm-up control valve (EFE)

Carbon canister

PCV valve

This schematic illustrates the major components of a typical 1975 GM emission control system

certification, of course, every certification car must pass. A margin of safety below the actual standard is necessary.

The groups meeting the statutory standards for HC and CO showed fuel economy that demonstrates the impact on fuel consumption of increasingly stringent emission standards. The average fuel economy of these 25 cars was 9.6 mi/gal, 21% below the average of similar vehicles calibrated for the 1975 interim Federal standards. Those vehicles set for the California 1975 standards, at 10.9 mi/gal, showed a 10% penalty (Table 3).

Economy effects

With technology at its present state, to meet standards much more severe than the present Federal standards would again require compromise of spark timing and air-fuel ratio. The condition would be particularly aggravated by a more stringent NO_x standard, which would require increased exhaust gas recirculation (EGR), since EGR shifts the maximum economy point toward a richer mixture and more advanced spark timing, and concomitantly increases HC emission (Figure 6). Both fuel economy and driveability would again be impaired by measures taken to improve emission control.

Last year, President Ford requested, by 1980, a voluntary 40% improvement in automobile fuel economy over the 1974 level. General Motors reacted affirmatively, projecting that, with decreases in car weights, increased power train efficiency, and improved aerodynamic qualities, as well as a refinement of the catalytic emission control system, GM could reach the President's goal of 18.7 mi/gal in 1980 models. This forecast was based on the premise that the exhaust emission regulations would not be made more severe than the 1975 standards for at least five years. Although President Ford's request was expressed as a 40% fuel economy increase over 1974, GM's improvement would have to be 53%, since the company's 1974 sales mix contained a greater proportion of larger cars than the average of the other companies in the industry (Table 4).

Sales-weighted fuel economy of 1975 GM cars is shown by the EPA city test schedule to be 28% greater than in 1974. The composite average fuel economy for 1975 GM cars is 15.5 mi/gal, computed on the basis of 55% city and 45% highway driving. This figure is expected to improve to about 16 mi/gal in 1976. With retention of 1975–76 emission standards, approximately 17 mi/gal is projected for 1977, and 18

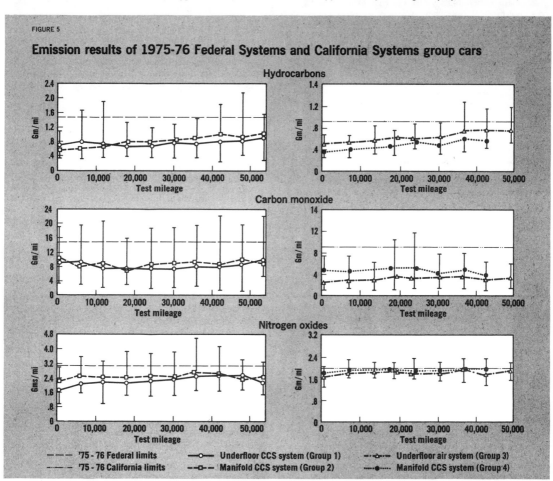

FIGURE 5

Emission results of 1975-76 Federal Systems and California Systems group cars

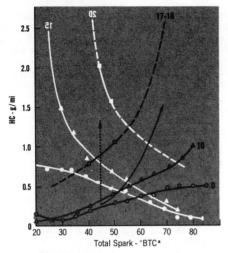

FIGURE 6

Emission and economy tradeoffs

at 55 mph road load condition
(Numbers on curves give percent EGR)

[a] Before top center

The bold solid line shows the increasing spark advance required to maintain best fuel economy with increasing EGR rates, and the resulting detrimental effect on HC emissions. The dashed line superimposed shows the effect of increasing the EGR rate while maintaining a constant spark setting

mi/gal may be exceeded in 1978. These projections, on the basis of experience in California and with experimental systems, will have to be discounted substantially if Congress does not see fit to order an extension of the 1975 standards.

Further problems

The catalytic converter's function is to oxidize unburned material in the exhaust. Unfortunately it also tends to further oxidize the small amount of sulfur present in gasoline (about 0.03% by weight as a national average). Some of the sulfur dioxide resulting from the burning of the sulfur is further oxidized to sulfur trioxide (which hydrolyzes to H_2SO_4) in the exhaust system. The magnitude of the resulting sulfate emissions is very small, and there should be no immediate health hazard. At the very least, several years are available for analysis of that possibility. During this period the benefits of decreased HC and CO will surely outweigh any adverse effects of sulfuric acid from automobile sources.

The most pessimistic outlook is that, after a substantial portion of the American automobile population consists of catalyst-equipped cars, a concentration of sulfuric acid potentially adverse to health could develop at a heavily traveled roadside under the worst possible conditions. The probability of such occurrence appears slight. Present projections are based on highly questionable atmospheric models. General Motors is, therefore, conducting a large-scale experiment to collect data intended to test and improve these models. In this program, which is being conducted in the fall of 1975 at the GM Proving Ground near Milford, Mich., more than 350 catalyst-equipped cars will be driven on a 4-lane track under schedules duplicating heavy traffic on a freeway. Samples of air will be taken from a matrix of locations in the area.

Another possible obstacle to the continued use of the catalyst, the cost of platinum, is not expected to be a controlling factor in future decisions regarding catalyst use, but the fact that it is imported from South Africa could present political awkwardness in the future.

Alternate powerplants

While alternative engines, under continuing evaluation by the auto industry, may ultimately include a candidate that can decrease air pollution while providing adequate transportation, none has yet been shown to have this capability. More in the realm of near-term probability would be the spark ignition

engines designed to burn extremely lean air-fuel mixtures efficiently. Both stratified charge engines (SCE) and the "lean-burn" engines that have been under investigation by the auto companies are in this group.

The stratified charge became a production reality in the 1975 model year. As with all lean-running engines, carbon monoxide emission is inherently low in the stratified charge engine (SCE), and NO_x is low relative to a conventional engine. While NO_x can be brought below 2.0 g/mi, apparently NO_x from an SCE car will have difficulty in meeting the 0.4 g/mi.

GM has achieved the statutory 1977 HC and CO standards in the laboratory with a stock engine modified for stratified charge, with fuel economy levels comparable to that of 1974 cars. Improved economy could be expected with this design if the engine were recalibrated for less stringent levels, but there presently appears to be little practical near-term prospect of meeting the standards for 1978, when all three statutory levels become mandatory.

The diesel engine appears to have the ability to meet the statutory HC and CO standards, along with an NO_x standard of about 2.0 g/mi. While it does have a fuel economy advantage over the gasoline engine, part of this gain is at the cost of performance, which is definitely poorer than that of the comparable gasoline engine.

The NO_x reduction measures used up to the present, exhaust gas recirculation and spark retard, have been effective at moderate control levels, but spark retard impairs fuel economy. Even to approach the 0.4 g/mi. level, however, catalytic reduction appears to be required. So far none of the experimental reducing catalysts or combinations of methods tried have met the statutory NO_x level for any sustained mileage. One of the systems under investigation, that offers some promise, entails simultaneous oxidation of HC and CO and reduction of NO_x by a single catalyst.

However successful the U.S. Congress and the automobile manufacturers are in establishing and meeting exhaust emission standards appropriate to the atmospheric needs of the nation, mass transit can be expected to play a more prominent role in the future. The individually owned car, nevertheless, will remain the dominant form of transportation for the near-term, but that car is expected to become smaller and lighter.

Additional reading

Stempel, R. C., and Martens, S. W., "Fuel Economy Trends and Catalytic Devices," Paper 740594, presented at SAE West Coast Meeting, August 1974.

Gumbleton, J. J., Bolton, R. A., and Lang, H. W., "Optimizing Engine Parameters with Exhaust Gas Recirculation," Paper 740104, presented at SAE Automotive Engineering Congress, February 1974.

Miles, D. L., Faix, L. J., Lyon, H. H., and Niepoth, G. W., "Catalytic Emission Control System Field Test Program," Paper No. 750179, presented at SAE Automotive Engineering Congress, February 1975.

Report by the Committee on Motor Vehicle Emissions, National Academy of Sciences, Commission on Socio-Technical Systems, National Research Council. Washington, D.C., November 1974.

Ernest S. Starkman *is vice president of General Motors in charge of the Environmental Activities Staff. He recently served on the White House Task Force on Air Pollution and on the Office of Science and Technology Ad Hoc Panel on Unconventional Automotive Vehicle Propulsion.*

Coordinated by LRE

Porsche stratified charge engine

This German engine uses less fuel and produces
less exhaust emissions than the conventional
four-stroke spark ignition engine does

T. Ken Garrett

Hemel Hempstead
England

Reprinted from ENVIRON. SCI. TECHNOL., **9**, 826 (September 1975)

The stratified charge type of engine shows the greatest
promise for solving both the problems of emissions and fuel
economy over at least the next decade. Porsche has been
engaged in basic research work in this field.

The Porsche SKS engine is similar to the Honda CVCC
unit; the essential feature of stratified charge, of course, is
that it can render the engine tolerant of a wide range of fuels
which, in the overall context of energy utilization, is a great
advantage in getting the utmost out of each barrel of crude oil
at the refinery is concerned.

At the outset, Porsche had to chose between an open and
a divided combustion chamber system. They chose the latter
because of the difficulties of stratifying the charge consistently in the open—or single chamber—combustion system.
These difficulties arise because of the effects of varying load
and speed.

Their studies were concentrated primarily on the four
phases of the combustion process:

- ignition
- development of the combustion kernel
- the main combustion phase
- afterburning.

For ultimate success, optimum conditions have to be established for each of these phases.

Phase 2—combustion kernel

For the initial ignition and then the development of a stable
kernel of combustion, it is necessary for the rate of generation of heat in a small volume of combustible mixture around
the spark to be greater than the rate of loss through its
boundaries. The velocity of turbulence also has to be relatively low to avoid a reduction in the rate of ionization—the essential process immediately prior to combustion. On the other
hand, once the combustion kernel is firmly established, a high
degree of turbulence is necessary for rapid and complete
combustion.

The Porsche SKS engine was designed with a main and
small auxiliary combustion chamber, Figure 1, next to which
is yet another, even smaller chamber, into which the sparking
plug is screwed. The main combustion chamber is supplied
with a very weak mixture through a conventional inlet valve.
A rich, yet combustible, mixture is supplied to the auxiliary
chamber either by direct injection of fuel or through a supplementary valve, according to whether injection or carburetion
is used for the main mixture supply.

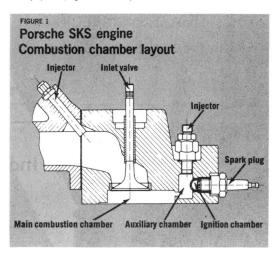

FIGURE 1

**Porsche SKS engine
Combustion chamber layout**

Injector Inlet valve

Injector

Spark plug

Main combustion chamber Auxiliary chamber Ignition chamber

The mixture flows freely from the auxiliary combustion
chamber into the tiny ignition chamber, but the connection
between the auxiliary and main combustion chambers comprises one or more nozzles, designed to generate swirl in the
auxiliary chamber and thus to stratify the charge in it during
the compression stroke and to direct jets of flaming gas out
into the main chamber during the firing stroke. Accordingly,
ignition is initiated in the relative stillness of the smallest
chamber and, once the kernel of flame has developed, it
spreads very rapidly throughout the turbulent auxiliary chamber, whence the flaming gases are forcibly ejected to spread
the combustion throughout the weak mixture in the main
chamber. Although the overall mixture strength can be considerably weaker than stoichiometric, combustion occurs regularly and consistently over a wide range of speed and load
conditions.

From the indicator diagram, Figure 2, it can be seen that,
after an extremely short ignition delay, the burning spreads
rapidly throughout both the auxiliary and main chambers. The
relative rates of increase of pressure in these two chambers
depend on the size and shape of the nozzles that intercon-
nect them. Good results are obtained when the velocity of the

burning gases passing through the jets is high enough to increase the turbulence in the main combustion chamber. In these circumstances, the thorough intermixing causes combustion to originate in and spread rapidly from many centers in the very weak mixture in the main combustion chamber.

Periodic fluctuations in gas pressure develop at a frequency determined by the relative volumes of the interconnected cavities—the main and auxiliary chambers. In contrast to the vibratory fluctuations experienced during knocking combustion, however, the fluctuations in gas pressure cannot be suppressed by increasing the octane number of the fuel or by retarding the ignition; moreover, variations in the physical and chemical properties of the fuel have little influence on the combustion process. On the other hand, the amplitudes are small, so there is no risk of mechanical damage. Similar vibrations are induced in the combustion chambers of gas turbines to encourage complete combustion.

Phase 3—ignition

Ignition will occur reliably in the auxiliary chamber at air:fuel ratios ranging from 0.4-1.2 times stoichiometric. As the mixture strength in this chamber becomes richer than 0.8, partly as well as completely burned products of combustion are ejected through the nozzles into the main chamber. The partly burned products include not only CO and C (partial products of combustion occurring instantly in the initial stages) but also H and OH radicals and many others. Because many of these products are highly active, they serve as a multiple of sources of ignition, setting up chain reactions throughout the main combustion chamber.

As the richness of the mixture in the auxiliary chamber increases, the temperature of the wall of this chamber becomes high. But, with a reduction in the strength of the mixture in the auxiliary chamber, the quantities of products of incomplete combustion decrease and the temperature of the gases expelled into the main chamber increases. At the same time, ignition and combustion in the main chamber become irregular. Obviously, therefore, it is not the temperature of the expelled gases but the quantities of products of partial combustion that are important for the development of chain reactions in the main chamber. The regularity of ignition and combustion in the auxiliary chambers of multicylinder engines is much better than in the combustion chambers of a conventional engine, as can be seen from Figure 3.

FIGURE 2

Indicator diagrams of a conventional and SKS engine

Conventional engine

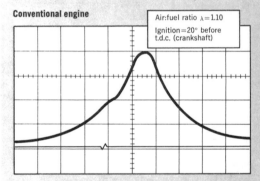

Air:fuel ratio $\lambda = 1.10$
Ignition = 20° before t.d.c. (crankshaft)

Speed = 2000 rev/min; specific load = 0.25 kJ/dm³; compression ratio = 8.65:1

Porsche SKS engine

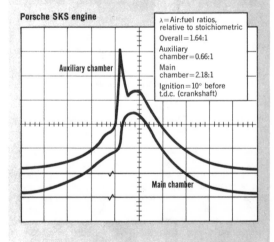

Auxiliary chamber

Main chamber

λ = Air:fuel ratios, relative to stoichiometric
Overall = 1.64:1
Auxiliary chamber = 0.66:1
Main chamber = 2.18:1
Ignition = 10° before t.d.c. (crankshaft)

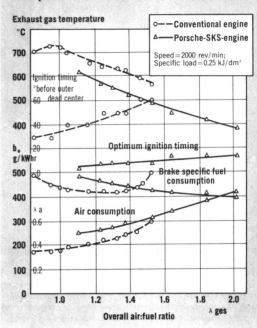

FIGURE 4

Comparison of combustion characteristics

Exhaust gas temperature

Speed = 2000 rev/min;
Specific load = 0.25 kJ/dm³

o – – Conventional engine
△ —— Porsche-SKS-engine

Optimum ignition timing

Brake specific fuel consumption

Air consumption

Overall air:fuel ratio

FIGURE 3

Variations in peak pressures in successive cycles in the conventional and Porsche SKS engines

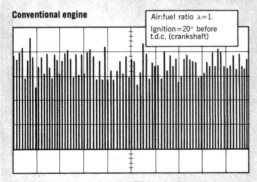

Conventional engine

Air:fuel ratio λ=1.
Ignition=20° before t.d.c. (crankshaft)

Speed=2000 rev/min; specific load=0.25 kJ/dm³;
compression ratio=8.65:1

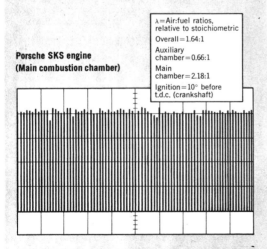

Porsche SKS engine
(Main combustion chamber)

λ=Air:fuel ratios, relative to stoichiometric
Overall=1.64:1
Auxiliary chamber=0.66:1
Main chamber=2.18:1
Ignition=10° before t.d.c. (crankshaft)

Performance and emissions

In Figure 4 is shown the variations in performance with changes in overall air:fuel ratio of a single cylinder research engine under part load. If the injection of fuel into the auxiliary chamber is regulated to give minimum fuel consumption, the SKS engine will run on overall air:fuel ratios as weak as 2.1 times stoichiometric. The limiting weakness is determined only by the ability of a wide-open throttle to pass sufficient air to maintain the required power output at that air:fuel ratio.

At lower power outputs, overall air:fuel ratios of 0.8–2.2 times stoichiometric can be used. As with conventional engines, on the other hand, maximum power is obtainable only at an air:fuel ratio of about 0.9.

In Figure 4, the engine with which the SKS unit is compared is of the most modern design. At their minimum values the brake specific fuel consumptions of the two do not differ much; however, the brake specific fuel consumption of a conventional engine increases as the mixture is weakened, in direct contrast to the characteristic of the SKS engine.

Because of the better conditions for ignition and subsequent burning in the SKS engine, its optimum ignition timing is later than in a conventional one. Moreover, provided the relationship between the volume of the auxiliary chamber and the dimensions of the nozzle is optimized, variations in ignition timing have much less effect on the performance and emissions of an SKS engine as compared with a conventional one.

Measurements of the exhaust concentrations, Figure 5, show that the constituents are qualitatively the same for both types of engine. However, quantitatively the NO_x content of the SKS engine exhaust is 90% lower than the highest values obtained with a conventional engine and still 60% of that of a

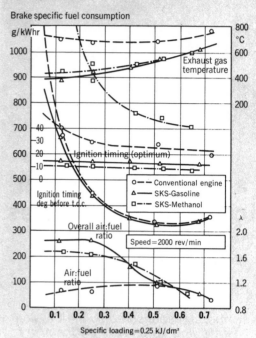

FIGURE 6

Relationship between performance data and load

Brake specific fuel consumption

FIGURE 5

Comparison of combustion products

conventional engine running on a very weak mixture. The CO emission is about the same as that of a conventional engine on weak mixture, while the HC output is lower. With another variant of the SKS engine, it is possible still further to reduce the HC emissions, and at the same time to lower the fuel consumption, but at the expense of increasing NOₓ output.

Control of the SKS engine

Because of the wide range of mixture strengths acceptable in the Porsche SKS engine, qualitative control, by varying the strength of the mixture supplied to the main combustion chamber, becomes possible. This method has decisive advantages over quantitative control, by means of a throttle valve, so long as the conditions remain favorable for combustion. As the mixture strength is reduced beyond a certain value, however, there is a risk of the rate of progress of combustion becoming too slow. At this point, conventional throttling of the ingoing mixture becomes necessary. According to Lange and Gruden (see additional reading) it has been proved that, with the SKS engine, qualitative control give best results when the mean effective pressure exceeds half the maximum attainable value. Below the half-way point, quantitative control should be introduced to reduce specific fuel consumption and exhaust emissions. This latter method of control maintains the exhaust gas temperature at a level conducive to afterburning of the CO and HC emissions. It also reduces the NOₓ omissions. With the variant of this engine the NOₓ output is only 10–20% of that of a conventional engine, while the CO and HC emissions are about the same for these two types of engines.

Research has shown that the performances obtained with methanol and gasoline as a fuel do not differ so widely in an

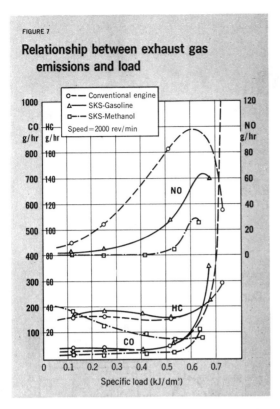

FIGURE 7

Relationship between exhaust gas emissions and load

Conventional engine
SKS-Gasoline
SKS-Methanol
Speed = 2000 rev/min

Specific load (kJ/dm³)

SKS engine as in a conventional one. This is because of the favorable method of propagation of combustion in the SKS unit. In Figures 6 and 7 the characteristics over the whole range of power output at a constant speed of 2,000 rev/min are compared. The greatest differences were in exhaust gas emissions, and the SKS engine run on methanol had approximately double the specific fuel consumption of either engine run on gasoline. The reason for the high fuel consumption was the lower calorific value of methanol. However, the thermal efficiency obtained with methanol was higher. Since the optimum ignition timing for methanol and gasoline in the SKS engine is approximately the same, obviously the rates of combustion are, similarly, virtually equal.

With methanol, the CO emission of the SKS engine is 50–70% lower than with gasoline. It was not possible to come to any valid conclusions regarding the actual reduction in HC emission—though it was obviously substantial—because the methods of measuring it in the exhaust gases are not directly comparable. Over a wide operating range, however, the NOₓ emission with methanol was almost zero.

Additional reading

Second Symposium on Low Pollution Power Systems Development, Dusseldorf, Nov. 1974. Paper by K. Lange and D. Gruden.

T. Ken Garrett *began his career in aeronautical engineering. In 1951, he was invited to the editorial staff of* **Automobile Engineer;** *by 1958 he was Editor of that publication where he remained until 1972. Since then he has been writing articles and broadcasting on technical and scientific subjects.*

An automotive engine that may be cleaner

Even without add-on emissions control equipment, the
stratified charge engine boasts low CO, HC, and NO$_x$

Reprinted from ENVIRON. SCI. TECHNOL., **7,** 688 (August 1973)

What's the likelihood that you'll be driving a car with a stratified charge engine in 1975? Not very likely. How about 1980? Chances are better. And 1985? An excellent possibility indeed. The stratified charge engine is looming larger in Detroit's game plan to control automobile emissions. The reason: Japan's Honda Corp. and Texaco, Inc., in the United States claim to have vehicles with stratified charge engines which meet the 1976 "Muskie numbers" for automobile emissions (see "Pollution Free Power for the Automobile," *ES&T*, June 1972, p 512).

A long time coming

The stratified charge engine is, of course, an internal combustion engine (ICE). What makes it different from the conventional reciprocating piston ICE is the pattern of fuel distribution within the cylinder. In the conventional ICE, the air–fuel mixture enters the cylinder through the intake valve as a pretty nearly uniform mixture. In the stratified charge engine, there may be zones which are fuel rich or fuel poor.

In both kinds of engine the object is the same—complete combustion of the fuel in an orderly manner for the greatest efficiency. In the conventional ICE the air–fuel mixture must be rich enough to be ignited by a spark. Once ignition has occurred, however, flame or "torch ignition" can ignite gas in a leaner mixture.

That's the advantage of stratified charge engine. While all the air–fuel mixture in the conventional ICE must be rich enough to be ignited by a spark, only a small fraction of the mixture in a stratified charge engine needs to be that rich. The remaining mixture can have an excess of air which, in turn, is good for fuel economy and pollution control.

In the conventional ICE, as the flame fans out from the point of spark ignition, there may be areas along the cylinder wall which are not hot enough to burn the mixture, re-sulting in unburned hydrocarbons which are then discharged through the tailpipe. Furthermore, in richer fuel–air mixtures, combustion tends to favor the formation of carbon monoxide rather than carbon dioxide. The stratified charge process, therefore, helps cut down hydrocarbon and carbon monoxide emissions by optimizing combustion parameters.

Many designs

The concept of stratified charge combustion has been around a long time, dating back to the early 1920's when Sir Harry Ricardo first described the system theoretically. Since that time, several different designs have been proposed. Today, the designs settle around two distinct configurations. The configuration chosen by Honda is an adaptation of a precombustion chamber design of the type used in the Russian Nilov engine of the 1950's. The second configuration, chosen by Texaco and Ford, uses direct fuel injection and a swirling within the cylinder to achieve the same results.

The Nilov engine had two separate carburetors—one for the rich fuel–air mixture in the precombustion chamber and one for the lean mixture in the main combustion chamber. In addition to the conventional intake and exhaust valves, the Nilov engine used a third valve for metering rich mixture into the precombustion chamber.

Honda's adaptation uses a double barrel carburetor, one barrel for the rich mixture and the other barrel for the lean mixture to admit fuel to the cylinder. The Honda engine also uses a third valve which admits the rich mixture to the precombustion chamber. Both this third valve and the conventional intake valve allowing the lean mixture to enter the cylinder are opened at the same time. The air–fuel mixtures then spiral in a sort of miniature whirlpool, giving rise to Honda's name for the engine the Compound Vortex Controlled Combustion (CVCC) engine. The engine has a conventional intake, compression, ignition, exhaust cycle sequence.

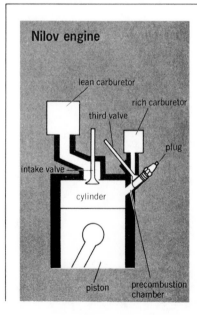

Nilov engine

lean carburetor

rich carburetor

third valve

plug

intake valve

cylinder

piston

precombustion chamber

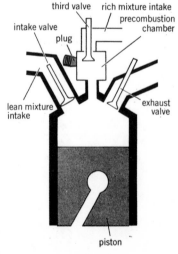

Honda engine

third valve

rich mixture intake

precombustion chamber

intake valve

plug

lean mixture intake

exhaust valve

piston

The engines developed by Texaco, the Texaco Controlled Combustion System (TCCS) and Ford, the Programmed Combustion (PROCO) use fuel injection and air swirl. The major difference between Ford's and Texaco's is the timing of fuel injection. In Texaco's engine, air is swirled into the cylinder during the intake stroke and compressed. Combustion occurs near the end of the compression stroke when fuel is injected into the cylinder and immediately ignited by a spark from the plug. The spark, therefore, ignites a fuel-rich mixture in the immediate vicinity of the plug where a stationary flame front is established. Additional fuel is burned as soon as it is injected, mixes with air, and enters the flame front. The excess of oxygen in the cylinder assures more complete combustion, reducing levels of carbon monoxide and unburned hydrocarbons (HC).

Ford's Proco system is similar in concept, except that fuel is injected into the cylinder over a longer period of time. Both Ford and Texaco have modified the piston to include a cup or depression in the top surface to improve combustion characteristics. The Texaco engine has no octane or cetane number requirement, hence the fuel can be made more efficiently in the refinery than conventional gasoline or diesel fuels.

Tailpipe emissions

How good are the stratified charge engines? By most accounts, pretty good. In tests certified by EPA on the Honda engine, the CVCC emitted 1.96 g/mile CO, 0.21 g/mile HC, 0.81 g/mile NO_x at low mileage.

For one car which accumulated 50,000 miles, the levels were 2.57 g/mile CO, 0.26 g/mile HC, and 0.98 g/mile NO_x. Those are figures for a stripped emission control system, without catalysts, exhaust gas recirculation (EGR), and the like.

Texaco's L-151 stratified charge engine developed for the Army tank command can produce similar numbers. Although Texaco's numbers are not certified by EPA and apply only to one test on one car at low mileage, the stripped engine with no additional pollution control equipment gave 3.34 g/mile CO, 1.89 g/mile HC, and 1.00 g/mile NO_x. The same vehicle with a full control package, including Texaco's electronic ignition system, exhaust gas recirculation, combustion event retard, and three catalytic converters, gave 0.36 g/mile CO, 0.28 g/mile HC, and 0.31 g/mile NO_x. After 50,000 miles, under level road, 60-mph conditions, the numbers were 1.31 g/mile CO, 0.34 g/mile HC, and 0.73 g/mile NO_x.

So the numbers are hovering right around the 1976 standards. But such levels of emission control appear to exact a heavy price in terms of fuel economy. Honda's CVCC engine for example, is estimated to consume about 25% more fuel than an uncontrolled 1973 model car. Although Honda has not published a great deal of technical information on the CVCC, knowledgeable industry spokesmen speculate that the excellent hydrocarbon emission characteristics shown by the engine can only be achieved by very lean running with a considerable fuel penalty.

Texaco's engine, on the other hand, installed in a jeep and tested on the CVS cycle got about 24 mpg of gasoline with no additional controls. On the full controlled model with pollution control accessories such as EGR and catalytic converters, the figure was 16 mpg. Texaco's vice-president for environmental protection, W. J. Coppoc, pointed out, however, that the same engine in its conventional configuration as a premixed charge gas burning ICE only got 13–14 mpg. What that means according to Coppoc is that if the U.S. were willing to accept slightly relaxed emission standards which would still protect the environment, stratified charge engines could give back a significant portion of the mileage which is being eaten up by emission control devices. As it now stands, however, the stratified charge engine will allow automakers in a sense to "run faster just to be able to stand still."

Do you want one?

Coppoc has no doubt that the stratified charge engine will make significant inroads within the next decade. Before then, however, the sheer logistics of tooling up to produce a radically new kind of engine will ensure that the conventional ICE remains on the road for a long time. Honda is planning to offer a limited number of its Civic model car in the U.S. in 1974, containing the CVCC engine. Honda says the Civic will cost about $2000 and will come with either a manual or automatic transmission. Beyond that the only plans are tentative ones.

Ford president Lee Iacocca recently told Congress that if the nitrogen–oxygen standards were relaxed from the proposed 0.4-g/mile level to 2.0 g/mile, that Ford could make about 500,000 stratified charge engine automobiles by 1977. That's still a drop in the bucket compared to the annual engine output of Ford which is something on the order of 4 million. Nevertheless, it's a commitment to a running start.   HMM

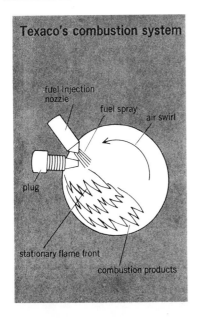

Texaco's combustion system

fuel injection nozzle

fuel spray

air swirl

plug

stationary flame front

combustion products

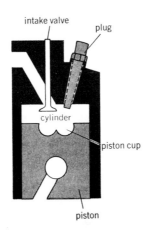

Cup combustion chamber

intake valve

plug

cylinder

piston cup

piston

Nilov engine achieved stratified charge with separate carburetors and a prechamber design. Third valve admitted rich mixture to precombustion chamber where it was ignited, in turn igniting lean mixture in cylinder. Honda engine achieves same result with double barrel carburetor, regulating flow of air-fuel mixture to both precombustion chamber and cylinder. Texaco's engine uses air swirl and a stationary flame front pattern to "torch ignite" lean mixture with burning rich mixture. With all stratified charge engines, excess air in cylinder cuts pollutant formation. Cup in Texaco's piston improves combustion characteristics.

Former chief scientist in one federal agency, and now a Washington, D.C.-based consultant on materials and energy, seriously questions how much we as a nation can afford to be . . .

Paying for new car emission controls

Reprinted from ENVIRON. SCI. TECHNOL., **8,** 807 (September 1974)

Earl T. Hayes

Consultant, Silver Spring, Md. 20901

For want of a nail the shoe was lost is a well-known story leading up to how a battle was lost. It could well be that we have a similar situation coming to resolution this year which might do irreparable damage to the whole environmental movement and gains of the past seven or eight years. The particular nail is the automobile emission standards which bid fair to be the most expensive experiment of all history. The cost could be as high as $20 billion a year and the fuel penalty translates to the loss of several hundred thousand jobs in a gasoline starved economy. There is no clear cut evidence as to why America should be called upon to pay such a price.

In the few years that have passed since the passage of this act there has been a radical change in the U.S. energy demand and supply picture. The price of petroleum has more than tripled and the whole economic structure of the country is facing drastic change in view of soaring petroleum prices.

The fuel penalty incurred in reaching the present standards is now known to be at least 500,000 barrels of crude oil per day and attaining the 1976 standards will raise the total penalty to something between 750,000 and 1,500,000 barrels per day. The Achilles heel of all these regulations is the one for NO_x for which there is no demonstrated need or supporting evidence to say that it should be removed to the levels now on the books. The American public has so far failed to grasp the principal problem of a permanent change in their way of life and certainly have been unable to adjust to the petroleum supply levels of 1973–74.

Times have changed since the passage of the first environmental laws when no one dared vote against any law with the word environment in it and when it was assumed that our affluent life would be able to achieve the goals of zero risk to the health of the American public at any cost. The thundering vote which passed the Alaska Pipeline Bill 380 to 4 shows that public sentiment can change drastically in a very short time especially when it affects the gasoline tanks of 100 million American drivers.

Assessing the petroleum situation

The U.S. is in a bind now and forever as regards petroleum. With a 16th of the world's population we consume over a third of its total energy. However, we possess only 5% of the world's petroleum reserves, and the Arabs and Middle Eastern nations have at least 53%. We are heavily dependent on a liquid and gaseous hydrocarbon economy since 44% of our total energy use is in the form of petroleum and 34% in natural gas. We are particularly vulnerable in petroleum because 25% of the national total energy is used for transportation, and there are simply no near-term ·alternatives for replacing the internal combustion engine which is used in more than 100 million automobiles in America today.

We have not been a prudent or forward planning nation in almost any sense when it came to determining what the automobile population ought to be in terms of a national resource economy. The insatiable petroleum appetite of the last 10 years, and the probable course for the next 6–10 years is shown (Table 1). Since 1920 the use of petroleum has compounded at a rate of 4% a year gradually growing to 6% in the last two years. What this means is that our demands for petroleum were growing each year at a rate of a million barrels per day. Also, since we passed the peak of production that we could ever hope to attain from this country's oil fields and those of Alaska and the Continental Shelf some years ago, all this increase had to come by way of imports.

We have the situation then that whereas our imports were about 20% of our petroleum requirements 6–10 years ago, they have risen to 6.7 million barrels per day or 38% of our total use. In terms of dollars the effect of the tightening of the noose by the Arabs on October 20, 1973, pulled the rug out from under all known economics. In the years just passed the average price of oil ranged from $3.00–$3.50 a barrel. Even last year, after prices started to rise, we had an oil import bill of only about $9 billion. However, in the year 1974 alone, just to buy us last year's imported oil will cost $18–25 billion.

The alternatives are quite clear. If we continue to import oil at the increasing rate of the last three years our oil import deficit will rise to more than $30 billion in 1976 and $40 billion in 1980. It is obvious, therefore, that there will be no growth in oil imports and that we are faced with a steady state economy, otherwise known as a zero gain of the gross national product.

The legal requirements

In the 1960's it became apparent that with the car population growing each year something had to be done about the tail pipe products of the internal combustion engine. Three gases—HC (hydrocarbons), CO (carbon monoxide), and NO_x (nitrogen oxides)—were identified as requiring control for reasons of health and air quality. A digest of these is shown in Table 2.

The original standards were set by arbitrary decision at 90–97% of pre-1968 values and have no positive chemical or physiological basis for the lower limits. They are so much tougher than the heretofore stringent California standards that the term "overkill" has been used. In a floor speech June 18, 1973, Sen. William Brock (R-Tenn.) said "...but as former EPA Administrator William Ruckelshaus has stated, it has become apparent that the data on which some of these decisions were made are either out of date, or were inaccurate in the first place."

He continued, "It is now obvious that it is not necessary to reduce automobile emissions of nitrogen oxide to hour air quality standard of 9 ppm CO more than once a year.... Present federal emission requirements of 0.41 gpm HC and 0.4 gpm NO_x seem more restrictive than need be by a factor of about three ... These conclusions suggest that the 90% reduction of CO and NO_x specified in Sec. 202 of the Clean Air Act may be more than is required to meet the present national air quality standards for CO, NO_x, and oxidant."

There was fairly rapid progress from 1968 to 1973 in lowering the pollutants by making various changes in the engine carburetion, compression ratio, and other modifications but other approaches are needed to meet the future lower standards. This takes the form of a platinum catalyst in the exhaust system to convert the pollutants to less noxious forms. Here we come to the crux of the whole automobile emission problem (aside from the indefensible low standards). Lead fouls up the catalyst rapidly and must be removed from gasoline. This is a real contradiction because lead in the form of a chemical additive has been the greatest conservation measure ever de-

Loans. "Implementation of the 1976 automotive emission standards will cost approximately $20 billion per year"

a level of 0.4 gram per mile in order to have a safe and healthful atmosphere. The figure is probably too low by a factor of three or four. Yet, by setting that figure in 1970, Congress has had an almost unimaginable effect on American industry. The result of that single figure has been that oil companies have found it necessary to divert a substantial portion of their production to no-lead gasoline, which requires an approximately 7% greater consumption of energy on their part. At the same time, this figure has severely restricted the options of the automobile makers in the kinds of antipollution devices they might use on their cars. It excluded, for example, the promising stratified charge engines, as well as thermal reactor systems. And now millions of dollars later, we are told the figure wasn't even right in the first place."

Along the same line Rowe in *The Wall Street Journal* of September 28, 1973, reported, "It's now almost a total certainty that Congress erred in setting the stringent auto emission standards of the 1970 Clean Air Act. And unless it amends the law this session, consumers will have to pay for the mistake for more than a decade."

There's no need for this, as testimony before the House subcommittee on public health and environment has revealed. Perhaps the only completely independent and objective source that has studied the issue told the panel that the federal emission standards are tougher than necessary.

Representing the National Academy of Sciences' Committee on Motor Vehicle Emissions, Prof. Arthur Stern of the University of North Carolina's School of Public Health testified, "An emission limit for CO approximately three times as high as that promulgated by EPA for 1975 vehicles would give assurance of not exceeding the eight-vised for giving increased engine performance to the American automobile over the last 50 years.

There is another reason for taking the lead out and that is the unsubstantiated fear that the fine particles discharged by the cars will eventually produce a health hazard. Over the long term this removal remains a desirable goal. Rapid technology development by reputable firms has developed filters in the last three years that can be installed on all cars both new and old. Even now an 80% recovery of the lead appears feasible and this can certainly be improved. There is the bonus that the leaded filters would be available for saving lead by recycling. True enough, they would cost $20–40 for a life installation but would not incur an energy penalty. The lead filter development appears to be far ahead of the catalytic converter development. In short, modern American technology has solved the lead problem in a much more satisfactory manner than EPA proposals or present regulations, both from the standpoint of technology and conservation.

The costs

It must be remembered that the 450,000–600,000 barrels of oil per day fuel penalty has only brought us to January 1974. Going on to meet the statutory standards of 1976 will result in an additional fuel penalty of the same magnitude. In the late 1970's then we have low estimates of the loss of 700,000–850,000 barrels of oil per day by EPA and high estimates of over 1,000,000 barrels per day by the Bureau of Mines and 1,750,000 by industry.

In a letter to *Science* February 15, 1974, Naumann gives the following estimate of these costs, "If we assume that 10 million cars are sold each year, the total cost for these devices (in 1975) will be $3.14 billion per year. Replacement of catalytic converters every 50,000 miles requires an amortized cost of $40 per year for each car. The total annual cost will thus be $4 billion. The fuel consumption of cars manufactured in 1973 is already 30% higher than that of 1970 models and we are nowhere near meeting the 1976 emission standards. Even if we assume that these standards can be met with no further sacrifice in economy, the cost of the 30% increase in fuel consumption is more than $12 billion at current prices. Thus, we can estimate that the implementation of the 1976 automotive emission standards will cost approximately $20 billion per year."

The real price of this fuel penalty is its effect on the Gross National Product and unemployment. The National Petroleum Council in a report of November 15, 1973, estimated that a loss of 2 million barrels a day of petroleum imports would cut the GNP $48 billion and raise unemployment by roughly a million people. The low estimates of fuel penalty for emission controls could mean the loss of 100,000–250,000 jobs, the high estimates are several

step in solving the energy problem is to inform society of the cost of environmental and ecological programs and allow the people to choose. If people want the end products of such programs they will have to pay the cost in higher energy prices. Without adequate information, society will not be able to decide which programs are worth the cost and which are not. . . Such a system seems preferable to allowing a bureaucrat to decide for them."

At this time there appears to be no change in the EPA objective to enforce the statutory standards in 1977. This position is holding in spite of the fact that numerous people have questioned the validity of the data on which the standards were based, that the assumptions made on the future car population are now known to be in serious error, that we are approaching a no-growth economy, that the cost of imported oil has risen from $7 billion to $18–25 billion between 1973 and 1974, and that the total cost of attaining the statutory standards is of the order of $20 billion a year with the accompanying unemployment of perhaps hundreds of thousands of people. We were

	TABLE 1		
	Petroleum statistics		
	(Millions of gallons per day)		
Year	Domestic supply, crude + natural gas liquids	Imports	Total
1964	8.8	2.3	11.2
1966	9.6	2.6	12.4
1968	10.6	2.8	13.8
1970	11.3	3.4	15.1
1972	11.2	4.8	16.4
1973	10.7	6.7[a]	17.3
1976	10.2	6.7	16.9
1980	11.8[b]	6.7	18.5

[a] No growth in imports. [b] Includes Alaska oil.

	TABLE 2		
	Automobile emission standards		
	HC	CO	NO$_x$
		(grams per mile)	
Pre 1968[a]	8.6	87.5	3.5
1973	3	28	3.1
1975 (original)	0.41	3.4	3.1
1975 (interim)	1.5	15	3.1
1976 (original)	0.41	3.4	0.4
1976 (interim)	0.41	3.4	2.0
1977	0.41	3.4	0.4

[a] No standards.

hundred thousand. This latter would be in the range of a $20–35 billion drop in the GNP.

The benefits

There is no clear cut case that the 1973 auto emission levels need to be decreased. In a March 29, 1973, floor speech Sen. Philip Hart (D-Mich.) originally a strong supporter of the Clear Air Act, called for its re-examination and stated that little time was given to cost-benefit questions when the laws were passed. He said, "There is no sound scientific evidence that the '75–'76 standards will do anything to improve health. If it can credibly be said four years from now that we have caused the expenditure of billions to no purpose or to questionable purpose, the clean air cause will be dealt a blow from which it will be difficult to recover."

Likewise, no direct evidence has ever been presented for taking lead out of gasoline. This was well stated by Dr. William A. Vogley, Deputy Assistant Secretary of the Interior, in a December 7, 1973, letter to Mr. Alvin L. Alm, Assistant Administrator for Planning and Management, EPA, "In the final analysis, these draft regulations continue to threaten a substantial impact on our Nation's limited oil resources. Considering that the conclusions relative to the health issue are largely judgmental and somewhat subjective and the health effect of airborne lead is a controversial question that is unlikely to be resolved satisfactorily one way or the other by available scientific evidence, we cannot agree that any impact on our inadequate fuel supply is justified. Analysis of these impacts and reasonable alternatives to the limiting of lead remain the same as stated in our previous review."

The confrontation

No one has ever really told the American public what these costs are all about. Phillip Gramm writing in The Wall Street Journal November 30, 1973, said, "Another

never that rich as a nation and in the developing priority battle in social, productivity, and environmental gains, a more realistic assessment of the true benefits will have to be made.

Up to now, the general public has been in good agreement with all the environmental causes. Clean water and sulfur dioxide removal efforts have no specific emotional effect, but when 100 million motorists are further curtailed in their driving they will rebel. An hysterical outburst could demand complete repeal of the Clean Air Act or at least the auto emission controls section.

Remember two things: The American public only has an attention span of a few years and witness the overwhelming vote on the Alaska Pipeline Bill. In this new ball game we are not a rich enough nation to pay $20 billion a year and have several hundred thousand out of work for future health standards which have no solid basis in fact.

Unless there is a rapid meeting of the minds on what this all costs and the American public is given a voice in the selection of auto emission standards, we will witness a tearing down of present laws. Total repeal would be disastrous, for no matter what the new way of life that we must adjust to in the coming years, the quality of that life will be just as important as today.

Earl T. Hayes is a consultant on materials and energy. Until recently he was chief scientist of the Bureau of Mines in the U.S. Department of the Interior where he spent most of his professional career. Dr. Hayes is a registered professional engineer and has been concerned with planning, evaluating, and directing research programs in the area of energy, mining, metallurgy, and mineral supply.

Flywheels: energy-saving way to go

Reprinted from ENVIRON. SCI. TECHNOL., **10,** 636 (July 1976)

In operation on subway cars today, flywheel technology
may be used on private cars tomorrow to help you get
to and from work and play in a less polluting way

Would you ride or drive a flywheel car?
Fly what? A kite. No, a car that has a
spinning wheel that stores mechanical
energy that can be put in and taken out,
much like water in a reservoir.

Such a car could use an electric motor
to store energy in a spinning flywheel. Or
it might store energy from a conventional
internal combustion engine in a spinning
flywheel; this energy would then be used
to propel the car. Using less gasoline, it
would be one way to effect a fuel savings
or substitution, in the case of the electric
battery car. Rapid use of this technology
is one step toward Project Independence,
relieving U.S. dependence on imported
oil.

To learn how such developments might
be practical for the real world of daily
urban commuting, the Energy Research
and Development Administration (ERDA)
will be spending nearly $1 million in R&D
studies in fiscal year 1977 that begins this
month. Earlier, ERDA started its R&D in
fiscal 1975; then in fiscal 1976 the federal
agency spent about $800 000 on flywheel
technology.

George F. Pezdirtz, director of the division of energy storage systems, explains
that the three-pronged goal of the ERDA
energy storage activities are:
- conservation of fuel
- fuel substitution
- improved environmental quality.

"What ever this energy storage system
division does it must satisfy these three
requirements," he says.

Organizationally, the division consists
of a chemical energy storage branch, a
thermal energy storage branch, and a
mechanical energy storage branch. The
division is part of the Office of Conservation, one of six main offices in the
ERDA.

Why the interest?

The sudden interest in flywheels stems
from a technical feasibility study on the
subject by Rockwell International. Funded
for $152 000 with ERDA fiscal year '75
funds, the 1-yr study was completed this
February. Referred to as a baseline study,
it answered both technical and economic
questions.

Last November, ERDA and the Lawrence Livermore Lab sponsored a 3-day
flywheel technology symposium (Berkeley, Calif.) that drew more than 150 specialists from the flywheel community.
Both the Rockwell report and the proceedings of the symposium are available
from the Government Printing Office.

What will the federal energy office be
doing with its funds—your tax dollars—to
help solve the world's need for a nonpol-
luting vehicle?

ERDA flywheel expert George C.
Chang, chief of the mechanical energy
storage branch, says that one of the major
emphasis of his branch is on transportation; another is on utility applications. He
says that there are several ways that the
energy from a car engine can be captured
and put to later use. All three are hybrid
concepts; they range from a minor dependence to a total dependence on the
flywheel. All three systems are being
funded; ERDA is putting nearly $1 million
on a number of developments.

The first concept is that of regenerative
braking; here the flywheel would play a
minor role. The basic power of the car
would come from a conventional internal
combustion engine or perhaps an electric
engine. "Here we are trying to design a
flywheel, and the accompanying transmission requirements, that would capture

Flywheel development today, at a glance

Technology application	Projects
Subway rapid transit car	UMTA,[a] Air Research Div. of Garrett Corp.
Electric trolley bus range extension	UMTA
Flywheel/internal combustion hybrid car	Ford Motor Co., ERDA at University of Wisconsin
Mine shuttle car	Bureau of Mines, Dept. of the Interior
Switching engine	Federal Railroad Administration
The urban car	Lear Motor Corp. and U.S. Flywheels use flywheels developed by Brobeck Associates of California
Aircraft landing gear power boost	Rockwell International program supported by the Air Force
Helicopter high-speed hoist	Lockheed program with Army support
High-powered mobile equipment	DOD classified project totaling $1.5 million
Utility peaking	Public Gas and Service Co., of New Jersey with Battelle; Brobeck Assoc. (Berkeley, Calif.)
Aircraft catapult	U.S. Navy program with all steel flywheels
Advanced flywheel technology	NSF, Applied Physics Lab., The Johns Hopkins University

[a] UMTA, Urban Mass Transit Administration in the Dept. of Transportation

Subway car. *Energy savings of 30% have been demonstrated with flywheel units*

ACT-1. *A new car from the wheels up, the advanced concept train uses flywheel unit*

the energy lost from a moving vehicle resulting from braking,'' Chang explains.

ERDA is working with the electric jeeps of the Post Office on the regenerative braking concept. Ultimately, an improved performance of these jeeps is anticipated by incorporating flywheel-related technology into the vehicles.

The second concept, a flywheel-heat engine hybrid, is intermediate in its use of flywheel technology. This hybrid would rely about 50% on flywheel technology; a gasoline engine would be used to charge up the flywheel.

''In this case,'' Chang says, ''one runs an engine such as a conventional internal combustion engine (ICE) under optimum conditions, to charge the flywheel.'' He explains that such a hybrid would get the best mileage from gasoline because the engine would be run at its best efficiency, which is somewhere between 45–50 miles/h—a condition under which no one wants to drive down today's highways.

Putting the energy equivalent of gasoline into perspective, Chang tells *ES&T* that today's ICE converts only 12% of the energy in gasoline to motive power. A dollars worth of gasoline gets 12 cents worth in energy. Chang estimates that flywheel hybrid development could up that energy efficiency to more than 20%, and theoretical efficiency is even higher than that. The ultimate third concept would rely 100% on the flywheel. Essentially, it would be a flywheel-only car.

Funding

With their fiscal year '76 funds, ERDA awarded a $200 000 contract to the Lear Motor Co. (Reno, Nev.) for development of flywheel technology for possible use in commuter cars. Already, Lear Motor Co. has teamed up with U.S. Flywheels, Inc. (San Juan Capistrano, Calif.). This development is aimed at the ultimate hybrid, the 100% flywheel car. Earlier, the Lear Co. was in the great steam car race (*ES&T*, June 1972, p 512).

Other awards include $125 000 to the Sandia Laboratory (Albuquerque, N. Mex.), which is looking at the heat engine hybrid vehicle, the second concept. Investigation started earlier with fiscal '76 funds. As

part of this contract, Sandia will be making an assembly of different rotors to help visualize the flywheel concept.

Materials

''Without any doubt the real advance in flywheels will be in materials and designs,'' Chang says. The Lawrence Livermore Lab ($50 000 in fiscal '75 and $270 000 in fiscal '76 funds) is looking at advanced composite flywheel rotors. ''In the past, nearly all flywheels were made of metal,'' he explains. ''The way to compare flywheel rotors is in terms of energy density—on the basis of watt-hours per pound (Wh/lb) of material.'' A typical state-of-the-art metal rotor would have an energy density of 3–14 Wh/lb. Within the next 10 years improvements in metal rotors can only increase the energy density to 5–15 Wh/lb.

''The real advance would be in the area of composite rotors made of plastics, fibers, and the like,'' Chang continues. ''Here the state-of-the-art is in the energy density area of 10–20 Wh/lb but within 5 years it could be up to 30–40 Wh/lb and within 10 years to 60–70 Wh/lb.''

A typical composite rotor is made of Kevlar, an aramid (aromatic polyamide) fiber made by Du Pont. It has, for example, a high volumetric density but is at the same time very light. Such a combination in rotors leads to more energy per pound of flywheel weight.

Other activity

All told, the federal government now has announced about a dozen RFP's (request for proposals) on flywheel technology totaling more than $10 million.

David Rabenhorst, a scientist at the Applied Physics Laboratory of The Johns Hopkins University, has been working in the flywheel area for the past eight years. Some six or seven basic patents on flywheel technology have been issued to APL/JHU. By now, he has made presentations to all major U.S. automobile manufacturers as well as the Lear Motor Company.

Rabenhorst's leadership in this area of technology is well known to every practioner of this art. Anyone researching the literature in this field finds the name Rabenhorst as a first source of information on the subject of flywheels. He is now writing a book on flywheels that will be published next year by McGraw-Hill.

Rabenhorst explains that only three things count in a flywheel—the strength

ERDA's Pezdirtz
''three requirements of energy storage''

120

APL's Rabenhorst
"flywheels from almost any material"

UMTA's Mora
"options with flywheel use"

UMTA's Silien
"returning energy to the third rail"

of the material, density of the material, and the configuration of the flywheel.

Kevlar, that wonder material used in belted automobile tires, has an ultimate tensile strength of 400 000 psi, Rabenhorst explains. Typical materials in the literature have tensile strength to 220 000 psi.

But Rabenhorst is interested in making more advanced flywheels less expensively. For example, he is interested in a bare-filament flywheel for which patent applications are pending; it is referred to as a string flywheel. It has no matrix yet a strength of 400 000 psi. "The string or filaments are tied at strategic places," he explains. When a flywheel is spun the centrifugal force causes a disc configuration to move as many as eight times sideways from its original position. In some cases, this movement causes the flywheel to break up; the filaments break away from the composite material in which they were imbedded.

Rabenhorst says that flywheels can be made from any monofilament polymer such as Dacron, Nylon, and even from such materials as cotton, steel wire, and wood fiber. He foresees flywheels of several types—super high performance, moderate performance, and low cost. One day, the latter might be made from paper, concrete—almost any material not costing more than a penny a pound.

Testing

"We have already demonstrated the feasibility of a flywheel having an energy density of nearly 100 Wh/lb in full size spin tests with a 30-in. flywheel," Rabenhorst says. "We did this by spining a 32-in. rod-shaped rotor in a 3-ft vacuum chamber." In this case the rod was made from Kevlar, the filament material called Flex Ten that Goodyear uses in automobile tires and for parachute tape. "Spinning this rod at 40 000 rpm in a vacuum chamber is equivalent to a flywheel rotor having a energy density of 60 Wh/lb," he says. "When the configuration of the rotor

is thin concentric rings, then the energy density would be 95 Wh/lb."

As part of the APL/JHU work funded by the National Science Foundation, the laboratory is building a flywheel assembly test rig. The APL/JHU rig will have an air turbine drive and a capacity to spin up to a 200-lb flywheel and up to 100 000 ppm, but obviously not each variable at the same time.

Moderate cost flywheels, those with an energy density of 20 Wh/lb and made from materials costing $4 per pound, have been spun at the laboratory, Rabenhorst says. He explains that the rotors were made of fiber glass-plywood and that the energy cost would be 5 Wh/dollar.

Rabenhorst says, "We can build flywheels having a configuration of 100 rings." For practical reasons, these flywheels are 30-in. in diameter. The 100-ring configuration would be cheaper to make than one with five rings because it allows a simple ring attachment method to be used, according to the APL scientist.

In this case, the flywheel takes advantage of the thin-shell, filament-winding technology that was perfected under the space program. Such filament-winding technology was used to make solid rocket motor cases, which need light weight and high tensile strength; both specifications are needed for flywheels also. The 100-ring flywheel could be made from a thin-walled tapered shell similar to a filament-wound rocket motor case simply by slicing the many rings from the shell.

Uses

Rabenhorst says that the Russians have flywheel buses and flywheel wind mills in operation today. In fact, the Soviets established a Mechanical Energy Recuperation Laboratory in 1965, that is currently under the direction of Dr. Nurbei Gulia. Flywheel technology could be an area of future environmental cooperation between the U.S. and the U.S.S.R. (*ES&T*, May 1976, p 414).

Rabenhorst says that the only thing that

will make the electric car attractive to the purchasing public is the incorporation of a flywheel into the battery car. He explains that the usefulness of a lead storage battery is dependent on how fast energy is drawn off the battery. The disadvantages of the present electric cars—limited range and poor acceleration—can be overcome. By using a flywheel in the car, the battery energy would only be drawn off slowly and intermittently. Such a flywheel-electric hybrid could use the lead battery conveniently. In this arrangement, the electric car would also have an infinitely variable transmission.

Putting flywheel technology in perspective, Rabenhorst says that earlier flywheels such as those used on buses in Switzerland and the Belgian Congo were in the operational energy density class of 1.25 Wh/lb. In Switzerland, a fleet of flywheel-powered buses were operated during a 17-yr period, ending around 1948. These buses were limited in distance between bus stops.

"When the public comes to realize that flywheels with an energy density of 40 Wh/lb are possible, then it will begin to appreciate the mechanical advantage of flywheels," Rabenhorst says.

Rail rapid transit

Energy storage or flywheel research and development projects of the U.S. Department of Transportation's Urban Mass Transportation Administration (UMTA) began in 1971 with the award of a grant to New York's Metropolitan Transportation Authority to equip two conventional rapid transit cars (R-32 type) with a unique energy storage propulsion system.

According to Joseph S. Silien and Jeffrey Mora of UMTA's Rail Technology Division, the energy storage system captures a portion of the vehicle's kinetic energy during braking. Normally, this energy is wasted as heat released through underfloor resistor grids, but in the energy storage system the energy is stored in spinning flywheels. The stored energy is then used to supplement third-rail energy requirements during vehicle acceleration. The primary objective of the project is to measure the energy consumption of the flywheel-equipped cars and compare the results with conventional systems.

Each of the two transit cars was retrofitted with a solid-state chopper control system, new separately excited traction motors (4 per car), and energy storage units (2 per car). Each energy storage unit contains four flywheel discs, a motor, and an alternator. The flywheels spin up to a maximum of 14 000 rpm.

In 1974, the cars were tested on UMTA's Rail Transit Test Track, a 9-mile electrified loop, at DOT's Transportation Test Center (Pueblo, Colorado). The cars were then tested in 1975 in non-revenue service on the New York City transit system, and went into revenue service on the BMT and IND Divisions January 29, 1976. The cars will operate for six months, and a final report will then be published on the results of the project. The total funding for the project was $1.8 million from UMTA, and $0.6 million from New York State's Department of Transportation and the Metropolitan Transportation Authority. UMTA spokesmen noted that energy savings of 30% have been recorded on test runs of the flywheel cars.

Why are transit authorities interested in flywheel systems? Mora explains that a transit authority pays for electricity based on its peak use with the maximum number of trains in service. If a line or a system is equipped with energy storage cars realizing a 30% reduction in energy consumption, there would be a reduction in power demand during the peak period. Therefore, both power consumption and power cost would be reduced.

The power bill of the New York City Transit Authority (UMTA) jumped from $30 to $80 million following the mid-1973 oil embargo. The NYCTA estimates its power costs could be reduced by $20 million annually if all of its nearly 7000 cars were equipped with the flywheel/energy storage system.

In addition to the New York project, UMTA is undertaking development of two rapid transit cars representing dramatic technological and design innovation. These cars, known as the Advanced Concept Train, ACT-1, will utilize an advanced energy storage/flywheel propulsion system. Garrett AiResearch (Torrance, Calif.) is the developer of the ACT-1 and New York flywheel systems.

The objectives of the ACT-1 project are to advance the state-of-the-art of rapid rail transit car design and construction, and to demonstrate the benefits of advanced technology when applied to existing and future rapid transit systems. In order to select a contractor for the ACT-1 cars, a design competition was held in 1972. Garrett was awarded the ACT-1 contract, largely as a result of its energy storage/flywheel concept and associated operating cost savings.

There are significant differences between the ACT-1 cars and the R-32 cars operating in New York City today. Silien explains, "a chopper control will not be used on the ACT-1 eliminating high-powered, circuit-controlling electronics. Armature circuits and control with low power fields driven by phase delay rectifiers will replace the chopper control."

The R-32 system consists of four 2-in.-thick steel alloy discs bolted together as contrasted to 27 thin laminated steel alloy discs that are pressure mounted on the rotors of the ACT-1 cars. At present, the flywheel units for ACT-1 are undergoing acceptance tests at Garrett, and the two cars are in the advanced stages of construction. At the end of this year, the ACT-1 cars will be delivered to the UMTA Test Track for test and debugging. In late 1977, the cars will be evaluated under revenue service operating conditions on some of the transit properties.

A number of options become available with extensive use of flywheel-equipped transit cars. Reduction of power consumption and cost is the most obvious. Alternatively, a transit authority with ACT-1 equipment could increase the number of trains, put more in service yet pay the same electricity bill. Also, the authority could air-condition presently non-air-conditioned cars at no extra energy penalty.

Nonflywheel activities

Another way to save on energy, analogous to automotive regenerative braking, involves regenerative choppers, Silien notes. These units would put energy back into the third rail, but the ability of the line to absorb the energy would only be possible when a second train was in the vicinity of the one with regenerative choppers, or when there was a load on the line. Otherwise the energy would go to a resistor grid where it would be converted to heat and lost, as is presently the case.

Checking it out

Built in 1970, the DOT test facility (Pueblo, Colorado) involves both the FRA (Federal Railroad Administration) and UMTA. UMTA's 9-mile Rail Transit Test Track has the capability of testing urban (or intercity) rail equipment at speeds up to 80 mph. The primary purpose of the track and the ancillary maintenance facilities is to serve as a reference for test and evaluation of urban rail vehicles.

A secondary purpose of the track is the test and evaluation of state-of-the-art and advanced track structures. For example, two different weights of running rail are used in the track (119RE and 100RE), and welded and bolted rail and both wood and concrete ties are used. The UMTA investment in track, maintenance, and administration facilities exceeds $26 million.

SSM

Control systems on municipal incinerators

Federal and state particulate matter emissions
standards dictate the use of electrostatic
precipitators, fabric filters, or scrubbers

Norman J. Weinstein and Richard F. Toro

*Recon Systems, Inc.
Princeton, N.J. 08540*

Reprinted from ENVIRON. SCI. TECHNOL., **10**, 545 (June 1976)

Even a modern well-designed and operated municipal incinerator cannot meet federal and state regulations for particulate matter emissions without an air pollution control system. Federal emission standards in effect require a control efficiency of at least 93% on a weight basis, but design bases are normally higher because of uncertainties and opacity requirements. To achieve the necessary efficiency, all particles larger than 1–3 μ must be removed, effectively eliminating the simple air pollution control systems traditionally used on incinerators.

Electrostatic precipitators, fabric filters, and certain types of scrubbers appear to be the only commercially available devices that have the capability to meet current emission standards for incinerators. Newer forms of these devices, including charged droplet scrubbers and high-velocity wet precipitators, may have advantages over more conventional devices, but these have not yet been commercially demonstrated for municipal incinerator applications.

Electrostatic precipitators

Not until 1969 were electrostatic precipitators applied to municipal incinerators in the U.S., although they had previously been used in Europe and Japan. Almost all new thermal processing facilities built since 1969, however, have used electrostatic precipitators for particulate emission control (Table 1). Although the design efficiencies for incinerators shown in the table are less than normally required for coal-fired steam boilers,

extraordinary attention to all design and construction details is still required to ensure continuing high-efficiency performance. Mechanical and electrical designs are as important to adequate electrostatic precipitator performance as the basic size parameters. With careful design and operation, efficiency requirements for "dry catch" particles (filterable at 121 °C) can be met.

Corrosion resulting from acidic gases can be a problem in precipitator operation. The gas temperature must be maintained high enough to avoid acid condensation on cold surfaces. Hot air purging and preheat burners to minimize acid-gas contact with cold surfaces during shutdown and startup, and sufficient insulation of metal surfaces exposed to outdoor conditions are especially important in corrosion prevention. Hopper heaters are useful in avoiding corrosion and bridging problems resulting from moisture deposition in flyash.

Scrubbers

Simple devices such as baffled-water-spray chambers traditionally used to protect duct, stack, and fan materials, are inadequate as scrubbers to meet modern particulate emission control requirements, but they can still be useful for flue gas cooling. Much more sophisticated devices are needed to efficiently remove particles in the important 1–5 μ size range.

Various techniques are used in scrubbing, but all rely on "wetting" the particle with water to increase its size and permit

TABLE 1. Some ESP installations at thermal processing facilities in the U.S. and Canada

Plant	Capacity, TPD	Furnace type[a]	Gas flow, ACFM	Gas temp, °F	Gas velocity, FPS	Residence time, s	Plate area, ACFM/ft²	Input, KVA	Pressure drop in. H₂O gage	Efficiency, wt %
Montreal	4 × 300	WW	112 000	536	5.5	3.3	6.2	35	0.5	95.0
Stamford	1 × 220	Special R	160 000	600	6.0	3.3	6.6	57	0.5	95.0
Stamford	1 × 360	R	225 000	600	3.6	5.0	4.5	225	0.5	95.0
Stamford	1 × 150	R	76 000	600	3.8	4.9	4.6	75	0.5	95.0
SW Brooklyn	1 × 250	R	131 000	550	4.4	3.2	6.7	47	2.5	94.3
So. Shore, N.Y.	1 × 250	R	136 000	600	5.6	3.3	6.8	33	0.5	95.0
Dade City, Fla.	1 × 300	R	286 000	570	3.9	4.0	5.7	48	0.4	95.6
Chicago, NW	4 × 400	WW	110 000	450	2.9	4.6	5.5	40	0.2	96.9
Braintree, Mass.	2 × 120	WW	32 000	600	3.1	4.5	5.5	19	0.4	93.0
Hamilton, Ont.	2 × 300	WW	81 000	585	3.5	5.4	3.9	70	0.5	98.5
Washington, D.C.	6 × 250	R	130 800	550	4.1	3.9	4.9	77	0.4	95.0
Eastman Kodak	1 × 300	WW	101 500	625	3.4	5.5	3.8	106	—	97.5
Harrisburg, Pa.	2 × 360	WW	100 000	410	3.5	5.1	5.0	40	0.2	96.8
Saugus, Mass.	2 × 750	WW	240 000	428	—	—	—	—	—	97.5

[a] R = refractory-lined; WW = waterwall.
Note: Except for capacity, data refer to design parameters for one precipitator; several may exist

easier removal from the gas stream. The efficiency of a particular type of scrubber on a given particle size can be related to the energy used to force the gas through the collector and to generate the water sprays. This energy usually is supplied by fans supplying pressure to the gas stream (gas motivated) in venturi or orifice scrubbers; or by pumps supplying pressure to the water stream (liquid motivated) in jet ejector or impact scrubbers. The energy required in either case represents a very significant incinerator operating cost.

Since scrubber water requirements are high, recirculation is usually practiced, both to minimize makeup water and the amount of wastewater to be treated. The ratio of recycle to makeup water is determined by the quantity of particles to be removed; by the tolerance of the scrubber design to the concentration of both soluble and insoluble materials in the water, which tend to buildup with increased recycling; and by the amount of water evaporated or otherwise lost.

Incinerator stack gases contain gases that dissolve during scrubbing and cause the water to become acidic. As a result, even stainless-steel scrubbers have been known to corrode. Therefore, pH control by the addition of alkali must be practiced. This has two other important effects: First, undesirable acidic gases such as hydrogen chloride, hydrogen fluoride, and sulfur oxides are removed to some degree; and second, some carbon dioxide (CO_2) is removed, which increases alkali consumption. The removal of CO_2 may also have an important regulatory effect. Since emission standards are based on a 12% CO_2 content, the lower CO_2 content exiting from a scrubber may require an even lower actual particulate emissions rate. The regulations are not clear on this matter.

Several venturi scrubbers have been applied to incinerators (for operating data see Figure 1). It would appear that a pressure drop of at least 22–32 mm Hg (12–17 in. H_2O) is required to achieve the Federal standard. A "clear stack" (0.07 g/scm or 0.03 gr/scf) may require more than 37 mm Hg (20 in. H_2O), although the scrubber water vapor plume tends to reduce this requirement by masking the opacity. The visible water vapor plume is exempt from opacity regulations.

Wastewater from the scrubbers can be used to quench the furnace residue prior to treatment or disposal, thereby reducing both water and treatment costs.

Fabric filters

Baghouses are widely used in industrial applications, but only a few have been built for refuse incineration in the U.S. and Europe. In this device, the particle-bearing gas stream is passed through a fabric-filter medium of woven or felted cloth that traps the particles and allows the gas to pass through the pores of the fabric. Although the pores are as large as 100 μ, sub-micron particles are captured because particles collect on the cloth to form a fragile porous layer that effectively decreases the pore size. For various economic and practical reasons, fabric filters are usually constructed in tubular form (bags) and several bags are housed together in a steel vessel, the baghouse.

The design of fabric filter baghouses depends on several parameters:
- choice of fabric (based on gas temperature, humidity, and particle characteristics)
- size-length, diameter, and number of bags (based on an empirically obtained air flow-to-cloth area ratio, and mechanical considerations)
- method of cleaning (based on particle characteristics and vendor preferences)
- method of precooling the gases to the operating temperature.

To operate continuously, the filter must be intermittently cleaned by manual or mechanical means or pneumatic shaking. The dislodged particles fall to a hopper where they are removed by screw or other types of conveyors.

For dry catch particles, there is no apparent reason why fabric filters will not easily meet any existing particulate matter standard. The lack of significant use in incinerators may be due to the filters' dramatic sensitivity to high and low temperature; large space requirements; difficult maintenance; and significant operating costs.

A pilot baghouse was operated with some success around 1959, and a recent commercial installation on a municipal incinerator has apparently been operating reasonably successfully (Table 2).

Selection of control systems

To choose among electrostatic precipitator, scrubber, and fabric filter systems for particulate removal from incinerator gases, several factors need to be considered. The first factor is initial cost, including those for the cooling systems, fans, stack, waste disposal, and other items dependent on the method of particulate matter control. Second, operating costs, including power, water, maintenance, labor, and waste disposal costs. A third consideration is reliability, which must take into account the best possible estimates for downtime, the effect of downtime on other operations, sensitivity to upsets and ranges of operating conditions, possible degradation of performance with age, and problems induced in associated equipment.

Finally, environmental and other considerations, including the ability to meet and exceed emissions standards; removal of non-particulate matter pollutants; the effect on the air quality of surrounding areas; the possibility of undesirable plumes; and the availability of facilities for waste disposal need to be assessed.

The initial cost for particulate matter control systems tends to be comparable when the complete system is considered, including startup heaters, insulation, gas coolers, hoppers, and conveyors for precipitators; alloy metal construction, alkali addition, water supply, wastewater disposal, water vapor plume control for scrubbers; and startup heaters, gas coolers, hoppers, conveyors, and pulse air supply for baghouses. However, definitive capital cost estimates are advised for system selection.

Energy requirements probably represent the single most important difference among systems. Because of low-pressure drop through an electrostatic precipitator, total energy requirement is low, even though power is required for the corona discharge and the rappers and heaters. Fabric-filter-pressure drops are higher, requiring more energy, but the energy requirement for scrubbers is by far the greatest of the three systems (Table 3).

Since all of these systems are highly automated, operating labor requirements are essentially comparable and low when the systems are operating properly. Maintenance material and labor requirements for particulate matter control systems may be very significant when design and preventive maintenance are

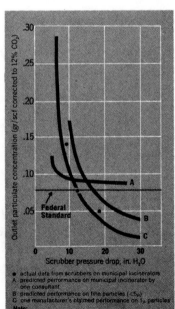

FIGURE 1
Performance of venturi scrubbers on incinerators

- ● actual data from scrubbers on municipal incinerators
- A predicted performance on municipal incinerator by one consultant
- B predicted performance on fine particles (<5μ)
- C one manufacturer's claimed performance on 1μ particles

Note:
grams/SCM = 2.29 × gr/scf
mm Hg = 1.87 × in. H_2O

TABLE 2. Operating and design parameters for a baghouse on municipal incinerator

Air flow, m³/min (CFM)	5090 (180 000)
Air temperature, °C (°F)	260 (500)
Fabric	glass fiber
Air/cloth ratio, m³/min/m² (CFM/ft²)	0.61 (2/1)
Bag size, diameter, m (in.) length, m (ft)	0.14 (5.5) 4.27 (14)
Number of bags (approx.)	4350
Method of cleaning	reverse air
Design pressure drop, mm Hg (in. H₂O)	3.7–5.6 (2–3)

TABLE 3. Comparison of energy requirements for control systems[a]

System	Gas motivated scrubber	Fabric filter	Electrostatic precipitator
Gas pressure drop mm Hg (in. H₂O)	28.0 (15.0)	9.3 (5.0)	1.9 (1.0)
kW per 1000 m³/min (hp/1000 CFM)			
Fan power	103.2 (3.92)	34.5 (1.31)	6.8 (0.26)
Pump power	2.1 (0.08)	— —	— —
Electro-static power	— —	— —	15.8 (0.6)
Total power	105.3 (4.00)	34.5 (1.31)	22.6 (0.86)

[a] This table is based on a hypothetical calculation for approximately equivalent particulate matter removal efficiency. It does not necessarily include sufficient fan power for all furnace and duct pressure drops. Fan efficiency is approximately 60%. Power for heating hoppers (electrostatic precipitators, baghouses) or for tracing water lines (scrubbers) is not included.

Dry flyash disposal, as usually practiced with electrostatic precipitators and baghouses, is considered advantageous; but, unless the flyash is carefully handled, a considerable amount of fugitive emissions can occur. The removal of solids in a slurry from scrubbers is less objectionable with incinerators than in other applications because this system can be integrated with the residue system for common water recycle and residue disposal facilities.

Scrubbers (or wet precipitators) have at least one major advantage over dry methods of particulate matter removal: The ability to simultaneously remove a significant portion of gaseous emissions. However, present regulations do not require this control; efficient, low-energy, second-stage, gas scrubbers can be added to dry systems at a later date, if extra space and static pressure allowance in the fans are provided.

The major disadvantage of scrubbers, in addition to the high-energy requirement, is the formation of visible moisture plumes and the possibility of icing and condensation problems in the surrounding area.

Additional reading

Niessen, W. R., et al., Systems Study of Air Pollution From Municipal Incineration. Volume I. Arthur D. Little, Inc. Cambridge, Mass. U.S. Department of Health, Education and Welfare. National Air Pollution Control Administration Contract No. CPA-22-69-23. NTIS Report PB 192 378. Springfield, Va., March 1970.

Stabenow, G., Performance of the New Chicago Northwest Incinerator. Proceedings, 1972 National Incinerator Conference. New York. June 4–7, 1972. American Society of Mechanical Engineers. pp 178–194.

Ensor, D. S., and Pilat, M. J., Calculation of Smoke Plume Opacity From Particulate Air Pollutant Properties. J. Air Pollut. Control Assoc. 21 (8), 496–501 (1971).

Fife, J. W., Techniques for Air Pollution Control in Municipal Incineration. Am. Inst. Chem. Eng. Symp. Ser. 70 (137), 465–473 (1974).

Hesketh, H. E., Fine Particle Collection Efficiency Related to Pressure Drop, Scrubbant and Particle Properties, and Contact Mechanism. J. Air Pollut. Control Assoc. 24 (10), 939–942 (1974).

This article was excerpted from *Thermal Processing of Municipal Solid Waste for Resource and Energy Recovery* by N. J. Weinstein and R. F. Toro, Ann Arbor Science Publishers, Inc., P. O. Box 1425, Ann Arbor, Mich. 48106. 1976. $20.

The work was performed pursuant to Contract No. 68-03-0293 with the U.S. Environmental Protection Agency.

inadequate. Thus, differences among systems may be less important than the care that is tendered to adequate design and operation.

Acid-dew-point corrosion of metal surfaces can be a problem in all systems, but is more likely to occur when gases are cooled with spray water rather than with steam boilers. Cooling with spray water is sometimes done ahead of either precipitators of baghouses, and is always an integral part of a scrubber operation. Increased pressure drop and plugging can be serious problems with scrubbers or baghouses, but seldom occur with precipitators. Problems of hopper operation, which can occur with baghouses and precipitators, are obviously not a part of scrubber operations. Moderate excursions of temperature may have relatively minor effects on precipitator and scrubber operations, but can have drastic effects on filter bags so that frequent changes involving significant labor and downtime may result.

The performance of all particulate matter control systems can deteriorate. Dust buildup on either discharge or collection electrodes will cause diminished precipitator performance. Discharge electrodes in precipitators are subject to deterioration and breakage, sometimes shorting out a section of the precipitator and reducing its effectiveness. Hopper bridging can cause problems in both precipitators and baghouses. Tears in filter bags can have drastic effects on baghouse performance. Deterioration in scrubbers can be caused by failure of spray nozzles, mist eliminators, and poor pump performance.

Norman J. Weinstein *is president of Recon Systems, Inc. He has been a consultant to industry and government for 10 years on environmental problems associated with chemical, petroleum, gasification, and combustion processes. Prior to consulting, Dr. Weinstein spent more than 10 years with Esso Research and Engineering Co. and played a key role in the development of the FIOR fluidized solids process for direct reduction of iron ore.*

Richard F. Toro *is vice president of Recon Systems, Inc., an environmental engineering consulting and testing firm. A chemical engineer by training, Mr. Toro is an industrial and governmental consultant on technical, economic and legal aspects of pollution control problems.*

Coordinated by LRE

Fluidized bed steam generators for utilities

Pope, Evans and Robbins' unit will burn a variety of U.S. coals; continuous operation at a West Viriginia utility starts next July

Reprinted from ENVIRON. SCI. TECHNOL., **8,** 968 (November 1974)

Fluidized beds have been used for many years in the chemical industry to enhance reaction rates, but their use in steam generators is a relatively new concept. More recent is the development of the fluidized bed combustion technique to produce utility boilers at a fraction of their present size. Such boilers would burn almost any fuel, including low-quality coals such as lignite. Although the basic principles of fludizied bed combustion of coal have been known for 50 years, little progress has been made on a commercial design.

Pope, Evans and Robbins (PER), consulting engineers, will complete installation of the first boiler at the Rivesville Power Station (Fairmont, W.Va.) in the first quarter of 1975. Continuous operation of the unit is scheduled to begin in July 1975. Mr. C. G. McKay, Vice-President—Operations of Allegheny Power Service Corp., says, "We were interested in the fluidized bed boiler project because it shows great promise of being able to use high-sulfur coals to meet the standards of power plant emissions that have been established by EPA and cut the overall cost of generating electricity without the use of stack gas scrubbers that still have not been perfected. Our Rivesville Station, operated by Monongahela Power Co., an Allegheny subsidiary, was selected as the site for this project because of the availability of local high-sulfur coal, appropriate operating and supportive facilities for this test, and steam conditions of

1270 psig, 925°F, with no reheat, that can use existing turbine generating equipment. Also, the 300,000 lb/hr packaged unit, the highest capacity factory-assembled boiler in transportable size, can be accommodated in space available in the present building."

Fluid bed combustion has the eyes and ears of high government officials. For example, in a February 15, 1974, letter to the President of the U.S., Secretary of the Interior Rogers C. B. Morton wrote, "We believe the fluidized bed combustion boilers offer the best prospects for providing large volumes of clean energy from coal at an early date. Current pollution controls allow only limited use of high-sulfur coal in conventional boilers. The early demonstration of a low-cost, pollution-free combustion technique would help alleviate this undesirable situation and stimulate the production and utilization of Eastern coal."

In 1965, PER began active development work on atmospheric fluidized bed boilers in their Alexandria, Va., laboratory using a unit capable of burning 100 lb of coal/hr. In 1967, a small boiler (5000 lb of steam generation/hr at 300 psig) was built by PER under contract to the Office of Coal Research (OCR) for development of the start-up technique, turndown method, and pollution control. By the end of 1973, about 5000 hr of test operation were performed by PER on this unit, using crushed limestone as bed material for sulfur cap-

ture. Combustion intensity of 1.2 million Btu/hr/ft^2 bed surface was achieved, equivalent to 480,000 Btu/hr/ft^3 within the oxidizing fluidized bed.

In fact, two general approaches are being taken in development work and in construction and operation of such boilers. In France and Czechoslovakia, fluidized bed furnaces have been developed for steam generators in which there is no direct contact between the inert fluidized particles and the heat transfer surfaces, the boiler tubes. On the other hand, work involving direct contact heat exchange has been actively pursued in England and in the U.S. In the boilers without direct contact, the hot off-gases generate all the steam in conventional fashion, and no appreciable size or cost advantages are attained.

How it works

Basic design for direct-contact heat transfer units involved inert granular material supported by fluidizing air from a distribution grid. The grid contains horizontal boiler tubes immersed in the fluidized bed and surrounded by heat exchange sur-

Allegheny Power's VP McVay
"to use high sulfur coals . . . without use of stackgas scrubbers"

face membrane walls. Combustion within the fluidized bed is very intense. The granular bed material is limestone or ash, and very high volumetric heat release and heat transfer rates are obtained. As a result, there is no need for a large space-wasting furnace, and the amount of surface required is reduced.

John E. Mesko, vice-president of PER, explains that start-up of an atmospheric fluidized bed boiler requires heating a portion of the bed to 800°F, hot enough to ignite bituminous coal. After coal ignition, the temperature of the bed rises rapidly until the system achieves thermal equilibrium. At equilibrium, the energy released in the bed equals the energy absorbed by the boiler tubes, plus the energy contained in the hot gases leaving the bed. The desired bed temperature is obtained by setting the proper ratio of heat transfer surface to heat release volume.

Operating characteristics of the bed dictate an optimum design temperature range of 1500–1600°F, with excess oxygen at about 3%. At these conditions, about 50% of the heat released by the burning coal is absorbed in the immersed tubes.

Coal burns rapidly in a fluidized bed, even at 1500°F. The rate is so high that at any point the bed is composed almost entirely of the inert particles that were added before combustion began. Typically, a sample of bed material would analyze at less than 1% carbon. Based on bed volume, the heat release is 300,000–400,000 Btu/hr per ft^3. Counting the open furnace space above the bed, the rate is 100,000–200,000 Btu/hr/ft^3. The wide range in this overall furnace release rate follows from the fact that the furnace volume is set by design considerations other than heat release—i.e., tube arrangement, access for maintenance, water circulation. In either way of calculating, the heat release rate is much higher, as much as a factor of 10 more than the release rate in conventional coal-fired furnaces.

The heat transfer coefficient is also higher, 50–60 Btu/hr/ft^2/°F. But the important advantage is in the uniformity of the heat flux. The peak and average fluxes are equal; there is no danger of burnout as long as the feedwater treatment is intelligently handled. A typical riser has a heat flux of 50,000 Btu/hr/ft^2.

Tubing length is reduced because most of the steam is generated in tubes heated on both sides. Actually, effective projected radiant surface (EPRS) is a concept not used in designing an atmospheric fluidized bed boiler. Superheater surface requirements are also reduced because of

PER-Foster Wheeler steam generator

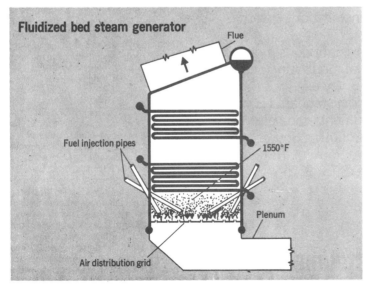

the high heat transfer coefficient and the fact that fireside deposits do not form.

Fireside corrosion is also avoided, apparently because the sodium, potassium, and vanadium in the coal are not released, or if released, are picked up by the bed particles instead of the tubes. The ratio of bed-particle-surface area to heat-transfer-surface area is about 15–1. Condensing vapors are thus more likely to deposit on a bed particle than a tube.

Unlimited coal type

Unlike a pulverized coal-fired boiler, ash properties are not significant to the design of an atmospheric fluidized bed boiler. The same basic design applies for all coals. The bed temperature is too low for the ash to soften or change chemically in any

detrimental way. All fuels are burned at a heat release rate equivalent to 110,000 Btu/hr/ft^2 of EPRS. In a pulverized coal-fired unit, this value may be as low as 40,000 with a high alkali coal.

The boiler designed and developed to date could not be used if the coal were very fine and dry, since the coal is difficult to feed uniformly, and successful operation depends on even fuel distribution. With current mining and transport methods, a ½–¾ in. top size is easy to get.

Keeping low emissions

Tests performed recently bv the Bureau of Mines, Department of the Interior, on the Pope, Evans and Robbins unit in the Alexandria, Va., laboratory established that both NO$_x$ and SO$_2$ emissions were held below EPA emission standards for new coal-fired plants. The conclusion reached by the Pittsburgh Energy Research Center of the Bureau of Mines indicates that "Test data obtained during the two 4-hr tests on the fluidized bed boiler showed emissions of 0.80–1.20 lb SO$_2$/10^6 Btu and 0.11–0.17 lb NO$_x$ (calculated as NO$_2$) per 10^6 Btu, compared with EPA Standards of 1.20 lb SO$_2$/10^6 Btu and 0.70 lb NO$_2$/10^6 Btu for new coal-fired plants," while burning coal with 4.5–4.8% sulfur content.

The coal fed to a fluidized bed boiler is crushed to a 1.4-in. or 3.4-in. top size, not pulverized. A good fraction of the ash stays in the fluidized bed, or if carried out with the products of combustion, is separable in a low-pressure drop cyclone dust collector. Little, if any, metal oxide or sulfate fume is formed, and the quantity of very fine particulate is relatively low. Sulfur trioxide has also not been detected when limestone is

Fluidized bed steam generator

Flue

Fuel injection pipes

1550°F

Plenum

Air distribution grid

used as the bed material. Plume opacity and particulate emissions may be low from a fluidized bed boiler without a precipitator. However, due to current EPA regulations, provisions for electrostatic precipitator are recommended.

Sulfur dioxide in the fluidized bed is actually controlled by the use of limestone as bed material, based on the following reactions:

$$CaCO_3 + heat \longrightarrow CaO + CO_2$$

$$CaO + SO_2 + \tfrac{1}{2}O_2 \longrightarrow CaSO_4$$

The limestone is kept reactive either by the addition of fresh limestone, in a once-through system, by regeneration to recover sulfur dioxide in useful concentrations,

$$CaSO_4 + CO \longrightarrow$$
$$CaO + SO_2 + CO_2$$

or by adding a low-cost catalyst (common salt) with the coal.

Current material balances indicate that for 2000 lb (1 ton) of 5.6% sulfur coal, 400 lb of limestone is required to remove over 90% of the sulfur. Studies are now being undertaken by EPA to develop alternate plans for calcium sulfate ($CaSO_4$) disposal. Since the temperature in the fluidized bed boiler is relatively low, NO_x emission is considerably below the value that EPA's rules would allow. A clear stack is readily achievable, inasmuch as there are relatively few micron-sized particles and there is virtually no sulfuric acid mist even at exhaust temperatures of 250°F.

Building the unit

The design of the 300,000-lb/hr unit, as a cell, for a 800-MW fluidized bed steam generator was developed by Foster Wheeler Corp. (Livingston, N.J.). Their design consists of four modules, each containing seven vertically stacked fluidized beds, called cells.

During the latter part of 1973, the 300,000-lb/hr capacity unit was released for fabrication. Based on competitive price bidding, the award went to Foster Wheeler. Final dimensions of the boiler are 12 ft wide, 25 ft high, and 38 ft long.

The atmospheric fluidized bed boiler is seen as the necessary first step on a long path toward better use of coal, the only major fossil fuel resource in the U.S. If and when the successful performance of the first 30-MW unit is demonstrated, then other utilities will commit funds for the installation of such boilers in the 200–800 MW capacity range.

Coal burns cleaner in a fluid bed...

... But it's always a good idea
to look at what pollutants may
remain. Here are a number of
possibilities

Reprinted from ENVIRON. SCI. TECHNOL., **11**, 244 (March 1977)

Paul F. Fennelly, Hans Klemm, and Robert R. Hall
GCA/Technology Division
Bedford, Mass. 01730

Donald F. Durocher
Kimberly-Clark Corp.
Lee, Mass. 01238

Fluidized-bed combustion systems operate at significantly lower temperature than conventional systems (about 1650 °F vs. 2700 °F, respectively). The solids are supported on a grid at the bottom of the boiler through which combustion air is passed at high velocities, typically 2 to 5 feet per second. The solids are held in suspension by the upward flow of the air and a quasi-fluid is created that contains many properties of a liquid. The most important liquid-like property to the boiler designer is the fact that the bed material is exceptionally well mixed and flows throughout the system without mechanical agitation.

Fluidized-bed combustion systems for the production of steam and/or electricity have several advantages over conventional combustion systems. These include:

• High heat transfer coefficients and volumetric heat release rates will reduce the boiler size by one-half to two-thirds or more compared to a conventional unit.
• Capital costs will be reduced because of the size reduction and the potential for shop fabrication instead of field construction.
• The use of limestone as bed material provides a means for in situ SO_2 removal.
• The high heat transfer coefficients permit lower operating temperatures (1550–1750 °F), which can potentially decrease NO_x emissions.

Questions have been raised concerning the emissions that could result at these lower operating temperatures. As a first step in answering these questions, one can conduct a "preliminary environmental assessment." In performing a preliminary environmental assessment, one's major role in a sense is to serve as a devil's advocate with respect to pollutant generation. The air is to focus attention on potential environmental problems as early in the development cycle as possible. This provides maximum lead time to gather the technical data on which decisions regarding control technology or process modifications (should they be needed) can be based.

It is widely known that fluidized-bed combustion (FBC) of coal results in low SO_2 and NO_x emissions. The idea in this feature article is to focus attention on the so-called "other" pollutants. These pollutants are divided into three generic classes: trace elements, organic compounds, and particulates. There are some limited experimental data available on trace elements in FBC; also, investigations of particulate size distribution and their chemical composition are just underway. Unfortunately, no data available on organic compounds that could be produced in coal-fired fluidized-bed combustion are yet available.

Numerous compounds can be included in an initial list of conceivable pollutants. Developing a list of conceivable pollutants is not necessarily a technically sophisticated task, but is an important effort. It establishes the scope of the environmental assessment; the more comprehensive the list, the less chance there is for unexpected pollutants to escape discovery.

Organic and other pollutants

Potential organic pollutants are those compounds that could form from the incomplete combustion of coal. One can simplistically view it as a sequential process. The particle vaporizes or volatilizes; the volatile compounds can react among themselves in a chemically reducing atmosphere. Next, they react with oxygen within the system in a diffusion flame. After devolatilization is completed, the char continues to burn. To specify the potential organic compounds that could form, the basic question is, what types of chemical species are produced during coal pyrolysis and to what extent will they survive in the reactive environment of a fluidized-bed combustor?

Only two classes of hydrocarbons should be of any significance in coal combustion: small hydrocarbons (less than about three carbon atoms) and polynuclear aromatic hydrocarbons. Both could form and survive within the bed at temperatures on the order of 1500 °F.

The chemical structure of coal can be viewed as a network of interconnected aromatic hydrocarbon compounds. Small hydrocarbons can form directly from cleavage of substituted alkyl groups. Polynuclear aromatic hydrocarbons can also form directly via bond cleavages in the structural network, or they can form through condensation reactions of various hydrocarbon decomposition products. Even though generally endothermic, at FBC temperatures of 1500 °F, these condensation reactions probably proceed at a significant rate. This belief is based on the fact that branched or cyclic hydrocarbons are seldom found as products of coal pyrolysis at similar temperatures.

Similar arguments apply to the generation of organic nitrogen and sulfur compounds. Species such as pyridine decompose at temperatures on the order of 1000 °F to form hydrogen cyanide (HCN) and small hydrocarbons. Thiophenes and mercaptans can also decompose to form small hydrocarbons and hydrogen sulfide (H_2S).

Concentration estimates

For a rough estimate of the concentrations at which some of the small hydrocarbons and reduced sulfur and nitrogen compounds might exist, one can use equilibrium calculations based on free energy minimization. An upper limit can be obtained from calculations performed in conjunction with coal gasification experiments where highly reducing, fuel-rich conditions exist. For example, even with only 60% stoichiometric oxygen present, concentrations of HCN, carbonyl sulfide (COS), carbon disulfide (CS_2), and the like, are less than 10 parts per million (ppm). Extrapolation of the calculations to typical operating conditions such as 20% excess air, in which case SO_2 and NO_x become the predominant sulfur and nitrogen compounds, indicates that compounds such as H_2S, HCN, COS, and cyanogen ($(CN)_2$) should be present in concentrations less than 1 ppm.

Free energy minimization calculations for the more complicated polynuclear aromatic hydrocarbons are impractical.

However, to estimate the concentrations at which these types of compounds might exist in the FBC flue gas, one can use empirical correlations between a compound such as benzo[α]-pyrene and methane (CH_4) concentrations from measurements in conventional coal-fired combustion systems.

Under normal operating conditions, about 3% O_2 in the flue gas (20% excess air), the concentration of hydrocarbons (as CH_4) is about 100 ppm (volume/volume or V/V). Although emissions can often vary between different fluidized-bed systems, 100 ppm provides a convenient average value. Previous measurements with conventional coal-fired systems indicate that the concentration of compounds such as benzo[α]pyrene is typically 10^{-5} times less than the concentration of total hydrocarbons as CH_4.

Thus, using a reference value of 100 ppm CH_4, one can infer that in a fluidized-bed system, polynuclear aromatic hydrocarbons (PAH) could exist in the flue gas at concentrations (V/V) on the order of 1 part ber billion (ppb). However, since flue gases are eventually diluted by roughly a factor of a thousand when they are emitted from the stack, ambient concentrations of PAH near FBC facilities would more likely be on the order of 1 part per trillion. This corresponds to about 0.6 ng/m^3, which is roughly comparable to the natural background concentration ranges found in rural areas. Accordingly, it seems that polynuclear aromatic hydrocarbon concentrations should not be high enough to cause problems.

Recently, some coal-fired flue gases have been tested for the presence of polychlorinated biphenyls (PCB) and trace concentrations have been reported. Experience in coal combustion

TABLE 1.

Assessment of trace element emissions from coal-fired fluidized-bed combustion

Elements of concern because they could be emitted in toxic concentrations (based on "worst case analyses.")	Be, As, U, Pb, Cr, V, Cl
Elements of concern because of possible enrichment on fine particles (<2 μm)	Pb, Cr, Se, Br, Hg

and coal pyrolysis indicates that at temperatures similar to that in coal-fired FBC, chlorine (when present) exists predominantly as hydrogen chloride (HCl). Accordingly, one would not expect significant concentrations of PCB's in coal-fired FBC. If present at all, they should be in concentrations less than those of the unsubstituted polynuclear aromatic hydrocarbons (1 ppb (V/

Trace elements

Trace elements and their compounds are of concern because some of these materials can vaporize and exit with the flue gas. Because they are in the gas phase, they are not captured by particle collection devices.

There is also concern about the enrichment of trace elements on fine particles in combustion processes. Studies have indicated that certain elements can concentrate in selected size ranges of particulates. For some elements, such as lead and cadmium, these sizes tend to be less than a few microns in diameter. Such small particles are of special environmental concern because they are difficult to remove from the flue gas, and once emitted, they can be readily embedded in the lung through normal breathing.

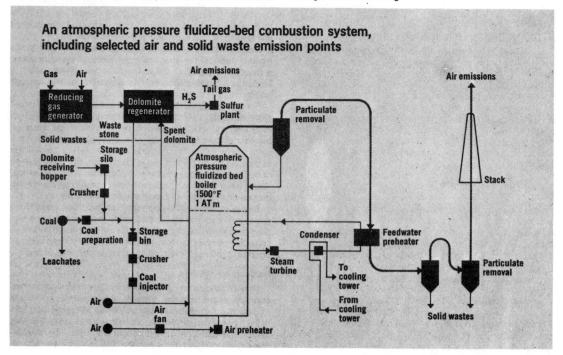

An atmospheric pressure fluidized-bed combustion system, including selected air and solid waste emission points

To assess the importance of trace element emissions in coal-fired fluidized-bed combustion, a "worst case analysis" approach was used. For both bituminous and lignite coals, ranges for the heat content and the concentration of trace elements were obtained. Under the assumption that all of the elemental material would exit with the flue gas as either a vapor or a particulate, a "worst case" emission factor was calculated (lbs/10^6 Btu), using the lowest heating value and highest trace element content of the coals. Based on this emission factor, stack gas concentrations were calculated and diluted by a factor of 10^3 to account for dispersion in the atmosphere. These "ambient" concentrations were then compared to industrial hygiene threshold limiting values (TLV).

Although industrial hygiene TLV's cannot be used to assess the absolute environmental impact of pollutants, they do provide a useful framework in which pollutants can be rank-ordered according to their toxicity. Any element whose predicted ambient concentration was within a factor of 100 of the industrial TLV was considered to be potentially harmful. This scaling factor of 100 was arbitrarily chosen to account conservatively for the effects of long-term exposure (industrial hygiene limits are usually based on exposures to healthy adults over an 8-hour period).

Table 1 lists those elements that could be of concern, based on the above "worst case analysis." It is important to emphasize that these worst case analyses are based on 100% emission of the various trace elements, which is most unlikely. Much of the trace element material will either be retained in the bed or be captured in particulate-control devices. Studies to determine trace element balances in FBC systems are currently underway, and preliminary results indicate that, in fact, significant advantages may result in controlling trace element emissions in fluidized-bed combustion. Compared with conventional combustion, fluidized-bed systems operate at lower temperatures and tend to reduce trace element vaporization.

Elements that could be enriched on fine particles are also listed in Table 1. The chemical composition of particulates as a function of particle size has not yet been investigated in fluidized-bed combustion. This investigation should receive high priority.

For the most accurate environmental assessment, one would prefer to identify the actual chemical compounds in which these trace elements exist. This is an extremely difficult task, because the concentrations of the elements are usually so low that the identification of specific compounds is beyond the capabilities of existing analytical technique. At present, the identification of trace element compounds is based primarily on speculation. Examples of exotic but highly toxic compounds that could form include nickel carbonyl, cobalt carbonyl and arsine.

By use of "worst case analyses" similar to those described previously, it can be shown that contributions to the trace element emissions from the limestone in an FBC system should be negligible compared to those of the coal. This estimate is based on a Ca/S ratio of 2 and coal sulfur content of 3%.

Particulate emissions

Only crude data on particulate emissions from fluidized-bed combustion are currently available. Preliminary data indicate that the mass median diameter of the flue gas particles (50% of the mass of the particles are above that size) is about 7 μ. This means that significant concentrations of troublesome fine particles could exist, but this is also the case in many conventional

Schedule for development of selected fluidized-bed combustion units

Unit size	Contractor and/or location	Approximate starting dates [a]		
		Design	Construction	Operation
0.5 MW [b]	Pope, Evans, & Robbins Alexandria, Va.	—	—	In operation
0.63 MW [c]	Exxon Research and Engineering Linden, N.J.	—	—	In operation
1.0 MW [c]	Combustion Power Co. Menlo Park, Calif.	—	—	In operation
3.0 MW [c]	Argonne National Lab. Argonne, Ill.	Mid '76	Mid '77	Late '78
6.0 MW [c]	International Energy Agency Grimesthorpe, U.K.	Late '76	Mid '77	Late '79
6.0 MW [b]	Morgantown Energy Research Center Morgantown, W. Va.	Late '76	Late '77	Late '79
13.0 MW [c]	Curtiss-Wright Woodridge, N.J.	Late '76	Late '77	Late '79
30.0 MW [b]	Pope, Evans, & Robbins Rivesville, W.Va.	—	In progress	Sept. 76
Industrial and institutional applications	Various	Late '76 [d]	Late '77 [d]	Late '79 [d]

[a] Approximate schedule as of May 1976; [b] Atmospheric pressure; [c] Pressurized combustor; [d] Earliest starting date of any projects.

TABLE 2.

Estimated concentration ranges of potential pollutants from coal-fired fluidized-bed combustion

Gas phase

One hundred parts/million:	CH_4, CO, HCl, SO_2, NO
Ten parts per million:	SO_3, C_2H_4, C_2H_6
One part per million:	HF, HCN, NH_3, $(CN)_2$, COS, H_2S, H_2SO_4, HNO_3, F, Na
One part per billion:	Diolefins, aromatic hydrocarbons, phenols, azoarenes, As, Pb, Hg, Br, Cr, Ni, Se, Cd, U, Be
One-tenth (0.1) part per billion:	Carboxylic acids, sulfonic acids, polychlorinated biphenyls, alkynes, cyclic hydrocarbons, amines, pyridines, pyroles, furans, ethers, esters, epoxides, alcohols, ozone, aldehydes, ketones, thiophenes, mercaptans

Solids

One part per million:	Al, Ca, Fe, K, Mg, Si, Ti, Cu, Zn, Ni, U, V
One part per billion:	Ba, Co, Mn, Rb, Sc, Sr, Cd, Sb, Se, Ca
One-tenth (0.1) part per billion:	Eu, Hf, La, Sn, Ta, Th

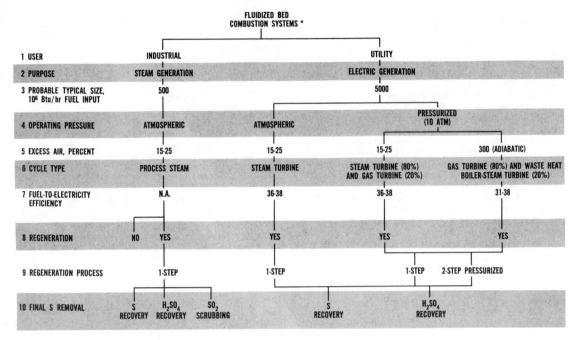

	INDUSTRIAL	UTILITY		
1 USER	INDUSTRIAL	UTILITY		
2 PURPOSE	STEAM GENERATION	ELECTRIC GENERATION		
3 PROBABLE TYPICAL SIZE, 10⁶ Btu/hr FUEL INPUT	500	5000		
4 OPERATING PRESSURE	ATMOSPHERIC	ATMOSPHERIC	PRESSURIZED (10 ATM)	
5 EXCESS AIR, PERCENT	15-25	15-25	15-25	300 (ADIABATIC)
6 CYCLE TYPE	PROCESS STEAM	STEAM TURBINE	STEAM TURBINE (80%) AND GAS TURBINE (20%)	GAS TURBINE (80%) AND WASTE HEAT BOILER-STEAM TURBINE (20%)
7 FUEL-TO-ELECTRICITY EFFICIENCY	N.A.	36-38	36-38	31-38
8 REGENERATION	NO YES	YES	YES	YES
9 REGENERATION PROCESS	1-STEP	1-STEP	1-STEP 2-STEP PRESSURIZED	
10 FINAL S REMOVAL	S RECOVERY H₂SO₄ RECOVERY SO₂ SCRUBBING	S RECOVERY	H₂SO₄ RECOVERY	

FLUIDIZED BED COMBUSTION SYSTEMS *

* PROBABLE Co : S MOLE RATIO IN FEED TO COMBUSTION IS 2:1. COAL SULFUR CONTENT MAY BE 4 PERCENT IN INITIAL SYSTEMS. PROBABLE OPERATING TEMPERATURE OF THE MAIN COMBUSTOR IS 850-950° C.

combustion systems. With suitable central devices, the problem can be ameliorated. Further experiments on particle size distributions and chemical composition as a function of particle size should receive high priority.

Table 2 summarizes concentration estimates for potential pollutants from coal-fired fluidized-bed combustion. These estimates are often based on limited data and simple assumptions; hence, they are probably good only to within an order of magnitude. The term "gas phase" includes gases, vapors, and very fine particulates ($<2\ \mu$); the term "solid phase" includes the agglomerated bed material and coarse particulates ($>2\ \mu$), which should be collected by conventional particle control devices.

Pollution control technology

The two areas of fluidized-bed coal combustion that will probably require pollution control are flue gas emissions and spent sorbent stone disposal. In FBC flue emissions, particulates will probably pose the most significant problem. Their presence will require add-on control devices; possible options include electrostatic precipitators, wet scrubbers and fabric filters. Fabric filters look especially promising, in view of the lower gas temperatures and SO_2 levels encountered in FBC flue gases.

For pressurized operations, gas purity for turbine requirements may be stricter than environmental regulations; hence, pollution control *per se* may be unnecessary. Moreover, high temperature (1500 °F) particle cleaning devices, for the most part, are unproven and not yet commercially available. Gravel bed filters and special cyclones are options under consideration.

Gaseous pollutants that may require control are CO, hydrocarbons and in some cases, perhaps HCl. CO and hydrocarbons can be controlled by increased excess air or the addition of secondary air in the freeboard zone. HCl can be removed via alkaline scrubbers in conjunction with particulate removal, if necessary.

Large amounts of solid waste consisting of spent stone and coal ash will be produced by FBC systems. The amount of solid waste related to sulfur removal could be greatly reduced by regeneration and recycling. Proper landfill site selection and management can minimize the potential for groundwater pollution. Besides disposal other options are also available for spent stone utilization, including use in agriculture, bricks, wallboard, and fill material for roads and concrete. However, utilization as a gypsum substitute (wallboard and other products) may be very limited because of transportation costs and a plentiful natural gypsum supply. Uses in products in which coal ash is utilized will also be limited, as evidenced by the fact that only 13% of coal ash is utilized. Control of pollutants from coal storage, cooling, and water treatment can be accomplished by practices currently in use at conventional power plants.

Although there is virtually no experimental information available concerning potential organic pollutants that could form in coal-fired fluidized-bed combustion systems, simple thermodynamic considerations, chemical experience and empirical correlations with conventional combustion systems indicate that no special problems should occur. Experimental verification, however, should receive high priority.

In comparison with conventional coal-fired systems, fluidized-bed combustion seems to offer significant potential for reducing trace element emissions. However, data are lacking in several areas, particularly with regard to the elements arsenic, beryllium, chlorine, chromium, lead, mercury, selenium, uranium, and vanadium, and their compounds. Nevertheless, experimental data indicate that particulate loadings should pose no special problems, provided conventional control devices such as cyclones, electrostatic precipitators, or fabric filters are used.

This paper is based on a report prepared by GCA for the U.S. Environmental Protection Agency. The authors would like to acknowledge many helpful discussions with Mr. D. Bruce Henschel who served as EPA's project officer. The opinions and conclusions are those of the authors and not necessarily those of the U.S. Environmental Protection Agency.

Additional reading

Sternling, C. V., and Wendt, J. O. L., Kinetic Mechanisms Governing the Fate of Chemically Bound Sulfur and Nitrogen in Combustion. U.S. Environmental Protection Agency. *Publication No. EPA-650/2-74-017.* August 1972.

Stinnett, S. J., Harrison, D. P., and Pike, R. W., Fuel Gasification: The Prediction of Sulfur Species Distribution by Free Energy Minimization. *Environ. Sci. Technol.,* **8,** 441 (1974).

Robinson, E. B., Bagnulo, A. H., Bishop, J. W., and Ehrlich, S., Characterization and Control of Gaseous Emissions From Coal-Fired Fluidized Bed Boilers. Pope, Evans, Robbins, Inc., Alexandria, Va. Report prepared for Division of Process Control Engineering, National Air Pollution Control Administration (now U.S. Environmental Protection Agency), 1970, p 103.

Hangebrauck, R. P., von Lehmden, D. J., and Meeker, J. E., Emissions of Polynuclear Aromatic Hydrocarbons and Other Pollutants From Heat Generation and Incineration Processes. *J. Air Pollut. Control Assoc.,* **14,** 267 (1964). See also: Hangebrauck, R. P., von Lehmden, D. J., and Meeker, J. E., Sources of Polynuclear Hydrocarbons in the Atmosphere. U.S. Department of Health, Education and Welfare. *Publication No. PHS 999-AP-33.* 1967.

Cowheard, C., Marcus, M., Guenther, C. M. and Spigarelli, J. L., Hazardous Emission Characteristics of Utility Boilers. U.S. Environmental Protection Agency. *Publication No. EPA-650/2-75-006.* July 1975.

Threshold Limit Values for Chemical Substances and Physical Agents in the Workroom Environment With Intended Changes for 1974. American Conference of Governmental Industrial Hygienists. Copyright 1974. Cincinnati, Ohio 45201 (P.O. Box 1937).

Swift, W. M., et al., Trace Element Material Balances Around a Bench-Scale Fluidized Bed Combustor. Paper presented at the Fourth International Conference on Fluidized Bed Combustion, sponsored by U.S. Energy Research and Development Agency/Fossil Energy. McLean, Va., December 1975.

Paul F. Fennelly *is a staff scientist at the GCA/Technology Division where he has served as program manager for projects investigating the environmental impact of new energy systems. Dr. Fennelly's interests also include atmospheric chemistry and the chemical kinetics of combustion.*

Hans Klemm *is a chemical engineer in GCA's Environmental Engineering Department. His recent experience includes experimental studies of fabric filtration and the design and evaluation of pollution control systems for the chemical process industry.*

Robert R. Hall *is a chemical engineer in GCA's Environmental Engineering Department. Mr. Hall has a wide range of experience in environmental science and engineering, having worked on projects involving boiler emissions, coke oven emissions, fabric filtration and SO_2 scrubbers.*

Donald F. Durocher *is now on the research staff at Kimberly-Clark Corp., in Lee, Massachusetts. While at GCA, Dr. Durocher worked on several projects investigating the environmental impact of new energy systems.*

Coordinated by JJ

Reprinted from ENVIRON. SCI. TECHNOL., **8**, 510 (June 1974)

Cleaner air and a new industry may both appear
if pilot tests check out the technology of . . . .

Cleaning coal by solvent refining

To scrub or not to scrub, that is the question.

Whether 'tis more economical and efficient to remove particulates and SO_2 from the end of the stack after the burning of coal,

Or to clean coal before it is consigned to the flames,

And thus, obviate the need for elaborate precipitator and scrubber installation . . .

So goes a major choice in the great coal burning debate. With enhanced technology, both principal routes may well be open. In any discussion of removal of potential air pollutants from coal, however, solvent refined coal (SRC) is receiving prominent consideration.

The principles of SRC were known since October 1932, when Alfred Pott and Hans Broche were granted a U.S. patent. However, it was not until 1964 that Pittsburgh and Midway Coal Co. (PAMCO, now a subsidiary of Gulf Oil), first started working with the U.S. Department of the Interior's Office of Coal Research (OCR) to determine the technical feasibility of SRC on a pilot scale.

Pilot plants

SRC work has progressed to the point at which different groups have built, or are completing pilot plants. One such group is PAMCO, under OCR contract, which began construction of a $18 million, 30–50-tpd pilot plant near Tacoma, Wash., in December 1972. The plant should start up in the summer of 1974; it was designed by Rust Engineering Co., a subsidiary of Wheelabrator-Frye, Inc. A second (all-industry) group consists of Southern Services, Inc. (SSI, Birmingham, Ala., a part of Southern Co.), and the Edison Electric Institute (EEI). The latter is now operating a 6-tpd pilot plant at Wilsonville, Ala., designed and built by Catalytic, Inc. EEI funding for this plant has been assumed by the Electric Power Research Institute (EPRI).

In the SSI process, now being "shaken down," Western Kentucky 14 coal containing about 3% sulfur (S) is slurried in an anthracene oil-type solvent. About 95% of the carbonaceous material is dissolved. Hydrogen (H_2) is added under pressure, and the slurry is heated to about 750–850°F. The material is then subjected to pressures of 1800 psig,

generally (although sometimes as high as 2800 psig), and H_2, hydrocarbon gases, water, and hydrogen sulfide (H_2S, made from organic sulfur in the coal), are separated. The H_2S is reacted with sodium hydroxide and water to make sodium sulfide; however, commercially, H_2S would be broken down to recyclable H_2 and marketable elemental S.

At this point, the pressure is decreased to 150 psi, at 600°F, to separate other gases, including remaining hydrocarbon values. After this step, the material contains solvent, dissolved coal, undissolved coal, and inorganics which would become SO_2 and ash if the raw coal were burned. Among these inorganics is pyritic sulfur, a major SO_2 source. The solid material is strained out through a Swiss "Funda" filter at 600°F so that low solvent viscosity and a high filtering rate may be maintained. Filtered solids are being stored for additional analytical tests.

Then the filtrate is warmed to 750°F in a vacuum column preheater, and the solvent is stripped off under 2 psia pressure and recycled to the plant. The distillation residue is SRC liquid; it is brought to 600°F and poured into water by a vibrating conveyor. The resulting solidified SRC, a clean, black, brittle material, minus virtually all ash and pyritic S, and over 60% of the organic S, is then drained.

Groundbreaking for the SSI pilot plant began in November 1972; construction was completed in September 1973. First runs were made this January. By mid-April, about 350 hr of operations were logged. The plant cost slightly less than $4 million.

Optimism

SSI project manager Everett Huffman is optimistic about the future for SRC. His optimism might not be without foundation.

First of all, SRC's heat value is about 16,000 Btu/lb; by comparison, raw coal's heat value generally ranges between 11,500–12,000 Btu/lb. Second, with solvent and filter treatment, coal of 2.5%, 5%, or even 7% S could be reduced to 0.9% S, or perhaps as low as 0.5% S; and from 8–20% ash to 0.166% ash. Such coal could meet EPA requirements for particulate and SO_2 emissions, especially in view of its higher Btu value.

Third, use of SRC might obviate or reduce needs for precipitators and scrubbers, which could have high capital costs, and, under the present state-of-the-art, a 5–8% energy penalty. Moreover, the level of technological advancement of such devices is a subject of spirited debate in many quarters, as is its potential for water or land pollution. On the other hand, purchase of SRC is an operational expense which can be "written off" in the year incurred, and use of SRC

Pilot plant. This year, this 6-tpd pilot plant in Alabama started operating . . .

should involve no installation and maintenance of elaborate capital equipment, according to one engineering and economic view recently expressed.

Huffman said that it is a bit premature to peg dollar figures on SRC economics at this time. He did, however, mention a few material figures. For example, it is expected that the SRC process will use 1.5–2.5% H_2 (of the total weight of coal). Liquefaction and gasification techniques require considerably more H_2, as well as more rigorous conditions of temperature and pressure. Therefore a material and energy saving could be realized.

The SRC process would spin off certain by-products which may be useful. One such by-product would be elemental S, recovered from organic and pyritic sources in the raw coal. A second would be hydrocarbon values which might be recovered and sold or recycled into the plant as an energy source. There may also be other useful mineral values in the solids removed; and the solvent itself, which can replace small losses sometimes incurred in the SRC process, is generated from the raw coal. Yet another benefit may be realized if the 5% of the coal, which remains undissolved in the solvent and is filtered out, could be economically gasified to produce energy for the SRC plant and its ancillary facilities.

Matters under study

One of the prime reasons for the SSI pilot effort was to determine the best method of removing solids (potential ash and particulates) from SRC. Filtering seems to remove enough solids so that as little as 0.1% solids may remain in the fil-

SSI's Everett Huffman

trate. However, hydrocycloning (hydrocloning), to be tried at the Tacoma, Wash., pilot plant could prove less expensive and more easily installable and maintainable than the filter scheme. Consolidation Coal Co. (Consol) tried hydrocloning on a small scale, but indications were that 1% solids may remain in what corresponds to a filtrate.

Another matter under study is the most rational way of handling the removed solids. One objective is to avoid having solid wastes; another aim is to prevent disposal in water that would cause a pollution problem similar to acid mine drainage. Under consideration is gasification of the carbonaceous material, during which H_2S would be produced from the iron pyrites. S would be recovered from the H_2S; H_2 would be recycled to the plant, as would the gas product. Other by-product values would be ascertained.

Finally, how should the SRC itself be handled? If a customer is very

near an SRC plant, the SRC could be supplied as a liquid. The other course is to solidify it, and front runners among approaches under consideration are the Sandvik belt (looked at by SSI) and the prilling tower (PAMCO/OCR/Rust). The Sandvik belt is of stainless steel and is cooled underneath by water. Liquid SRC would freeze on the belt and flake off. In the prilling tower, liquid SRC falls from top to bottom and solidifies on the way down.

Prospectives

If the results of the SSI and PAMCO pilot tests justify the present optimism, Old Ben Coal (subsidiary of Standard Oil of Ohio), together with Toledo Edison would start a $75-million, 5½-year project for a 900-tpd demonstration plant. The plant would be designed and built by Catalytic, Inc. Some components of the plant, however, would work under the more vigorous temperatures and pressures associated with liquefaction. Thus, although the plant would be an SRC facility, it would also be a two-pronged enterprise involving coal liquefaction.

For the future, Catalytic, Inc. looks to the possibility of a 20,000-tpd commercial SRC plant. Cost of such a plant is estimated at $200 million, but clean coal and by-product values, as well as solvent, hydrogen, and hydrocarbon recycling potential, and expected operational economy could be redeeming features.

SSI's Huffman cautions that while SRC shows real promise, it should not be regarded as "the answer to a maiden's prayer"—nor should any other technique be so regarded. From the standpoint of using coal and safeguarding environmental quality, a whole technical "bag of tricks" will have to be evolved. This bag of tricks will include improved coal liquefaction and gasification methods, as well as better gas cleaning at the end of the pipe; either approach would not only be more economical and efficient, but would be a source of useful by-products also. The evolution of these parallel approaches will proceed apace with the development of SRC. This is certainly understandable, since the demand for gas and liquid products and fuels will remain high.

In the long term, SRC could offer new prospects for those electric utilities needing to burn coal and meet clean air standards, and to avoid use of petroleum products or natural gas. Indeed, if SRC technology measures up to expectations, perhaps a new clean fuel and chemical industry will appear on the American scene—a coal refining industry.　　　　JJ

Here is the schematic diagram which shows how it works

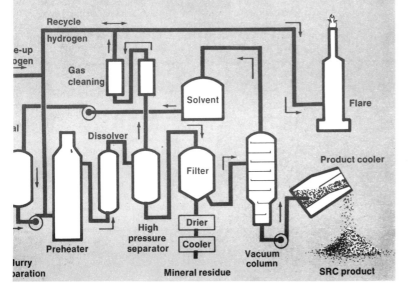

Three on-line processes offer economic and
environmental options that must be evaluated
before converting a dirty fuel to a clean fuel

Status of coal gasification

David A. Tillman

Materials Associates
Washington, D.C. 20037

Reprinted from ENVIRON. SCI. TECHNOL., **10,** 34 (January 1976)

Koppers-Totzek. *Modderfontein plant in South Africa is 13th since 1952; 4 others are under construction*

The decision to gasify coal and then burn the gas implies that an environmental/economic choice has been made. Coal gasification followed by gas combustion is inherently less efficient than the direct burning of coal. More land must be disturbed in mining for coal; more water may be consumed in fuel preparation. Yet gasification provides a cleaner burning fuel, and reduces air pollution. Within the available gasification schemes, limited environmental tradeoffs regarding operating efficiency and water consumption occur. These suggest certain approaches to gasification system optimization: either to minimize coal consumption and mining or water consumption. These system optimization criteria also provide certain guides to operating cost parameters. They emphasize capital cost and system reliability on the economic side, while equally encouraging surface mine rehabilitation and water pollution abatement strategies on the environmental side.

Near-universal agreement now exists: Coal must replace oil and natural gas as the primary energy source in the U.S. Coal reserves exceed 3,000 billion tons, a quantity larger than the combined resources of oil, natural gas, oil shale and bituminous sandstone. Oil, gas, oil shale, coal and current nuclear fuels offer 35,800 quadrillion Btu's of energy potential; coal, alone, offers 32,000 quads. Further, the coal industry has embarked upon an expansion program. Peabody Coal brought four new mines into production in 1974, with a combined annual capacity of 14.5 million tons. AMAX Coal projects a doubling of its capacity by 1978. Consolidation Coal recently completed three new mines, and has projects underway to open two more and expand five existing operations. Other companies have similar programs.

Several studies indicate that the most serious constraints to the increased use of coal are the environmental problems associated with its mining and combustion. Sulfur and ash make coal a dirty fuel. The failure of scrubbing technologies to reach an environmental and economic viability, plus the limited acceptability of tall stacks as a remedy, inhibit coal demand. Coal gasification facilitates sulfur and ash removal.

Environmental issues also exist with coal conversion pro-

Environmental impacts of coal gasification systems

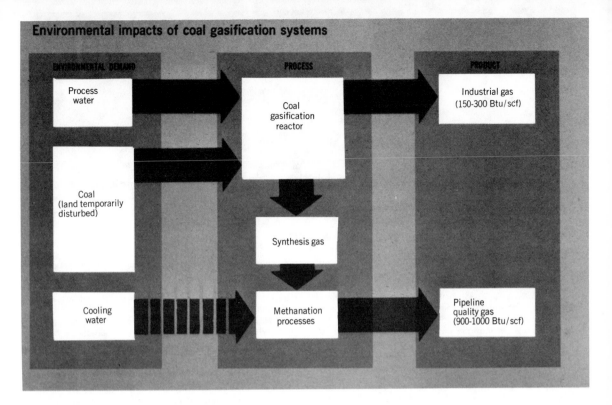

cesses. Two problems are of sufficient significance to be major factors in defining economic limitations to this technology. These factors are the efficiency with which any system converts coal to gas, and the rate at which conversion systems use water. The efficiency issue delineates the amount of land required to support gasification; it suggests the acreage needed for coal mining. Water consumption affects the degree to which mine-mouth gasification can proceed in the arid and semi-arid climates of the west. Those concerns, then, impose constraints on the overall environmental/economic acceptability of coal conversion.

Fuel markets

Popular conceptions of coal gasification largely involve producing substitute natural gas (SNG); many current programs are aimed at improving that technology. That is, however, but one market for coal gasification. Industrial applications using low Btu gas (100–200 Btu/scf) and medium Btu gas (300–500 Btu/scf) must be considered as well. They may, in fact, emerge as the larger market for coal conversion because the industrial sector will lose its natural gas supplies sooner than residential customers.

The market characteristics between industrial gas and SNG differ significantly. The industrial gas users must transport the solid coal to sites relatively close to the point of fuel utilization before gasification proceeds; low and medium Btu gases are not economically transportable over long distances. Further, industrial users must change their burner ports and feed apparatus. Where natural gas combustion requires a mixture of ten parts of air to one part of methane, low Btu gas utilization needs a more equally balanced mix. Finally, industrial users will tend to size their gasification plants to their specific requirements, rather than build the largest unit possible for optimized economies of scale. SNG users, on the other hand, receive and use the coal-derived fuel in the same manner as natural gas. They pay more for fuel, substituting higher operating costs for the expense of combustion equipment modification.

Process technologies

Three gasification units merit comparison in terms of environmental issues: the Wellman-Galusha, the Koppers-Totzek, and the Lurgi gasifiers. All of these technologies are now used in commercial applications. The Glen-Gery Corp. of Shoemaker, Pa., uses Wellman-Galusha gasifiers to supply fuel for the manufacture of bricks. Seventeen Koppers-Totzek units are operating around the world in Finland, East Germany and South Africa; none operate in the U.S. Lurgi units (licensed to Wheelabrator-Frye for U.S. distribution) have long been used by Sasol, Ltd., of South Africa. They will be installed in Farmington, New Mexico, to produce SNG for El Paso Natural Gas Co. Additionally, a major installation operates in Westfield, Scotland. These three systems have passed through the pilot and demonstration phases that the new Synthane, Hi-Gas, and CO$_2$-Acceptor processes have recently entered.

The three basic commercial process technologies differ in size, approach to gasification, and end-product fuels. These distinctions imply process-specific potentials and limitations on user markets and geographical applicability.

The Wellman-Galusha reactor, now produced by McDowell-Wellman Engineering Co., is a relatively small unit; its maximum capacity is 85 tons of coal per day. Such a size economically precludes its being used in SNG operations; it is an industrial system. In the Wellman process, coal is fed into the top of a vertical reactor. Drying occurs first followed by pyrolysis and coking as the coal proceeds downward. Finally, the reacted coal reaches the combustion zone where it is fired with minimal quantities of air. In this countercurrent scheme, hot gases rise upward against the flow of solid coal. This configuration maximizes thermal efficiency, but generates large volumes of tars and reduces the rate of gasification.

The Wellman system is air blown and works at atmospheric pressures; a stirring device facilitates the handling of both caking and non-caking coals. It is cooled by a water jacket that surrounds the reactor; process water is drawn from the cooling jacket into the reactor to supply hydrogen.

Wellman-Galusha. *Fourteen unit gasification system in Cuba; unit can use about 85 tons of coal per day*

Low Btu gas, containing 150–170 Btu/scf, leaves the reactor at 1000–1100°F. The composition of the gas is 2.7% methane, 28.6% carbon monoxide, 15.0% hydrogen, 3.4% carbon dioxide and 50.3% nitrogen when gasifying bituminous coal. The Koppers-Totzek reactor, a suspension system, also takes an atmospheric approach, although this system is much larger than the Wellman. It handles all types of coal, dried and pulverized to a 200-mesh size. These particles along with pure oxygen and steam are fed into the gasification chamber where the conversion reactions occur almost simultaneously. These systems, which range in size from 450–850 tons per day of bituminous coal, require auxiliary support systems for feeding oxygen and steam. They are less self-contained than the Wellman units.

The fuel gas leaves the Koppers-Totzek unit at 3300–3500°F. It is a medium Btu gas with a heat value of 290–300 Btu/scf. It contains the following constituents: carbon monoxide, 55.9%; hydrogen, 35.4%; carbon dioxide, 7.2%; nitrogen, 1.1%; hydrogen sulfide, 0.3%; and other materials, 0.1%.

Lurgi systems can use either air or oxygen along with steam as feeds to the coal reactor. Both the air- and oxygen-fed units are modified fixed-bed-pressurized reactors. Operationally, air has been preferred for industrial applications while oxygen has been chosen when the reacted gas is to be methanated. Commonwealth Edison will use air-fed units at its Powerton, Illinois, combined-cycle station; El Paso Natural Gas will use oxygen-fed reactors at its Farmington operations. While Lurgi units traditionally run on non-caking coals, the experience in Westfield, Scotland, indicates that Illinois and Pennsylvania bituminous coals, which are moderately to highly caking, can be gasified successfully. All coals must be sized, and the fines must be removed.

In the Lurgi system, coal is fed into a lock hopper at the top, and from there distributed into the vessel. While coal enters from the top, steam plus air or oxygen are fed through the bottom of the chamber. These reactors, which may be as large as 950 tons per day, gasify coal at 300–450 psig.

The air-fed Lurgi units produce a gas containing 110–180 Btu/scf. The gas contains 3% methane, 11% carbon monoxide, 17% nitrogen, 11% carbon dioxide, 31% nitrogen, 26% moisture and 0.5% hydrogen sulfide. Fuel values from the oxygen-fed Lurgi reactors are in the 500 Btu/scf range. Like Koppers-Totzek units, oxygen-based Lurgi reactors produce a synthesis gas devoid of nitrogen. Unlike the former system, however, the Lurgi produces a synthesis gas with 10% methane.

System efficiencies

System efficiencies determine the amount of coal that will have to be mined to obtain the required quantities of energy to fuel electric power plants, manufacturing installations, and residential dwellings. Efficiency implies the amount of Btu's delivered to the customer in gas form expressed as a percentage of the incoming energy contained in the coal. Additionally, this calculation includes by-product energy exports and additional energy imports for gasifier operations.

Within that framework; the systems were analyzed for the production of industrial gas; Koppers-Totzek and Lurgi units were also evaluated in terms of methane production. Low and medium Btu gas calculations, for all systems, were made by including sensible heat, tars, and volatiles in the coal gas energy; these efficiencies were also developed utilizing only the calorific content of the gas. Methanation schemes included only the calorific value of the gas, since synthesis gas must be cooled and cleaned before undergoing the shift reaction and methanation.

The Wellman reactor operates at one of the highest efficiencies in the industry. According to Wallace Hamilton, senior consultant to McDowell-Wellman, when the sensible heat plus the tars and volatiles are employed by the gas user, a 92% efficiency can be achieved. R. W. Damon, vice president of Glen-Gery, confirms that their reactors achieve a 90% efficiency operating on anthracite coal. When gas pro-

Lurgi. *Long used by Sasol, Ltd. of South Africa, process will be used in Arizona and Scotland*

duced by the Wellman unit is cooled and cleaned prior to combustion, the efficiency drops to 75%.

Unlike the Wellman unit, the Koppers-Totzek reactor requires additional energy inputs. A ton of eastern coal, containing 25,392,000 Btu's is accompanied by 57,000 Btu's of sensible heat in the dried coal feed; additionally, the oxygen carries 57,240 Btu's and the steam brings 667,290 Btu's to the reactor. These added energy inputs help to account for the higher coal throughput of the unit, while they do not impede its overall efficiency seriously.

The Koppers-Totzek reactor, operating in an industrial gas production mode, offers a higher heating value efficiency—ranging from 85% when operated on Illinois coal to 90.3% when gasifying eastern bituminous coal. The efficiency, when operating on western coal, is 88.2%. The efficiency range is from 83–89% when the calorific value plus the sensible heat is calculated, but the energy contained in the water vapor element of the product gas is discarded. When only the calorific value plus the sensible heat gleaned in a recuperator (waste heat recovery system) is available for utilization, the efficiencies are as follows: eastern bituminous coal, 75.2%; Illinois coal, 74%; and western coal, 75.5%. These are the efficiencies of burning cooled, cleaned gas.

Lurgi reactors, operating in an air-blown mode, offer a hot gas efficiency of 79.6% (exclusive of the energy contained in the gas pressure). The lower heating value, including some sensible heat regained through heat exchangers, is 70.2% according to B. J. Kristensen, senior process engineer for Rust Engineering. Operating the Lurgi air-fed reactor in connection with a combined-cycle-electricity-generating unit, however, capitalizes on the energy inherent in the gas pressure. J. Agosta and associates note that the volume of low Btu gas that will produce 493 MW from a conventional electricity plant will generate 534 MW from a combined-cycle operation. This represents an end-use efficiency increase of 14%; some of this increase comes from the electricity plant's ability to use the pressure, and the rest comes from the energy inherent in the gas pressure. Thus, apparent lower efficiencies of the industrial Lurgi gasifier are offset when the unit is linked to a power station that incorporates advanced power generation designs.

Both Koppers-Totzek and Lurgi gasifiers can generate raw gas for synthesis into methane. In such a situation, the Lurgi unit also uses an oxygen feed system. In methanation schemes, the Koppers-Totzek unit offers a net energy effi-

ciency of 64.7% before the gas is pressurized for transmission. Lurgi efficiencies, including the export of by-products, are 70.1%, according to Mr. Kristensen; that rate is based on the El Paso Natural Gas project. Much of the apparent advantage comes from the Lurgi unit's production of methane in the generation of synthesis gas, plus the pressure inherent in the fuel.

These efficiency data can be utilized to determine the amount of coal required by each technology to produce 10^{15} Btu's of energy useful to the U.S. economy. Based on the system efficiency calculations presented previously, a network of Wellman-Galusha reactors feeding hot gas to boilers would require 56,800,000 tons of high-grade western sub-bituminous coal containing 9,500 Btu's per pound. If these same reactors fed cold gas to the economy, they would consume 70,200,000 tons of coal. If methanation becomes the exclusive mode of operation, a Koppers-Totzek system would consume some 81,300,000 tons of coal, while a Lurgi-based network would require 75,100,000 tons.

Economically, these ranges transfer to the energy feedstock costs of the systems. If eastern coal is used at a cost of $25/ton, the use of hot Wellman-Galusha gas gains a $457,500,000 advantage in the cost of coal when compared to a Lurgi-based methane production network (in terms of producing 10^{15} Btu's in gas form). If western coal is used at a cost of $8/ton, the coal cost differential is some $146,400,000.

Environmentally, these efficiency ranges imply the number of acres temporarily disturbed in the mining operations required to support gasification systems. As an example, calculations can be made by supporting the annual production of 10^{15} Btu's in gas form from western coal deposits averaging 10 ft in thickness. Such seams are common in Arizona, Colorado and Utah. Using this example, and assuming that all of this energy could be generated by using Wellman-Galusha reactors producing hot gas, only about 4,500 acres would be disturbed annually by the associated coal mining activities. If, however, Koppers-Totzek-based methanation systems were used, approximately 6,500 acres annually would be temporarily disrupted. Underground mining, used to support the gasification network, would increase the acreage involved substantially. If 5-ft thick seams are assumed, the hot gas Wellman system will consume coal from 18,000 acres annually; the Koppers-Totzek methanation system will require 26,000 acres annually in the support of its gasification activities.

Factors to consider in calculating system efficiencies

$$\frac{\text{Coal gas energy} + \text{by-product energy}}{\text{Feed coal energy} + \text{auxiliary system energy}}$$

Cross comparison of coal gasification systems

Gasification system	Maximum size	Gas quality	Industrial gas efficiency[a]	Potential use in methane production	Methanation efficiency
Wellman-Galusha	85 TPD	160 Btu/scf	92%	No	
Koppers-Totzek	850 TPD	300 Btu/scf	88.2%[d]	Yes	64.7%
Lurgi	950 TPD	150 Btu/scf[b] 500 Btu/scf[c]	79.6%[e]	Yes	70.1%

[a] Industrial gas efficiency is based on utilizing the higher heating values contained in the gas immediately after production. [b] This quality of gas is produced by a Lurgi reactor fed with air. [c] This quality of gas is produced by a Lurgi reactor fed with oxygen. [d] This efficiency is based on western coal. The Koppers-Totzek reactor, fed with eastern bituminous coal, achieves efficiencies of 90.2%; when fed with Illinois coal, its efficiency drops to 85%. [e] This efficiency does not include the energy inherent in the gas pressure. Lurgi reactors, feeding combined-cycle electrical generating stations, do offer somewhat higher efficiencies.

Gasifiers. *Six units are used in the Modderfontein plant (Koppers-Totzek) to convert coal into ammonia*

These differences in environmental impact result from the mining methods involved. Western coals can be strip mined with 80% recovery rates. Upon completion of mining activities, the surface and land structure must be restored. Underground mines achieve 40–50% recovery rates, and they leave at least as much coal in the ground as they remove. In conventional underground mining, rooms eventually collapse after the coal has been extracted and this may cause subsidence on the surface. Subsidence is most prevalent if the coal seam is not particularly deep. Long-term environmental degradation, from either form of mining, can usually be avoided, but such protection necessarily implies an economic cost.

Water consumption

Data available on water consumption remain less precise than that for system efficiency. Water provides hydrogen for gasification reactions; it can also be used to moderate the reactions, and to carry away waste heat. Water consumption for hydrogen supply cannot be altered substantially. Water used as a reaction control agent can be modified to a limited extent. Cooling, by using water, can be adjusted substantially; it can vary from 15–100% of the cooling requirement depending on the availability of water. Air can be used as an alternative medium for waste heat removal. Finally, water requirements can be supplied by rivers or other traditional sources, from the varying quantities of moisture in the coal itself, and from the methanation reaction that produces water as a by-product. Thus, the number and complexity of variables in water consumption far exceed those in system efficiency calculations.

Variations in the data available make precise quantification of water consumption trying at best. The U.S. Geological Survey estimates that, for the production of pipeline quality gas, water requirements will range from 37–150 gal per 10^6 Btu's produced. Although Probstein and associates place the net water requirements at lower levels, their ranges are equally wide. Yet, certain conclusions remain.

Power or industrial gas requires less input water than synthesized methane. The low and medium Btu gas systems operate at higher temperatures. Further, the hydrogen requirements of industrial gas remain lower than those of synthetic natural gas.

Water demands for methanation schemes are higher, principally because of the requirement for hydrogen and cooling. The shift reaction that precedes methanation (CO & $H_2O \rightarrow CO_2$ & H_2) is exothermic, as is the methanation reaction itself. Since cooling can be accomplished by using either water or air, and since methanation produces water, Probstein and as-

sociates found no substantial difference between the net water requirements of industrial and SNG gas systems when both are optimized around this parameter.

Finally, a ranking of systems indicates that the Koppers-Totzek reactor requires the least water and the Wellman-Galusha consumes the most water. The differences are relatively small. Thus, water consumption data available indicate a far different optimization than system efficiency.

A summation

Both coal and water cost money. Mine rehabilitation and water purification require considerable effort and expenditure. Thus, the environmental issues of land use and water consumption contain certain strong economic implications.

To maximize coal utilization, and minimize land disturbance, industrial systems offer a preferable approach. Within that category, relatively small energy requirements appear to be best met through application of the Wellman-Galusha reactor system if water is not scarce. Large users may opt for a Koppers-Totzek or Lurgi system. If electrical power generation is desired, the Lurgi enjoys an advantage; its pressure enhances the economics of combined-cycle plants. If methanation becomes the dominant mode because of an existing transportation network, the Lurgi units profit from their somewhat higher efficiencies; on the other hand, Koppers-Totzek units require less water in operation.

Coal gasification does provide one route to the clean combustion of this fossil fuel. It is less efficient than direct coal combustion. Opting for coal conversion implies that one environmental choice has been made. Within the choice of coal gasification, certain reactors do provide limited optimization of either process efficiency or water consumption. Such performance maximization is application-specific and site-specific. The choice of gasification systems depends, in large measure, on the requirements and locations of the end-use markets. Such a selection, based on the environmental criteria of land and water consumption, can aid in determining the most economically viable unit for the project under consideration.

Additional reading

Bhutani, J., et al., *An Analysis of Constraints on Increased Coal Production.* Mitre Corporation. January, 1975. And The Hudson Institute. *Policy Analysis for Coal Development at a Wartime Urgency Level to Meet the Goals of Project Independence.* The Office of Coal Research, R&D Report No. 87. February, 1974.

Agosta, J., Illian, H. F., Lundberg, R. M., Tranby, O. G., Ahner, D. J., and Sheldon, R. C., "The Future of Low Btu Gas in Power Generation." April, 1973.

National Academy of Sciences. *The Rehabilitation Potential of Western Coal Lands.* Ford Foundation Energy Project. 1973.

Davis, G. H., and Wood, L. A., "Water Demands for Expanding Energy Development." U.S. Geological Survey Circular 703. 1974, p. 12.

Probstein, R. F., Goldstein, D. J., Gold, H., and Shen, J., "Water Needs for Fuel-to-Fuel Conversion Processes." AIChE Annual Meeting, Dec. 3, 1974.

David A. Tillman *is vice president of Materials Associates, Inc. Washington, D.C., and contributing editor to* Area Development *magazine. Prior to his present position, he was editor of* Vermont Business World. *Mr. Tillman is writing a book on industrial and municipal waste recycling for McGraw-Hill Book Co.*

Coordinated by LRE

A new "clean image" for coal

How can "America's Ace in the Hole" be played
as a winning card in the energy game?
The Louisville Conference/Expo offered some tips

Reprinted from ENVIRON. SCI. TECHNOL., **11,** 1148 (December 1977)

Choral songs sung by the Harlan (Ky.) Boys Choir greeted the 2700 attendees at the Coal Conference and Expo IV, sponsored by the National Coal Association (NCA) and Bituminous Coal Research, Inc., and held at Louisville, Ky., in October. NCA chairman Otes Bennett, Jr., remarked, "If we could mine, transport, and use coal as well as those boys sing, the U.S. would have no energy problem!"

Unfortunately, the U.S. has many difficulties with mining, transporting, and using coal, although recent activities would tend to mitigate these difficulties. One had but to hear or read the more than 150 technical papers, and view the 180 exhibits to be reassured that this is indeed the case.

Why all this feverish activity? The conference/expo's theme, "Coal: America's Ace in the Hole", may explain why. But, as Secretary of the Interior Cecil Andrus observed in his keynote address, an "ace in the hole" aimed at winning a poker or energy game has to be played in the correct, timely, and technical manner. Otherwise, it will not help to produce the desired outcome of the game in which it is played. Thus, if mining, transportation, and environmental problems of coal cannot be solved in an aggressive, systematic manner, the ace in the hole would remain in the hole, and not have a chance to be played.

Coal cleaning plant

Under present and forthcoming laws and regulations, coal, in order to be used, must have its pollutants removed before, during, or after burning. At present, the EPA still says that the use of flue gas desulfurization (FGD) scrubbers, the after-burning option, is "the best means" of protecting public health. But the EPA position is not as rigid as it was only two years ago, when former EPA administrator Russell Train characterized non-scrubber desulfurization as viable options for the 1980s, and perhaps later. For example, the EPA has co-sponsored meetings and activities in coal cleaning (the pre-burning option, *ES&T,* August 1977, p 750). And fluidized-bed combustion, the during-burning option, is rapidly gaining acceptance (*ES&T,* November 1977, p 1048).

One large-scale coal preparation facility is now under construction at Homer City, Pa., under the auspices of Pennsylvania Electric Co. (Pennelec), and New York State Electric & Gas Corp. This facility will process 5.2 million tpy of coal, which is then expected to meet existing federal and state emission limits for SO_2 of 4.0 lb/10^6 Btu, at two existing 600-MW units. The product coal will be medium-ash (less than 18%), and medium-S (2.25%), which will satisfy these limits.

Interior Secretary Andrus
told how to make the "ace" playable

Later, a new 650-MW unit will have prepared for it a lower ash (2.8%), lower-S (0.88%) product coal that will meet new source SO_2 standards of 1.2 lb/10^6 Btu.

The units will not need stack gas scrubbing, according to Pennelec. The company also says that a high-ash, high-sulfur refuse product will be deposited in a refuse area about one mile from the coal preparation facility.

More on pre-cleaning

One technology for cleaning coal, available since 1916, but now being upgraded, involves pneumatic cleaning of fine coal for sulfur reduction, as described by Robert Llewellyn, vice president of Roberts & Schaefer Co. (Pittsburgh, Pa.). For a time, this technology suffered reverses, but modernization, including improved dust collection, appears to give it new promise. This promise grows out of a "Super Airflow" unit, which acts on fine coal in such a way that heavy sulfur-laden particles fall out faster than others. They are ultimately collected through a cyclone dust collector-air filter baghouse system. This collected material, known as "hutch", is very high in pyritic sulfur.

At one plant with a feed rate of 150 tph, ash was reduced from 15.89% to 9.78%, and sulfur from 5.89% to 3.15%. At another (100 tph) plant, ash was reduced from 44.4% to 17.3%, and S from 5.25% to 1.98%. Plant costs, based on 500 000 t of clean coal, are pegged at less than 40¢/t. As long as the coal is crushed to minus $\frac{3}{4}$ in., or finer, no water is needed for this process. Llewellyn characterized this technology as "proven."

Sometimes, removal of pyritic sulfur brings coal to a point of "cleanliness" at which it will meet existing, or even new standards for SO_x. In other cases, much of the sulfur in the coal is organic; that is, tied into the coal's molecular matrix, and harder to remove. However, in that regard, there may be a ray of hope. Results of work at Iowa State University (Ames) suggest that crushed and ground coal, leached with an oxygen-rich basic, rather than acidic solution, can remove at least some organic sulfur at temperatures as mild as 130 °C. Pyrites are also more efficiently attacked, and sulfur in sulfate form is virtually eliminated.

Sodium carbonate seems to be the best leaching agent. Partial pressure of oxygen need not exceed 3 atm. The use of such a base does increase the ash content, to be sure, but a dilute acid wash easily handles that problem. The Ames researchers have proposed an industrial process for this alkali technique.

A water connection

There is a major trend toward increased opencast mining of coal, as well as augmented need to dispose properly of captured fly ash, and FGD scrubber sludge. These factors lead to requirements for improved knowledge of the effects of disposed-of ash, surface mine overburden, and the like, on surface water and groundwater.

Before these effects can be properly assessed, a thorough familiarity with the hydrology of mine spoil banks, disposal areas, and other such facilities is necessary. To meet this need, Ernest Crosby, Donald Overton, and *ES&T* advisory board member Roger Minear, all of the University of Tennessee (Knoxville), have evolved a realistic model of the fate of water involved with a spoil bank. They used the Tennessee New River Basin to help construct their model, and looked at all aspects of underground seepage, infiltration, rainfall and runoff, water tables, overland flow, evapotranspiration, and the waters of a spoil bank pond.

One principal aim was to work out a scheme of how pollutants—mainly dissolved material, in this instance—travel. The "bottom line" appeared to be that these pollutants travel mainly in surface runoff, or in small seepages where ponding occurs. Presumably, this modeling will lead to optimized designs for disposal areas and spoil banks, which would minimize stream and groundwater contamination.

Polish-American cooperation

A large portion of the pollutants that affect surface water and groundwater, as well as soils, consist of toxic sulfides in the form of pyrites, marcasite, and other such minerals. These minerals are abundant in surface mine spoil, and not only produce metal toxicity, but also acid pH's ranging from a relatively mild 5.0 to a harsh 2.8. These factors can very adversely affect water quality and future land reclamation, Kazimierz Bauman of the Central Research and Design Institute for Opencast Mining (Poltegor, Wrocław, Poland) told the conference. He noted that in Poland, where a mammoth surface coal mining program will be under way over the next 20 years, these problems must be surmounted, if mandated water quality and land reclamation goals are to be achieved.

Thus, Poltegor and the U.S. EPA joined forces during 1973 to try to work out means of attacking and solving these problems. Financing was arranged through the Special Foreign Currency Program of Public Law 480. The object was to find ways of neutralizing toxicity/acidity, and of providing proper soil fertilization to bring about agricultural and forest reclamation. Lime and magnesia lime were used for neutralization at two sites in Poland. Fertilizers were then applied. Despite diverse overburden soil types, some erosion problems, and adverse weather, the Poltegor EPA approach showed considerable promise over the past several years, and up to the present time.

The Poltegor/EPA project coordinator, Jacek Libicki, reminded the conference that in the U.S., Poland, and elsewhere, expanded coal mining and use will produce much refuse and fly ash, both of which could well be consigned to abandoned or worked-out opencast mines. However, he said that such disposal might seriously pollute groundwater, ultimately over long distances. But how, to what extent, and in what time frame?

To find answers to these questions, Libicki is coordinating a practical study that will determine the influence of refuse and fly ash disposal on groundwater quality. There is a systematic, expanding monitoring set-up—perhaps the world's first such one—that will tell what contaminants are entering groundwater at what rate, and where they are going. For example, it was found that for a refuse pile 20 m thick, groundwater contaminants appeared at a 60-m distance after about 15 months. From these and similar data, Libicki expects the Poltegor/EPA project to come up with really effective means of counteracting groundwater contamination effects during refuse/fly ash storage and eventual land reclamation.

For the years ahead

All that was heard and seen at the conference and expo offers an idea of the almost staggering problems involved in bringing coal back as a primary fuel and chemical source during the years ahead. But realistic approaches to tackling these problems were also seen and discussed. The environmentally sound development and expanded use of coal will certainly tax the technical and financial ingenuity, resources, and skills of many. But if what was witnessed and heard at the conference is any indication of things to come, one can be optimistic concerning a successful response to all of these challenges.　　JJ

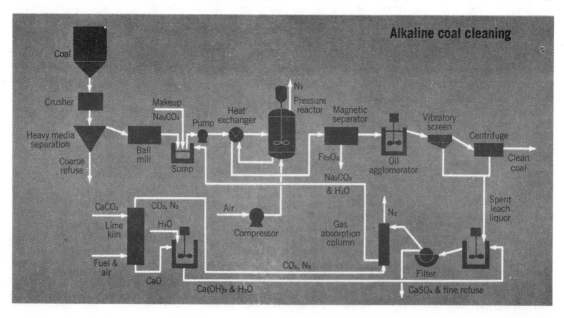

Alkaline coal cleaning

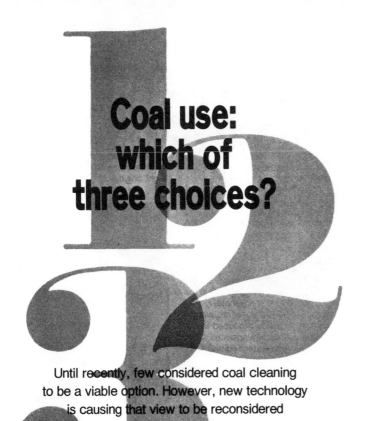

Coal use: which of three choices?

Until recently, few considered coal cleaning to be a viable option. However, new technology is causing that view to be reconsidered

Reprinted from ENVIRON. SCI. TECHNOL., **11,** 750 (August 1977)

Remember the good old days? You could see those factory or power plant chimneys belching out thick, black smoke. Your heart would swell with pride at this tangible expression of the nation's industrial prowess. In those times, much of the industrial might was powered by the burning of raw, or minimally cleaned coal. Then, pollution became a national issue; legislation was passed. Black smoke and sulfur fumes became a "no-no", and cheap oil and gas came to the rescue, for they were abundant.

The trauma of the 1973–1974 oil embargo rudely brought home the knowledge that petroleum products' abundance and inexpensiveness were things of the past. Suddenly, coal looked good again. But the problem was, and still is, how does one use coal without generating that no-longer-desired black smoke, and other products that people are better off without?

Three ways to go

There are three basic ways to go. These comprise cleaning coal before burning; combustion methods that reduce pollution (fluidized-bed, for example); or "scrubbing" stack gas before discharge.

The first alternative was discussed at the conference, "Coal Cleaning: An Option for Increased Coal Utilization?" (Arlington, Va). Cosponsored by the U.S. EPA and Battelle's Columbus (Ohio) Laboratories, the conference attracted 200 people.

About two years ago, top EPA spokesmen, including former administrator Russell Train, maintained that scrubbers were generally the most immediate, realistic approach to air pollution control necessitated by the burning of coal. At the conference, however, EPA's acting assistant administrator for research and development, Stephen Gage, noted that his agency is taking a second look at coal cleaning possibilities. Now EPA is helping to surmount the non-technical and technical barriers that have inhibited coal cleaning to date.

Gage also noted President Carter's goal of increasing coal production to 1 billion tons by 1985, from 660 million tons last year. He said, however, "There can be no compromise on the environment. Energy production and use must be controlled to minimize the cost to property and public health. In these matters, the environment–energy dichotomy is to be swept aside." Thus, stringent cleanup measures

must be taken, and in this regard, Gage observed that coal cleaning, for instance, could be an advantageous, economic approach, instead of scrubbing.

Physical approaches

At present, the most economical methods of cleaning coal are physical, according to EPA's Frank Princiotta. By "physical" is meant breaking or crushing coal, and removing pyritic sulfur, ash-producing material, and certain other non-combustibles by non-chemical techniques. Often, the physical techniques comprise washing, and its cost and the amount of undesirable matter removed will vary with the amount of crushing, screening, separation from waste, and actual washing done. What might the cost be? Battelle's Joseph Oxley provided an estimate. For example, for coal ground to "top size"—about $1\frac{1}{2}$ in.—costs could be $1.20/t. On the other hand, if the coal is pulverized ("deep cleaning"), and all coarse and fine material is processed, costs could be as much as $11.80/t. However, the degree of cleaning is then greatly enhanced.

EPA's Gage
taking a second look

Princiotta indicated that coal cleaning alone can now be considered an effective control approach for certain coals. He also suggested that coal cleaning/gas scrubbing mixes could be considered as possible low-cost alternatives to scrubbers alone, for many future applications.

Some chemical routes

How simple life would be if all of the world's coal were of identical composition! One general coal cleaning process could then be optimized, and would suffice. Unfortunately, as James Kilgroe, manager of the EPA's coal cleaning programs (Research Triangle Park, N.C.), reminded the meeting, coals and their "cleanability" vary not only from region to region but from bed to bed, and even within a given bed.

One salient example of this variability is found in a coal with a sulfur content that ranged from 0.2–7%. It may be pyritic or organic; proportions of these can vary widely. To remove this sulfur, or part of it, the task is eased when the sulfur is inorganic in the form of pyritic sulfur. But when the sulfur is organic, and bound up in the molecular structure of the coal itself, chemical means must be employed, and the job becomes more expensive. Indeed, Richard Balzhiser of the Electric Power Research Institute (EPRI, Palo Alto, Calif.) estimated that removing the organic sulfur from coals can more than double, or even triple coal cleaning costs, in some cases.

For coals, the bulk of whose sulfur is pyritic, the Meyers Chemical process (TRW, Inc.), which uses iron (III) sulfate as a chemical leaching agent, may be in the most advanced development stage, at present, of the approaches discussed at the conference. With EPA support, TRW has constructed a 667-lb/h desulfurization reactor test unit to prove the process out. It has removed up to 95% of pyritic sulfur in laboratory tests.

Another approach to getting the sulfur out before burning is Arco's catalytic method, now in laboratory stage. EPA's Kilgroe said that this method is slated to remove up to 95% of pyritic sulfur, and 40% of organic sulfur. Also, Battelle is testing a caustic leach process in a 0.25-tpd miniplant. That process, like that of Arco, is expected to remove 95% of the pyritic, and 40% of the organic sulfur.

On the other hand, ERDA is seeking designs for continuous units for an "oxi-desulfurization" air-steam system. Now in the laboratory stage, it should take care of 99% of the pyritic, and 40% of the organic sulfur.

Kilgroe said that 23 chemical coal cleaning processes have been documented as of late May. Of these, 7 or 8 are considered to be feasible, he pointed out. He also mentioned that in most cases, Appalachian coals are more "cleanable" than are Midwestern coals. Illinois coals, for example, not only have a high sulfur content, but much of that sulfur is organic.

SRC on the horizon?

For a yet newer generation of coal cleaning or conversion techniques—gasification, liquefaction—EPRI's Balzhiser saw solvent-refined coal (SRC) as a possibly viable option coming "down the pike". SRC removes pyritic, and a large portion of organic sulfur (ES&T, June 1974, p 510), as well as mineral matter from coal. The coal is essentially "dissolved" in its own organic solvents under conditions of hydrogenation, temperature, and pressure that are mild in comparison to those of gasification or liquefaction.

Coal cleaning. *This is how facility at Homer City, Pa., will appear*

The refined coal, with enhanced Btu value, is recovered in either solid (SRC I) or liquid (SRC II) form.

Balzhiser said that his organization plans to find out if the product from the SRC II process can be used in place of oil, even though it will still have some sulfur, and about 10% more mineral matter than would oil. He also said that EPRI is participating in a project to generate 20 MW of electricity at Georgia Power, by use of SRC. This use of SRC is not expected to worsen the NO$_x$ problem. He also noted, however, that SRC from most coals could generally meet the existing new source performance standard, 12 lbs of SO$_2$/10^6 Btu.

Much of the coal cleaning and SRC talk one hears concerns sulfur and SO$_2$ standards. But, as EPA's Kilgroe reminded the conference, mineral matter or ash content must not be overlooked, and that matter could comprise 5–40% of a given coal. The larger amount will almost certainly be a problem, especially if and when a fine particulate emission standard comes into effect. Indeed, one might do well to anticipate a particulate standard of 0.05 lb/10^6 Btu by no later than 1985, according to this EPA spokesman. Also, coal cleaning should reduce air emissions and solid waste from combustion, as well as eliminate scrubber sludge production. On the other hand, it could lead to a disposal problem for liquid and solid wastes from coal preparation.

Dollar estimates

Still, coal cleaning may have cost advantages. To illustrate this point, Lawrence Hoffman, president of Hoffman-Muntner Corp. (Silver Spring, Md.), compared figures of deep coal cleaning by physical means with scrubbing and found the latter more expensive. He assumed

that a Pennsylvania coal (2–3% S) would be used at Tonawanda, N.Y., with a 15-year plant payoff, 2% for insurance and taxes, unfavorable interest rates, and grinding costs at 50¢/ton. His "bottom line" figures for cleaning coal came to $3.55/t; comparative scrubbing costs were $5.56/t. In terms of 10^6 Btu, Hoffman estimated coal cleaning/use to run 10–12¢, while flue gas scrubbing might be 22¢. He listed other advantages for coal cleaning, such as less need for land space, no need to reheat flue gas, and more net power to sell.

One prediction was that coal cleaning might take place at the mine mouth. However, as Joseph Mullan, a vice-president of the National Coal Association said, such coal cleaning, there or elsewhere, could virtually call for development of a new industry with needs for new personnel skills. He added, "How many people know how to run a coal-cleaning plant? Or, for that matter, how to design one?" He also said that if some proposed more strict air standards come into force, coal cleaning or cleaning/scrubbing combinations may not meet these standards. "How do you tell that to potential investors in plants that would clean or burn coal?" he asked.

Perhaps, with the right incentives and technological advances, coal cleaning will advance to the stage at which coal refinery complexes, analogous to today's oil refinery complexes, will be developed. These complexes would not only provide clean coal but perhaps marketable sulfur and other chemicals, as well. However, there is a long way to go. Numerous technical, economic, and regulatory uncertainties will have to be resolved before coal cleaning becomes a factor in providing an environmentally acceptable fossil-fueled energy supply. **JJ**

Conversion's the name of the game!

FBC; energy storage; fuel alcohol—all are aimed at
putting fuels to work in a clean, efficient way. How to
do it was seen at Rivesville, and discussed in Washington

Reprinted from ENVIRON. SCI. TECHNOL., **11,** 1048 (November 1977)

It was a hot, hazy day at Rivesville, W. Va., on the banks of the Monongahela River. However, the heat did not deter numerous dignitaries, including Senate majority leader Robert Byrd (D, W. Va.), Sen. Jennings Randolph (D, W. Va.), and Congressman Austin Murphy (D, Pa.) from attending a rather unique dedication ceremony held there. What was unique was that the ceremony dedicated a 30-MW coal-fired fluidized-bed combustion (FBC) boiler at Monongahela Power Co. Now normally, dedications of boilers do not attract national leaders. Then why did they attend, and speak at this one?

One reason was because this boiler is the first of its kind at that scale. It opens up possibilities for burning coal without the need for SO_x scrubbers. The FBC boiler system was designed by Pope, Evans and Robbins, Inc. (PER, New York, N.Y.), built by Foster Wheeler (Livingston, N.J.), and funded to the tune of $23 million by ERDA. PER president Michael Pope told the attendees that the FBC concept "is now proved," especially because of its successful operation since early April.

Pope said that this system can remove up to 95 % of SO_2 emissions from coal— "more than enough to meet federal standards." He also noted that, in fact, virtually any fuel can be burned in a fluidized bed. FBC systems will also require the development of a new industry base to produce equipment. Pope announced that his firm, together with Foster Wheeler, has received a TVA contract to come up with preliminary designs for a 200-MW FBC-based central generating station at a site to be determined. He also predicted that 25 % of the electricity generating capacity of American Industry will use fluidized beds by the year 2000.

More on FBC

This Rivesville boiler also came up for discussion at the 12th Intersociety Energy Conversion Engineering Conference (12th IECEC), held at Washington, D.C., late

August, and attended by about 750 people. The conference heard PER's John Mesko describe the workings of the Rivesville unit which makes 300 000 lb/h of 1300-psi steam at 925 °F. He discussed types, problems, and prospects of FBC in general, and what is being done in the field. Mesko noted that the lower temperatures of FBC inhibit NO_x formation. They also eliminate slag problems, since combustion temperatures are below the 2100–2200 °F of ash fusion. By contrast, conventional boilers could run at 2500–2700 °F.

The Rivesville combustor/boiler is an

PER president Pope
"ready for scale-up"

atmospheric FBC (AFBC) unit whose sorbent "stones" go through only once. But FBC can be under pressures of up to 10 atm. S. Moskowitz of Curtiss-Wright Corp. (C-W, Wood-Ridge, N.J.) explained that pressurized FBC (PFBC) can offer higher combustion and SO_x retention efficiency, reduced reactor size and NO_x emissions, and, very significantly, the ability to power a gas turbine. Curtiss-Wright with Dorr-Oliver is now constructing a 13-MW pilot plant (3% S coal with dolomite sorbent, 1650 °F) to demonstrate the PFBC/gas turbine concept. One-third of the turbine compression air is to fluidize the bed and support combustion; the remainder will control bed temperature, and extract heat through an in-bed gas-to-air heat exchanger. For the future, a 500-MW commercial PFBC power plant is envisioned.

Another aim is to regenerate the FBC sorbent "stones", such as limestone or dolomite. One regeneration scheme is being developed at Exxon Research and Engineering Co. (Linden, N.J.). Exxon's L. Ruth explained how spent sorbent, much of which is calcium sulfate, could be regenerated under 7–10 atm pressure at 1100 °C, in a carbon monoxide-hydrogen atmosphere. Calcium oxide is the product. Ruth estimated that make-up sorbent requirements could be reduced by a factor of at least 4, in comparison with a once-through system for sorbent.

The sun shines bright . . .

FBC was only one of many energy conversion topics brought up at this energy conversion conference. Another was solar energy, and what its problems, prospectives, and markets·might be.

One near-term market for solar photovoltaic (SPV) electric systems could be the National Park Service, as Phillip Jarvinen of MIT/Lincoln Laboratory (Lexington, Mass.) explained. His organization did a market study for ERDA. Jarvinen gave a figure of 2765 kWh/d as representative of all typical electricity needs in a national park. This would include the visitor center,

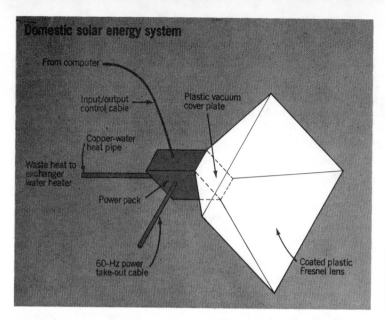

Domestic solar energy system

From computer

Input/output control cable

Plastic vacuum cover plate

Copper-water heat pipe

Waste heat to exchanger water heater

Power pack

60-Hz power take-out cable

Coated plastic Fresnel lens

and unheated homes are uncommon. Indeed, the energy crisis is of such low profile that, as Bishop remarked to the conference, a Gallup poll, taken in July, showed that 51% of the people sampled did not know that the U.S. imported petroleum!

Bishop warned that unchecked, exponential energy demand growth, such as is acutally occurring (though perhaps a bit slowed since 1973), cannot be sustained much longer. However, he noted that some years do remain before the real crunch would come. In those few years, vigorous efforts toward conservation, and developing and putting in place many different energy sources and systems, must be undertaken.

Energy advisor Bishop
shed light on "the invisible crisis"

ranger stations, sanitary needs, and other requirements. Probably the most promising site for an SPV system would be Natural Bridges National Monument (Utah). Jarvinen said that the site is well located, 6500 ft in altitude, and in a place for which the cost of bringing in commercial power (from 37 mi away) is unrealistic.

In Brazil, availability of low-cost land, especially in remote regions, could make solar energy economically attractive, D. Magnoli of Centro de Technologia Promon (Rio de Janeiro) told the conference. Magnoli said that 1000 collectors, each 3 m², provided 10 tpd of steam in trials. A special application of solar-derived low-pressure steam could be in the Brazilian program aimed at producing 5 billion L/y of fuel alcohol by the early 1980s, from cassava (manioc or tapioca) grown in the hinterlands, and other sources.

In solar and certain solar-derived energy, a weak link, at present, is heat or electricity storage during night hours and cloudy days. Energy storage approaches discussed included flywheels, compressed air, and underground pumped hydro (Sandia Laboratories and ERDA); chemical (Lawrence Berkeley Laboratory); thermal (Westinghouse); and other means.

One intriguing presentation covered a miniature solar-electric system based on ferroelectric heat engines and heat pipes. For example, it would help to power and heat a house, and weigh only 3 kg. There would be no dynamo required to convert mechanical to electrical energy. Moreover, the system would be built of light materials, and have a long (about 20-y) lifetime. Its waste heat can be used for some domestic applications, including washing, or swimming pool heating.

This system was described by James Drummond of Maxwell Laboratories, Inc.

(San Diego, Calif.; he is now with Power Conversion Technology, La Jolla, Calif.). It consists of a 4 X 5 m plastic fresnel lens, vacuum cover plate, household computer, a molten salt, power pack tuned to 60 Hz, control cable, heat pipe, and waste heat exchanger.

Impact assessment

Efforts to improve the energy conservation and supply situation will have environmental impacts, and the EPA has, as part of its mission, the job of minimizing these impacts, Harry Bostian of the EPA (Cincinnati, Ohio) told the conference. Other objectives are to develop cost-effective controls for any new pollutants generated by energy processes, and to see that energy conserved by new technologies is not consumed by required pollution controls.

Funding for the necessary impact studies is devoted to conservation assessment and development; waste-energy utilization; advanced power systems; geothermal and solar energy; and special studies. These tasks are being handled by EPA's Power Technology and Conservation Branch in Cincinnati, of which Bostian is head. Total fiscal year 1977 funding was about $1.1 million.

"Invisible crisis"

The foregoing represents some examples of what is hopefully an accelerating effort to mitigate the energy situation in which the nation presently finds itself, with due regard for environmental protection and improvement. This situation was aptly characterized by James Bishop, Jr., of President Carter's Energy Policy and Planning Office, as "the invisible crisis." After all, fuel, with the possible exception of natural gas, seems plentiful, if somewhat expensive. Gasoline lines

But sources, of themselves, are of little avail, unless they can be optimally converted to useful work. For instance, while solar energy is abundant and "free", it is of little help until it can be cost-effectively concentrated and changed into heat, electricity, or motion. Cleaner ways must be found to obtain thermal, chemical, and mechanical energy from coal, lignite, and western and Devonian oil shales. One can go on listing problems and needs of energy conversion *ad infinitum*.

However, cataloging needs and problems does not meet or solve them. That takes "1% inspiration and 99% perspiration" on the part of many skilled and not-so-skilled people who realize that economic and political survival may be at stake. As Bishop put it, "solutions will not come from Washington, but from millions of Americans." He warned that reducing economic and political vulnerability and maintaining economic growth, together with environmental protection, would be a monumental task. But he also said, "it is not necessary to shed bitter tears, for these goals are attainable." They will come closer to attainment as the technology of energy conversion progresses.   JJ

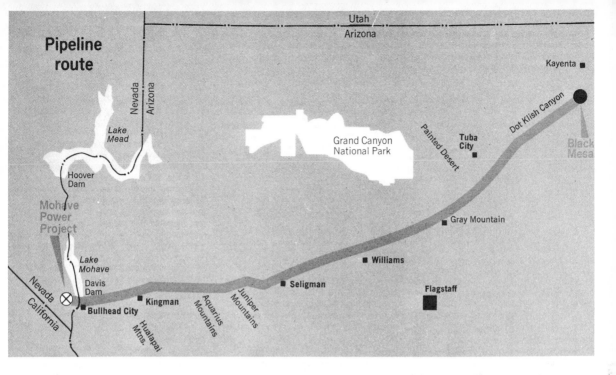

Pipeline route

Utah / Arizona

Nevada / Arizona

Lake Mead

Hoover Dam

Mohave Power Project

Lake Mohave

Davis Dam

Nevada / California

Bullhead City

Kingman

Hualapai Mtns.

Aquarius Mountains

Juniper Mountains

Seligman

Williams

Grand Canyon National Park

Flagstaff

Gray Mountain

Painted Desert

Tuba City

Dot Klish Canyon

Kayenta

Black Mesa

How the coal slurry pipeline in Arizona is working

The only operating one in the U.S., it is performing very smoothly so far

Reprinted from ENVIRON. SCI. TECHNOL.,

10, 1086 (November 1976)

Suppose a large electric power generating station is established at point "A," to be fueled by coal to be mined at point "B," which is almost 300 mi away. Suppose, further, that the terrain over which the coal must be transported is largely remote and rugged. Finally, suppose engineering/economic studies show that road or rail shipment of the coal is infeasible. Yet the generator must go on line, and must get coal to do so. What do you do?

You build a coal slurry pipeline, just like the Black Mesa Pipeline (Flagstaff, Ariz.) did to transport coal from the Peabody Coal Black Mesa Mine near Kayenta, Ariz., to the 1500-MW Southern California Edison (SCE) Mohave Power Station (*ES&T,* June 1976, p 532) near the southern tip of Nevada. After all, the coal slurry pipeline concept existed since late last century; and from 1957 to 1963, a 10-in. inner diameter (i.d.) slurry line moved 7 million tons of coal over 108 mi in Ohio for Consolidation Coal Co. That pipeline closed down because of rail rate reductions for large tonnages, not because of any technical problems.

Easier said than done

Building a coal slurry pipeline, however, is easier to talk about than to do, even if it is to be installed in a fairly gentle terrain, such as is found in Ohio. The Black Mesa Pipeline, by contrast, is 278 mi long, and crosses the rugged terrain of northern Arizona. Moreover, it must provide smooth operation, high availability, and steady coal delivery, with extremely minimal environmental disruption or pollution in the fragile desert area through

which it runs. In addition, this pipeline, when it was planned, would be the longest of its kind in the world—it still is—and involve much pioneering in its planning, engineering, construction, and use.

These and other considerations were what the Black Mesa Pipeline had to take into account when it was organized as an operating entity in 1966, as a wholly-owned subsidiary of Southern Pacific Pipelines, in turn 100% owned by the Southern Pacific Co. By January 1967, the

pipeline was deemed to be a "can-do" situation, and a 35-y contract was signed with Peabody Coal Co., to design and operate it. The system cost about $39 million to build, and has been in operation since August 1970. Buried at least 3 ft deep, and protected against the sharp volcanic rocks that surround it, the line took less than three years to build.

John Montfort, vice president and general manager of Black Mesa, told *ES&T* that the pipeline's operations since 1970 have been quite satisfactory. To be sure, there were some problems, as one would expect in any new system. For example, in the spring of 1971, a 40-ft plug, caused by too coarse a slurry, "consist", developed, partly because of slurry restart troubles after power failures. The plug was broken by a proprietary technique Black Mesa developed, and Montfort said that no such problem has been experienced since.

How it works

Perhaps the smooth performance of the Black Mesa Pipeline can best be ascribed to the most painstaking planning, engineering, and development of operating methodology that went into it. For instance, not only did difficult northern Arizona topography have to be taken into account, but so did avoidance of pipeline breaks, slurry spillages, and inner surface

> The solid material carried by the slurry is what are called "consists" (pronounced "*con*sists").

abrasion, along with smooth restart after interruption. Yet these needs are being met. Water needs are also met with approximately 3000-ft wells near Kayenta. These wells are sealed from their surroundings with steel casings.

After proper preparation at Black Mesa near Kayenta, the coal is so crushed that slurry consists will be no larger than "$\frac{3}{16}$ by zero," as Montfort expressed it—that is, $\frac{3}{16}$ in., wet-ground. At Kayenta, the slurry is pumped into the pipe at a flow rate of about 4000 gpm. The flow rate and consist size assure that a non-moving layer of liquid hugging the pipe's inner surface will inhibit abrasion. Corrosion protection is handled by proprietary techniques. Since there is pressure and flow loss along the way, Black Mesa maintains pumping stations at Gray Mountain, Williams, and Seligman, Ariz. (see map). Along the last 12 mi to Mohave, the 18-in. i.d. line is reduced to 12 in., because of a 3000-ft drop in those last 12 mi.

The slurry (48% solids) completes its trip in 67 h, and enters 7.8 million-gal tanks at Mohave. Here, the slurry is stirred with 500-hp agitators, then withdrawn from tankage by centrifugal pump to 40 centrifuges. There, 75% of the water is removed, and what was a slurry is now a wet cake that is dried, further pulverized, and burned. About 1400 gpm of water are salvaged from the slurry, and this water is of good enough quality to become part of the 15 000 gpm needed for make-up water for Mohave's cooling towers. No known pollutants are in the recovered slurry water.

Montfort told *ES&T* that the actual pipeline availability has been more than 99½%. In 1975, however, the pipeline was on slurry 75% of the time; on water, for flushing purposes, 7% of the time; and shut down 18% of the time. The reason was that while the line could deliver 5 million tpy of coal, only 4 million tons were to be delivered last year; consequently water flushing of applicable sections and shutdown were necessary to prevent overdelivery of coal to Mohave.

The water debate

Montford characterized the Black Mesa Pipeline as "an economic and technical success" after more than 5 years of operation. Might the slurry method of moving coal not apply to other routes? Montfort sees no technical reason why not, given the necessary coal, water, and engineering quality. Indeed, Energy Transportation Systems Inc. (ETSI), a joint venture of Lehman Bros. (New York, N.Y.), Bechtel Corp. (San Francisco, Calif.), and Kansas-Nebraska Natural Gas Co. wants to build a $750 million, 1036-mi pipeline from coal fields near Gillette, Wyo., to a 1400-MW power plant at White Bluff, Ark., being built by a subsidiary of Middle South Utilities, Inc.

The terminal at White Bluff is a barge-

head from which barges would deliver coal to a number of plants. A 1400-MW plant would require only 20% of ETSI's capacity.

This proposal has stirred up lively debate, to say the least, even on Capitol Hill. One keyword of the debate is water, and Rep. Teno Roncalio (D-Wyo.) is concerned that the slurry's potential use of up to 6.5 billion gal/y from his water-poor state. He indicated that he might relax his opposition if he could have solid assurances that at least 30% of the water used could be recovered and recycled in Wyoming. However, Rep. Joe Skubitz (R-Kan.) characterizes slurry pipelines as "voracious" users of the West's scarce water. This June, he proposed that the answer to any potential pipeline problems is, "Do not build the pipelines." Numerous environmentalist organizations seem to subscribe to Rep. Skubitz's view.

Bitter opposition also comes from the railroads. For example, Burlington Northern Inc. (BN) warns that the ETSI line would cost BN's railroad $120 million/y in revenues, and could jeopardize a $450 million program to increase coal-hauling volume by a factor of five by 1980.

Fears are also expressed concerning line breaks and spillages, with tons of polluted coal-carrying water released into the environment. In the case of Black Mesa, there were dire warnings given, for instance, about contamination of the Dot Klish Canyon near Kayenta, or of other areas.

Montfort explained why these fears need not be entertained, especially with respect to Black Mesa. First of all, there are dump ponds along the line, at the pump stations, and these have mostly stood empty. Secondly, the line is carefully monitored, since Black Mesa is vitally interested not only in "keeping things clean," but also in not incurring a loss of coal. Montfort said that what little has been released at the booster stations went to the ponds, and, at Kayenta, to open-top sump tanks for water and coal recovery. Finally, he said that if the very worst somehow happened, at least in Arizona, soil absorptivity and rainfall lack would allow very little runoff; also, a system of highly sophisticated controls would instantly shut the line down. Montfort feels, however, that the Black Mesa Line's engineering and maintenance would prevent a major spillage.

Perhaps more answers will be forthcoming after a coal slurry pipeline study approved by the Congress' Office of Technology Assessment (OTA) in July. To be completed by next May this study will entail:

• technological, energy, and legal issues
• environmental effects, especially where water, land use, and community planning are concerned
  • evaluation of costs and returns
  • impacts on the railroads.

Black Mesa manager Montfort
"a technical and economic success"

For the future

The economic incentive to build coal slurry pipelines is strong. For the case of Black Mesa, for instance, Montfort said that there was no way the rail shipment could have been cheaper since the rail route would have been so much longer. Moreover, in ETSI's case, 80% of the costs could be fairly stable capital, and 20% very escalatable operational and maintenance (O&M), although Montfort does not necessarily agree with these exact numbers. He does say, however, that a high capital cost, plus building in of sophisticated, automated supervisory controls—such as the GETAC system built for Black Mesa by General Electric—would provide a strong hedge against inflationary cost escalation, especially that brought about by rising labor costs. Montfort recommended that even in figuring capital costs, one should account for dollar value depreciation, even if heavy labor cost increase impacts need not be considered too much.

Be that as it may, Montfort said that "there is definitely a future for coal slurry pipelines in the U.S." Perhaps opposition will decrease somewhat if it can be conclusively shown—indications are favorable—that brackish water from deep wells, provided large-volume use does not disturb other aquifers, can be a transport medium for the crushed coal. This aspect is important, because Montfort sees future slurry lines as carrying western coals, although he "knows of no reason" why they could not apply to eastern coals. He also sees good possibilities that if and as coal slurry lines are built, the expertise in the field that Black Mesa has developed under very trying circumstances will be called upon to help put such new lines in place, and anticipate and solve any economic, environmental, and engineering difficulties before they can become real problems. JJ

Reprinted from ENVIRON. SCI. TECHNOL., **10**, 1200 (December 1976)

Coal for energy self-sufficiency

With solutions to problems of acid mine drainage,
land reclamation, and clean use, those dark nuggets
are America's nearer-term trump card

Brandeis vice president Bryant
"can meet equipment needs"

In Louisville, the Kentucky Derby is an old tradition. Now, a new tradition is rapidly establishing itself there. This tradition may have results that could affect the entire U.S. economy for a long time to come, while Derby win, place, and show statistics become only a matter of historical equestrian interest, and a few filled, but many emptied pockets. This new tradition is the annual NCA/BCR Coal Conference and Expo, sponsored by the National Coal Association (Washington, D.C.) and Bituminous Coal Research, Inc. (Pittsburgh, Pa.). The theme of the third conclave, held in October, was "Coal—Energy for Independence"; indeed, every American has a vital stake in what the approximately 3000 attendees discussed and learned there, and in what was displayed in 196 exhibits.

At a press conference, Interior Secretary Thomas Kleppe said that he expected all 50 states to have mandated surface-mined land reclamation by 1980. Noting that the EPA and Council on Environmental Quality have already endorsed proposed federal legislation, he noted that tougher surface mining laws are not needed, and that states can impose stiffer rules, if they see fit. Secretary Kleppe predicted that "there will be a better environment than before, with use of coal." He also foresaw that coal slurry pipelines (*ES&T,* November 1976, p 1086) will be supplemental to, and not a replacement for rail transportation, at least until the problem of water necessary to move coal slurries is solved.

Land restoration

Despite what the Secretary of the Interior said, one might be hasty if one discounts all possibilities of increased federal surface mine reclamation rules. For example, a differing view was given to *ES&T* by Milo Bryant, executive vice president of Brandeis Machinery and Supply Corp. (Louisville), and Joe Hammell of the same organization. They predicted some additional federal legislation, but said that Brandeis, and perhaps certain other companies in the industry, would be able to meet the needs for the equipment that would do the heavy reclamation work.

As for reclamation work itself, there are many approaches to the removal of overburden, separation of various soil horizons, and replacement in the worked-out mine (*ES&T,* July 1976, p 642). Still, there are other nitty-gritty factors involved in reclamation, and some of these involve soil build-up and re-vegetation.

One attempt at fast soil build-up and revegetation might also contribute to residential municipal solid waste problem solutions. Real Earth, Ltd. (Beverly Hills, Calif.), a conference exhibitor, announced the availability of a thermophilic aerobic bacterial conversion system that can change a combination of residential refuse and sewage sludge into compost in as quickly as six days. Pathogens are destroyed by conversion temperatures of 168 °F. The material is ground to about ⅛ in. in diameter, and contains about 3% each of nitrogen, phosphorus, and potassium, as well as 23 trace elements that plants need. Real Earth suggests that their compost can be applied to rapid improvement of replaced mine overburden, and provide a measure of relief to communities facing severe solid waste disposal problems. A competing rapid composting process is the German BRIKOLLARE process (*ES&T,* October 1976, p 971).

It is often heard that surface mining will irreparably damage land. True, a surface mine is a wound of sorts, and there is no such thing as instant reclamation. True, also, there are some areas that would be extremely difficult to reclaim; but generally, present industrial practice calls for avoiding such areas, where at all possible. However, well-planned, sophisticated, modern techniques of mining and soil engineering should, over the years, normally lead to restored land and associated water of a quality difficult to distinguish from that of the presurface-mined land and water.

Mine drainage

Mine drainage, especially acid mine drainage (AMD) is one of the principal environmental problems associated with surface mines, and even some deep mines. Thus, a better understanding on how AMD comes about, and on specific approaches to the abatement of its pollution, is necessary if this monumental problem is to be tackled without losing too much time.

Surprisingly enough, not all pyritic materials associated with coal necessarily cause AMD, even though pyritic materials can often constitute about half the sulfur content in coal. Some pyrites are stable in water. However, a real culprit appears to be the reactive fine-grained framboidal pyrite that changes to the iron (II) ion and sulfate ion. During this process, the hydrogen ion (H^+ or acid) is also formed.

Iron (II) then oxidizes to iron (III)—partially catalyzed by three types of iron-oxidizing bacteria—and the iron (III) combines with OH^- ions from water to precipitate iron (III) hydroxide, also known as "yellow boy." Thus, H^+ ions accumulate. Moreover, what iron (III) dissolves in AMD water will react further with pyrites to form more iron (II), sulfate ion, and H^+, and the cycle continues.

Still, oxidation to iron (III) is normally desirable, because that form of iron can be insolubilized. Indeed, at The Pennsylvania State University (Penn State), microbial oxidation is a technique proposed to help remove iron pollution from AMD. Penn State is testing this technique at Hollywood, Pa., and estimates operating costs at 4¢/1000 gal treated.

(continued on page 1202)

149

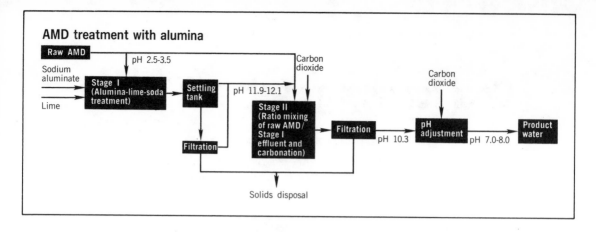

AMD treatment with alumina

Another treatment approach suggested by James Kennedy of EPA's Rivesville Field Site (W.Va.) involves the use of lime/limestone when iron (II) content is not very high. Advantages he listed involve a low-volume, high-density sludge; cheaper raw-materials; and, in some cases, reduced usage costs. However, such treatment must be tailored to a given AMD situation; and, in cases of high iron (II) content, this approach would be infeasible, Kennedy told a conference technical session.

John Nebgen of the Midwest Research Institute (Kansas City, Mo.) described an alumina-lime-soda process for AMD aimed at recovering potable water. His approach was originally conceived as a brackish water pretreatment prior to desalination by reverse osmosis. After treatment, no aluminum, iron, or manganese are found in the water. The treatment reagents are carbon dioxide (CO_2), sodium aluminate, and hydrated lime. He estimated costs of 79¢/1000 gal for a 5-mgd multi-stage treatment system.

Russell Klingensmith of Gannett Fleming Corddry and Carpenter, Inc., consulting engineers of Harrisburg, Pa., suggested at-source AMD control. Among the steps he listed were reconstruction of stream channels, restoration of surface mines, closing deep mine entries, chemical measures, and various disposal schemes. Klingensmith gave some examples of stream-channel reconstruction approaches in the Susquehanna River basin, which worked well. For instance, the reconstruction of Little Sandy Run near the Beach Creek surface mine, in Pennsylvania, completed in September 1974, cost just under $96 000. The restoration of the Beech Creek surface mine itself, completed at the same time, came to about $71 600. Klingensmith reminded the conference that in 1968, Pennsylvania allocated a total of $150 million for 10 years for AMD amelioration. He also listed other Pennsylvania streams and mine sites, such as Tioga River and Catawissa Creek, that were satisfactorily reconstructed and restored.

High-sulfur coal use

One principal strategy of the effort to achieve energy independence is to find better ways of using high-sulfur, high-ash coal in an environmentally acceptable manner. At the last conference/expo, EPA Administrator Russell Train said that flue gas desulfurization (FGD or scrubbing) would be the only technology that would make high-sulfur coal use possible in the near future (*ES&T*, January 1976, p 16).

Nevertheless, other approaches are being examined at a fast pace. Apart from use of naturally occurring low-sulfur coal, these approaches consist of coal conversion to clean fuels by gasification or liquefaction, physical and chemical sulfur and ash extraction before burning, and, perhaps, fluidized-bed combustion. Combinations of coal pre-cleaning and FGD are also under construction.

One coal-cleaning process described at the conference was the Meyers Process, which TRW Systems and Energy (Redondo Beach, Calif.) is developing for the EPA. This process is expected to remove 90–95% of pyritic sulfur from the coal matrix by chemical leaching with aqueous iron (III) sulfate at 90–130 °C, and would apply mainly to various Appalachian coals rich in pyrites, rather than organic sulfur. TRW estimates that such coal's sulfur content could be cut from 3%, for instance, to 0.6–0.9%, and that elemental sulfur, iron sulfate, and, in some cases, gypsum by-products can be made in large quantities.

PolyGulf Associates (New York, N.Y.), a joint venture of Gulf & Western Industries, Inc., and Polymer Research Corp. of America, exhibited "soluble coal." This form of coal is made from bituminous coal by a proprietary "chemical grafting" process, and can be dissolved in a small amount of plasticizing solvent, if "liquid coal" is desired. Production of soluble coal is at atmospheric pressure at about 140 °F, and the product contains less than 0.7% sulfur and less than 0.5% ash, according to PolyGulf. Experimental quantities are available on order.

It will be recalled that solvent-refined coal (SRC) has most of its pyritic sulfur and a generous portion of its organic sulfur removed (*ES&T*, June 1974, p 510). One company working on SRC process development with ERDA is The Pittsburg and Midway Coal Mining Co. (P&M, Denver, Colo.); its 50-tpd pilot plant is near Tacoma, Wash. P&M executive vice president Richard Holsten told *ES&T* that he is "optimistic on SRC," and that, in his view, "it makes more sense than does gasification." The plant's operating expenses are about $1 million/mo, according to an ERDA spokesman.

Still, gasification and liquefaction are going ahead full-blast. For gasification, the SYNTHANE pilot plant (*ES&T,* January 1976, p 17) is one salient example. Another is the ERDA/Continental Oil CO_2 acceptor plant—40 tpd—at Rapid City, S.D., in operation since 1972, and now scheduled to operate through next June. Also, the Institute of Gas Technology is heavily involved with Hi-Gas, Bi-Gas, and other efforts. Moreover, Frost & Sullivan, Inc. (New York, N.Y.) predicted substitute natural gas production at 1 trillion ft^3/y by 1990. As for liquefaction, a pilot plant for 600 tpd of coal, using the proprietary H-Coal process, will be built at Catlettsburg, Ky., pursuant to a $178 million contract to Ashland Oil, Inc., and others, of which ERDA will furnish $142 million.

Not too soon

So the effort to use cleaner coal, and diversify the nature of its preparation and use goes on at full speed, and will accelerate. This effort comes not a moment too soon; it will not come cheaply. On the other hand, the prospect of abject dependence on certain foreign sources for fuel, with all of its geopolitical ramifications, a possible severe displacement of the U.S. industrial economy, or some horrendous combination of these would be a far worse price to pay. By comparison, the cost of a stepped-up, concerted effort to achieve a greater degree of energy self-sufficiency through clean mining, preparation, and use of coal could be said to be reasonable, indeed.   JJ

Hazardous chemicals from coal conversion processes?

Reprinted from ENVIRON. SCI. TECHNOL., **10**, 1104 (November 1976)

Maybe. Because unexpected chemicals harmful to health and the environment may be produced, these methods and their products should be carefully studied now

David W. Koppenaal and Stanley E. Manahan

*Department of Chemistry
University of Missouri
Columbia, Mo. 65201*

The use of coal for energy is certain to increase in the U.S. during the next several decades. Utilization of domestically abundant coal for a majority of U.S. energy needs will require mining, transportation and processing operations on a scale not previously approached for any mineral resource. Environmental, occupational, sociological and economic changes resulting from the projected uses of coal will be substantial, and proper account must be taken of them in developing this resource. This is especially true if conversion of coal to hydrocarbons develops on a large scale.

The largest synthetic fuel industry to date was developed in Germany during World War II. In 1944 peak production reached 100 000 bbl/day—the equivalent of the crude oil consumption of one moderately large petroleum refinery. It is dwarfed by current U.S. consumption of about 20 million barrels a day.

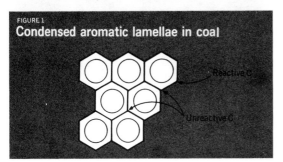

FIGURE 1
Condensed aromatic lamellae in coal

Reactive C

Unreactive C

On an enormous scale

The commonly accepted figure for the most efficient size of a synthetic natural gas (SNG) plant is 250 million ft³/day. Approximately 176 sites have been identified in the U.S. that have sufficient coal and water to support a plant of this size for at least 20 yr. Typically one of these plants would require about 15 000 tpd of coal. Approximately the same quantity of steam would be required. Oxygen would be consumed at the rate of almost 3000 tpd. Solid wastes from the gasifier and from coal pretreatment would amount to approximately 2500 tpd.

If a typical high-sulfur Eastern coal is used for feedstock, approximately 600 tpd of sulfur would be produced. Assuming this coal has a heat value of 11 000 Btu/lb and a waste heat loss of approximately 35% during processing, 1.1×10^{11} Btu/day would be released to the environment. Gaseous waste effluents would be produced in appreciable quantities. Some processes would require as much as 3 million gal/day of water to quench and scrub the raw gas product. This water would pick up sub-

stantially more than 1000 ppm each of phenols, ammonia, and COD and appreciable levels of suspended solids, thiocyanate, and cyanide. It must be treated before it is used as cooling tower makeup water or it is released to receiving waters.

Production of the order of 5×10^9 scf/day of SNG is reasonably feasible by 1990. This would require 20 standard-size SNG plants. The amount of coal required would be 110 million tons/year.

A total of 11 pilot plants for various coal-conversion processes are operating or have been demonstrated in the U.S. The latest of these is a 70 tpd Synthane plant dedicated by the U.S. Bureau of Mines in November 1975. During that same month a $237 million demonstration plant to be constructed by Coalcon was authorized for Athens, Ill. If constructed this plant, which will process 2600 tpd of coal and produce 22 million scf/day of SNG and 3900 bbl/day of liquid hydrocarbons, will be the first commercial operation for synfuels production in the U.S.

The production of "exotic chemicals"—species which in trace amounts have profound environmental and health effects—has received only scant attention in studies of coal-conversion processes. (Much of what is known has been inferred from studies of occupational health and environmental impact in the coking industry). The chemical nature of coal and the mineral elements contained in it are conducive to the formation of a variety of toxic substances under conditions that are attained during typical coal-conversion processes. If coal conversion is developed on the mammoth scale required to satisfy an appreciable fraction of domestic demand for hydrocarbon fuels, the environmental and occupational health problems will affect vast areas of the country and substantial numbers of people.

Chemical nature of coal

The organic portion of coal is largely composed of polycyclic aromatic structures and, in addition to carbon and hydrogen, contains significant quantities of organic oxygen, nitrogen and sulfur, residual biological compounds modified through the coalification process, and organically bound metals. Functional groupings include methoxy, hydroxyl, carbonyl, and carboxyl. Nitrogen, oxygen, and sulfur may exist within relatively nonreactive cyclic configurations. Aliphatic carbon and hydrogen are also present to some extent. Coal mineral matter contains appreciable quantities of practically all of the elements in the periodic table.

Physically, coal has a widely varying porous structure. The physical behavior of heated coal is especially important to its reactivity and handling in a conversion process. Eastern U.S. coal resources swell and cake when heated. These coals cannot be used with the Lurgi gasification process.

It is generally conceded that coal is made up largely of condensed aromatic hydrocarbon lamellae (Figure 1) held together physically or linked through chemical bridging groups. The molecular weights of coal species are thought to fall mostly within the range of 2000–12 000. These molecular weights are relatively low compared to those of tars and residual fuel oils (whose molecular weights may range into the millions) so that coal is a more reactive feedstock for conversion processes than petroleum residues.

Functional groups containing oxygen, nitrogen and sulfur may have a pronounced influence on coal reactivity and products. They lead to the formation of contaminants including phenols, aromatic nitrogen compounds and catalyst-poisoning sulfur compounds in coal liquids.

The functional groups of oxygen are understood to a relatively greater extent than those of nitrogen or sulfur. The oxygen in coal is primarily incorporated into hydroxyl (phenoxy) and carbonyl groups. A small amount is incorporated as heterocyclic oxygen and, particularly in low rank coals, carboxyl groups. As noted above, oxygen in coal may lead to formation of various phenolic compounds in coal liquids. Oxygen functional groups are also involved in the binding of metal ions.

Nitrogenous functional groups in coal are not so well characterized. Nitrogen may exist in concentrations up to 5%, although it is generally present from 1–2% by weight. Nitrogen in coal is probably the major source of NO_x compounds, which are produced when coal is burned. The same is true of nitrogen in coal liquids.

Sulfur presents several technological and environmental problems. Therefore, all major conversion processes provide for the elimination of sulfur from coal. The sulfur in coal may be either organic or inorganic. The inorganic portion exists primarily as pyritic iron sulfide, FeS_2, which is fairly easy to remove. A small portion also exists as sulfate ion. Most of the organic sulfur is incorporated into heterocyclic configurations. As such, it is somewhat harder to remove, and ultimately contributes to air pollution. Sulfur concentrations in coal may range between 0–10%. There has been surprisingly little work on the analysis of the various functional groups of sulfur in coal. In view of the problems that sulfur presents, this should be an area of research priority.

Coal-conversion processes

The primary coal-conversion processes that are planned for commercial development in the U.S. are:
- coal gasification
- coal liquefaction
- solvent-refined coal (SRC).

Figure 2 illustrates basic flow diagrams for each of these conversion processes. Of these, the coal gasification and SRC processes are receiving the most attention and several commercial plants are being planned.

In the gasification process, the coal is ground, dried, and sent to a preheater where it is mixed with O_2 and steam. It is then gasified at temperatures of about 1000 °C and pressures of 500–1000 psi. The gas product is "scrubbed" to remove tar, and then sent to the "shift" reactor where the H_2/CO ratio is optimized. Acidic gases are removed, and the product is methanated over a catalyst. The gas is then dehydrated and compressed to form pipeline gas.

In the SRC process (ES&T, June 1974, p 510), the pulverized coal is mixed with a coal-derived aromatic solvent. This coal slurry is then reacted with hydrogen in a dissolver at 300–400 °C and 1000–2000 psig. It is believed that coal minerals catalyze the hydrogenation. While in the dissolver, most of the coal (95%) is dissolved, almost half of the organic sulfur is converted to H_2S, and hydrogen, amounting to 1–2% of the weight of the coal, is consumed. Liquid, gaseous and solid phases are then separated. Effective removal of H_2S, CO_2 and other hydrocarbon gases is necessary. These gases are consequently treated and/or recycled. The mineral ash is removed by centrifugation or filtration. Removal of up to 99.9% of the ash is possible. The solvent is

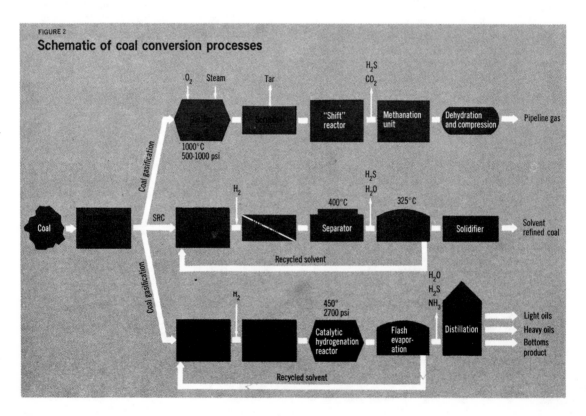

FIGURE 2

Schematic of coal conversion processes

Mean analytical values for chemical elements in coal

Constituent	Mean, ppm	Std. dev.	Value in ash-free, solvent-refined Kentucky No. 9 coal [a]
As	14.02	17.70	1
B	102.21	54.65	
Be	1.61	0.82	
Br	15.42	5.92	6.2
Cd	2.52	7.60	
Co	9.57	7.26	0.8
Cr	13.75	7.26	
Cu	15.16	8.12	
F	60.94	20.99	
Ga	3.12	1.06	
Ge	6.59	6.71	
Hg	0.20	0.20	
Mn	49.40	40.15	1.7
Mo	7.54	5.96	
Ni	21.07	12.35	
P	71.10	72.81	
Pb	34.78	43.69	
Sb	1.26	1.32	
Se	2.08	1.10	
Sn	4.79	6.15	
V	32.71	12.03	2.8
Zn	272.29	694.23	
Zr	72.46	57.78	
Al	12 900	4500	65
Ca	7700	5500	
Cl	1400	1400	167
Fe	19 200	7900	
K	1600	600	
Mg	500	400	196
Na	500	400	2.6
Si	24 900	8000	
Ti	700	200	96
moisture	9.05%	5.05%	0.00%
ash	11.44%	2.89%	0.051%

Source: Illinois State Geological Survey [a] Data obtained by Manahan's research group, primarily by neutron activation analysis.

removed by flash evaporation and recycled. The remaining liquid product is then solidified to form a brittle, hard, clean-burning solid. Or, if desired, the SRC liquid may be further hydrogenated over a catalyst to produce a liquid hydrocarbon fuel of high-Btu value.

The coal-liquefaction process closely resembles that of the SRC process, except that the former has a catalytic hydrogenation unit. The coal is hydrogenated over a catalyst at high temperatures and pressures. Hydrogen is supplied to the unit continuously. Solids and gases are then separated, and the resulting liquid undergoes a distillation and flash-evaporation process like that of the SRC process. The primary product of the liquefaction process is a synthetic crude oil that can be refined to produce light oils and gasoline.

These conversion processes take a dirty, bulky, inconvenient solid fuel and transform it into a clean, easily transportable solid, liquid or gaseous fuel. The heat content per unit weight of the product normally is higher than that of the original fuel. Overall, the products are more acceptable environmentally, especially with respect to air pollutants. However, there is some potential for the formation of hazardous organometallic and/or carcinogenic chemicals.

Formation of organometallics

The chemical nature of coal combined with the conditions under which it is liquefied or gasified are conducive to the formation of a variety of organometallic compounds consisting of metals bonded to organic groups or ligands. Coal liquefaction in particular may produce organometallic compounds.

Elemental analyses of a variety of coals (101) has revealed the presence of well over half of the elements in the periodic table (Table 1). Many metal or metalloid elements that might be expected to form some types of organometallic compounds or stable organic complexes occur at relatively high levels in these coals. By way of comparison, some values are given for elements analyzed in an ash-free, solvent-refined coal product prepared from Kentucky No. 9 coal. The solvent refining process yields impressive reductions in trace element content; however, a fraction of the metals remaining may be organically bound. Some organometallic compounds may result from the chemical interaction of trace metals in the ash with the organic portion of coal.

Surprisingly, very little work has been done on elucidating the exact nature and effects of organometallic compounds that may be produced in coal-conversion processes. Some idea of the various types of organometallic compounds that may be formed can be inferred from the structure of coal, the biological origin of coal, and the bonding tendencies of known organometallic species. Some of the major types of organometallic compounds that may be formed include:

• Metal-porphyrin compounds. The biological origin of coal makes it probable that porphyrin-type compounds are common components. Porphyrins are capable of binding metal atoms, and are known to be important carriers of vanadium and nickel in crude oil. It is possible that various metal porphyrins indigenous to coal may survive the processing intact or in an altered form.

• Metal-carbonyls. The relatively high-partial pressures of carbon monoxide during various stages of the conversion process may lead to the formation of metal carbonyls. Carbon monoxide may react with most of the transition series metals under certain conditions of temperature and pressure to form carbonyls. Nickel, iron and cobalt carbonyls are the most significant carbonyls in the petroleum industry and may be expected to arise from coal-conversion processes. Although the extreme temperatures will normally preclude the persistence of large amounts of these compounds, trace amounts may escape during particular phases of the process. All of the metal carbonyls are reactive and several are acutely toxic.

• Metallocenes. Also known as π complexes, these compounds consist of aromatic rings bounded to metal atoms. Usually, the metal atom is "sandwiched" between two aromatic systems—π bonds linking the metal to the carbon atoms of the rings. Iron, nickel, chromium, vanadium, tantalum, molybdenum and tungsten metallocenes have been observed. Although the iron cyclopentadienyl compound (ferrocene) is quite stable, other metallocenes are relatively less stable. Introduction of various functional groups on the ring portion of the compound may have significant effects on the stability of these compounds, however. The possibility of the formation of quite stable metallocenes must not be overlooked.

• Arene carbonyls. Arene carbonyls are organometallic species in which the metal atom is bonded to both an aromatic ring system and to carbon monoxide molecules. Certain arene carbonyls may be more stable than their metallocene counterparts. Many transition metal arene carbonyls have been synthesized. Because of the high pressures of carbon monoxide in many conversion processes, the aromatic nature of coal, the presence of metals that form arene carbonyls, and the stability of these compounds, they are one of the most likely types of organometallics to be found in coal liquids.

• Metal alkyls. Many transition metal alkyls are known. In addition, lead, tin, aluminum and silicon may form metal-alkyl combinations. Because of the limited stability of these compounds, it is not anticipated that they will be common constituents of the final conversion product. However, as they may be

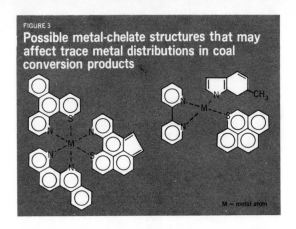

FIGURE 3

Possible metal-chelate structures that may affect trace metal distributions in coal conversion products

M = metal atom

involved in the transport and mobilization of heavy metals in coal liquids, their presence must not be completely discounted.

• Organo hydrides. Several metal and metalloid elements are known to form relatively stable organo hydrides of the general formula R_nMH_{4-n}. (R represents an organic group and M a metal or metalloid.) Organo hydrides of Pb, Sn, Ge, and Si may form in the reducing atmosphere of the conversion process. Their relative stability makes their persistence likely.

• Metal chelates. Coal contains many chelating structures (phenolic OH, carboxylic acid groups, and amino groups) that can effectively bind metal atoms. These structures may have various effects on the mobilization and release of many metals. Figure 3 shows hypothetical chelates that may have an influence on metal distributions in coal-conversion products.

In addition to the above mentioned possibilities, other unpredictable compounds involving metal interactions with the organic portion of coal should be considered. Although ash removal is very efficient in most conversion processes, small, but important quantities of many metals may be left organically bound in the conversion product. It is important that the organometallic portion of coal conversion products be well characterized. These compounds may be toxic—carcinogenic in some cases—and detrimental to the use of coal liquids as a fuel or feedstock for further refining processes. The release of such compounds to the atmosphere or water systems will pose significant pollution problems. In addition, such compounds may foul catalysts and interfere with the conversion process. On the other hand, such compounds may be found to have a catalytic effect, or prove to be a valuable source of some metals.

Formation of PAH's

Considering the aromatic nature of coal, it is not surprising that the polycyclic aromatic hydrocarbon (PAH) content of coal-conversion products is high. This is of concern because of the carcinogenic activity of many PAH's. It has been known since the early eighteenth century that a carcinogenic hazard is present when working with coal tar or its by-products. Skin cancer in particular was a common occupational hazard to chimney sweeps, dyestuff workers, and coke oven workers.

The first chemical agent in coal tar separated and tested for its carcinogenic activity was 3,4-benzopyrene. It was found to be extremely carcinogenic to experimental animals and is now known to be carcinogenic to humans. It is often analyzed as an indicator of possible carcinogenic hazards. Many other PAH's have been found to be carcinogenic.

The formation of PAH's in coal-conversion processes is well known. There have been studies on the carcinogenic hazard associated with coal liquefaction processes. The aromatic content of coal liquefaction products is generally in the range of 50–70%, depending on the boiling point range considered. The aromatic content generally increases with boiling point.

Certain aromatic amines are also known to be carcinogenic. Polycyclic aromatic hydrocarbons and aromatic amines are thermally stable, and are present in coal-conversion products.

In addition, PAH's have been found in atmospheric particulate matter around coal liquefaction plants. Therefore, the release of hazardous compounds during process failure and shutdown may be of considerable importance. These compounds may also be found in scrub waters, which are eventually released to the environment.

Contact with coal-conversion products may be hazardous. Worker contact with these products may be frequent, particularly when considering the problems that occur during the initial start-up of a new industry.

The production and release of carcinogenic compounds, in addition to the hazard of personal contact with the conversion product, makes the consideration of these chemicals as a carcinogenic risk very important. A closer look at the occupational hazards of the industry is certainly warranted.

Associated hazards

It is evident that some health hazards are to be expected with the advent of a large-scale coal-conversion industry. In particular, the hazards associated with carcinogenic polycyclic aromatic hydrocarbons, aromatic amines, toxic metals and organometallic compounds need to be investigated.

It is important that these processes and their products be methodically examined for these hazards, especially in view of the possibility of a massive coal-conversion industry in the U.S. Although these processes look promising with respect to the removal of "traditional pollutants," it is likely that they will produce chemicals that have not even been identified or characterized with respect to their health and environmental effects.

The development of any new technology requires investigation of toxicological and environmental impacts of that technology. This is especially true of coal-conversion processes because of the diverse nature of the raw material. In order to prevent any delay in the large-scale development of this industry, a major effort should be made now, before developments proceed beyond the pilot-plant stage, to eliminate potential hazards that may be imposed upon the environment and the health of personnel.

Additional reading

"Carcinogenic Potential of Coal and Coal Conversion Products," A Batelle Energy Program Report, R. I. Freudenthal, G. A. Lutz, and R. I. Mitchell, 1975.

"Coal Structure and Reactivity," A Batelle Energy Program Report, G. L. Tingley, and J. R. Morrey, 1973.

"Industrial Carcinogens," R. E. Eckhardt, Grune & Stratton, New York, N.Y., 1959.

David W. Koppenaal is working on his Ph.D. degree in analytical chemistry at the University of Missouri. He is interested in new energy systems and their environmental effects, analytical instrumentation, and the occurrence and distribution of chelating agents in natural waters.

Stanley E. Manahan is currently associate professor of chemistry at the University of Missouri where he directs a research program in environmental and analytical chemistry. Dr. Manahan's recent studies have focused on coal humic acids as they relate to pollution control and the fate of trace elements in solvent-refined coal. He is the author of some 30 technical publications and one textbook (Environmental Chemistry, 2nd ed., Willard Grant Press, 1975).

Coordinated by LRE

ERDA's fossil energy activities

Synfuels from coal and enhanced methods to recover oil, natural gas and oil shale may be competitive and socially acceptable options to alternative energy sources in the not-too-distant future

Philip C. White

Energy Research and Development Administration Washington, D.C. 20545

Reprinted from ENVIRON. SCI. TECHNOL., **10,** 746 (August 1976)

The Energy Research and Development Administration (ERDA), established January 19, 1975, inherited the coal, petroleum, natural gas, and oil shale programs that principally originated in the Department of the Interior's Office of Coal Research and the Bureau of Mines. The newly created agency greatly expanded its horizons for fossil energy technology in 1975–76. During this year it moved toward the implementation of large demonstration projects for converting coal to clean fuels.

Significant progress also was achieved in developing the Fossil Energy Research Program. Here ERDA was able to enlist university participation and to secure industrial contributions to cost-sharing projects for the enhanced recovery of petroleum and natural gas.

The new Fossil Energy organization, headed by an assistant administrator and a deputy assistant administrator, consists of five program divisions:

- coal conversion and utilization
- fossil energy research
- fossil demonstration plants
- MHD (magnetohydrodynamics)
- oil, gas and shale technology.

Staff offices include Program Planning and Analysis, an Administrative Office, and the Senior Staff for special assistance and counsel.

The conversion and utilization of coal, the Nation's most abundant fossil fuel, was given high priority in ERDA's National Energy Plan. An updated version of the Plan, issued in April 1976, placed more emphasis on conservation. Light-water-nuclear reactors and the enhanced recovery of oil and gas were also ranked as fuel sources that could be tapped to bridge the energy gap between now and the year 2000. After these energy sources, ERDA foresees the successful development of more inexhaustible supplies of energy such as the breeder reactor, solar, electric, and fusion.

Fossil energy activities

In essence, the Fossil Energy program seeks to develop and demonstrate, in conjunction with industry, the technology necessary for establishing a synthetic fuels-from-coal industry, as well as for improving methods to recover petroleum, natural gas, and oil shale. These objectives have the potential for being economically competitive with alternate energy sources and environmentally and socially acceptable in the near-term (by 1985) and mid-term (by 2000).

To achieve its objectives, the Fossil Energy program has grown substantially, as reflected by the dramatic increases in budgetary funds. In fiscal year 1975, expenditures totaled $204 million, nearly triple the $76 million expended in fiscal 1974. In fiscal 1976, the budget increased to just under $350 million and, for fiscal 1977, funding requests amount to $442 million (Table 1). These figures do not reflect cost-sharing funds put up by contractors.

Industry cost sharing is concentrated in the more advanced phases of coal conversion and in the enhanced recovery of oil and gas, although there is some agency cofunding in earlier developmental stages. A typical development sequence from

Highlights of the direction of ERDA's activities

- A request to industry and others for technical proposals to design, construct and operate a demonstration plant for converting coal to pipeline-quality (high Btu) gas. ERDA received five proposals that were being evaluated. Such a plant would supply data for scale-up to commercial-sized facilities producing gas from coal and/or lignite.
- A request for proposals (RFP) to design, construct and operate demonstration projects for converting coal into clean fuel (low Btu) gas for power generation and on large- and small-scale industrial users. Fourteen firms responded to the RFP.
- Award of contracts to develop engineering information on magnetohydrodynamics (MHD) generators and systems for testing in a component development and integration facility to be built near Butte, Mont.
- RFP's to perform studies of high-temperature gas-turbine-combined cycle development program operating on coal or coal-derived fuels. Four large manufacturers were selected to perform the first phase of a six-year program at the initial cost of approximately $9 million.
- Selection of the Curtiss-Wright Co. (Wood-Ridge, N.J.) to negotiate a contract to build and operate a pilot electric generating plant using advanced technology to burn high-sulfur coal. The plant would use a coal-fired, pressurized fluidized-bed combustion technique to power steam and gas turbines in a combined cycle operation.
- Initiation of site preparation work for the H-Coal pilot plant to be built at Catlettsburg, Ky. The plant, the largest of its kind, would process up to 600 tons of coal daily in the production of clean liquids.
- Issuance of a program opportunity notice (PON) for proposals to develop and demonstrate a fluidized-bed concept for clean burning of high-sulfur coal and coal waste in industrial boilers and heaters. Thirteen proposals were received from industry and cost-sharing contracts are expected to be awarded by the end of 1976.
- RFP's to design, develop, test and evaluate valves to be compatible with operating conditions encountered in various coal gasification plants.
- Issuance of two RFP's to demonstrate methods for recovering oil from tar sands.
- Issuance of two RFP's for field demonstration projects with use of a micellar-polymer chemical flooding process to recover crude oil. One of the RFP's asks for small pilot projects, the other for one large-scale project, using the recovery technique.
- RFP's for projects to stimulate natural gas production by hydraulic and chemical explosive fracturing methods that could release the vast amounts of fuel now locked in tight geological formations.
- RFP's for field demonstration projects with use of carbon dioxide injection methods to recover crude oil from shallow underground reservoirs.
- RFP's for field demonstration projects with use of thermal processes to recover crude oil.
- The award of millions of dollars in cost-sharing contracts with industry to test improved methods for the enhanced (secondary and tertiary) recovery of oil and gas. By 1985, ERDA expects this technology to add 2 billion barrels of oil and 10 trillion cubic feet of natural gas to U.S. reserves. This would mean an increase in production of 500 000 barrels of oil per day and 3 billion cubic feet of natural gas per day.
- Distribution of a PON seeking proposals for the development of technology to recover shale oil in place.
- Launching of a new program to stimulate the production of natural gas from Devonian shale in the eastern U.S. Trillions of cubic feet of gas, locked in tight formations, cannot be developed economically with current technology.

156

Table 1. ERDA's 1977 request to Congress for fossil energy programs

Coal	$ million
Coal liquefaction	73.9
High-Btu gasification	45.0
Low-Btu gasification	33.0
Advanced power systems	22.5
Advanced research and supporting technology	37.0
Direct combustion	52.4
Demonstration plants	53.0
Magnetohydrodynamics	37.4
Total coal program	354.2
Petroleum & natural gas	36.9
In-situ (oil from shale and coal gasification)	30.6
Total operating expenses	$422.0

concept to pilot and demonstration projects requires from 15–20 years, but overlapping sequences can reduce this time to a degree. This sequence of scaling up engineering data to demonstration projects will provide information for cost estimates and design of commercial plants.

To date, ERDA has defined several demonstration projects for translating advanced concepts into commercial use, with one—a clean boiler fuel plant—already underway. The others include a pipeline quality gas plant aimed at residential and commercial heating, and a fuel gas plant for electric power utilities or industrial uses.

In conjunction with this essential scaling up of second-generation technologies, President Ford has proposed a Synthetic Fuels Commercialization Program directed at constructing a limited number of plants in the 1980's. By using present day processes to produce synthetic fuels for the marketplace, these plants could yield valuable economic, environmental, regulatory, and institutional data needed for commercializing both first- and second-generation technologies.

The transfer of pilot-plant technology to the demonstration stage gained impetus in January 1975, with the award of a $237 million contract to the Coalcon Co. (New York, N.Y.). The firm proposes to design, build and operate a coal-to-clean-boiler-fuels demonstration plant on a site near New Athens, Ill. For other highlights, all laying the foundation for technical advances on a larger scale, see box material.

The Fossil Energy research program expanded steadily in 1975–76, particularly in the area of university support and materials research. At the academic level, the annual rate of expenditures increased from $2.9 million in fiscal 1974 to $8.6 million in fiscal 1975 and $14 million in fiscal 1976. The number of projects increased from 48 in fiscal 1975 to about 80 in 1976.

Second-generation technology

The coal conversion and utilization effort is directed toward demonstrating second-generation technology on a near-commercial scale in the early 1980's. A variety of processes is being developed to convert eastern and western coal to liquids and gases. Although some processes may appear to be similar, technical differences in key aspects of the processes make them suited for a particular rank of coal, type of product, or plant site.

Coal conversion processes are being developed to convert this fossil fuel into products that substitute for those derived from oil and natural gas. These substitutes will include: crude oil, fuel oil, and distillates; chemical feedstock; pipeline quality (high Btu) and fuel (low and intermediate Btu) gas; and other by-products such as char that may be useful in energy production. The liquefaction, high Btu, and low Btu sub-programs are designed to develop these products and their use in the market.

Coal utilization programs are developing processes that would permit increased use of coal by direct combustion in utility, industrial/institutional boilers, and process heaters, as well as primary fuel for electric power generation. These objectives may be attained through improved direct combustion systems, advanced power systems with gas turbines, and MHD electric power.

Liquefaction. Products derived from coal liquefaction compete in two distinct markets. One market uses low-ash, low-sulfur boiler fuels suitable for clean electric power generation and industrial steam generation; the other uses high-grade fuels such as gasoline, methanol, diesel oil, heating oil, and chemical feedstocks.

Several problem areas have been identified; for example, solid/liquid separation and utilization of remaining solids for the generation of hydrogen. As part of the solution, the revamped Cresap, W. Va., Test Center will permit pilot testing of advanced solid/liquid separation concepts, and the evaluation of valves, pumps, and other auxiliary equipment. The information collected will be fed into the pilot-plant program (such as the Solvent Refined Coal plant at Tacoma, Wash.) and into current and future demonstration projects.

High-Btu gasification. Improved gasification processes should produce a substitute natural gas (SNG) with a heating value of approximately 950 Btu's/ft^3 with combustion characteristics similar to natural gas but essentially free of sulfur and other pollutants.

The processes are distinctive in that each represents a different approach to the technique of high-Btu gas production. The first-generation Lurgi gasification process that incorporates a methanation step is being used as a guide to determine if any of the processes under development are, in fact, improvements over existing technology. Improvements are measured primarily in terms of comparative capital and operating costs, and the ability to operate successfully with American caking coals.

Four gasification processes are relatively advanced in their development. The HYGAS and CO_2-Acceptor pilot plants are currently operating. The Synthane pilot plant is being readied for operation, and construction is completed on the Bi-Gas pilot plant. A fifth plant, steam-iron, produces hydrogen from char for use in other coal gasification or liquefaction processes. A sixth project, a process development unit (PDU) that uses the self-agglomerating ash process, is nearing completion. The program also includes the pilot-plant study of alternate techniques for methanating intermediate-Btu synthetic gas, including a new concept called liquid-phase-methanation.

Low-Btu gasification. The three most likely markets for low-Btu gas are electric power generation, industrial heating, and chemical feedstocks.

The major milestones for low-Btu projects are the operation of an atmospheric, entrained-bed PDU; operation of a two-stage pressurized, fluidized-bed gasifier PDU with sulfur removal by

Coal gasification. *The Bi-Gas pilot plant recently completed at Homer City, Pa.*

limestone; the design of a combined cycle pilot plant with a pressurized, entrained bed; and the operation of a molten salt PDU.

In addition to the conventional above-ground gasification projects, underground or in-situ gasification processes are being developed by ERDA. The in-situ method potentially offers economic and environmental benefits over conventional mining and gasification. In such processes, coal deposits are chemically reacted in the ground and low- to intermediate-Btu gas is extracted in a manner similar to that used with natural gas.

The major effort to date consists of field tests at Hanna, Wyo., a technology based on vertical wells for generating a low-Btu gas to be used for power generation. Another approach to in-situ gasification involves evaluating directional wells drilled in thin eastern coal seams. A third project, using the packed-bed process, is planned to gasify thick western coal seams to produce an intermediate-Btu gas; this gas may be upgraded to a synthetic high-Btu gas by surface processing. Pilot projects for these last two approaches are being planned.

Advanced power systems. These systems are directed toward the use of coal-derived fuels in an environmentally acceptable manner at lower cost than current electric power generating systems. Specifically, developmental efforts are directed toward improved gas turbines for combined cycle systems and topping cycles in order to use high combustion temperatures more effectively. The two major categories of advanced power systems—open and closed cycle—are undergoing comparative studies.

Energy Conversion Alternative Studies (ECAS) are being undertaken to determine which systems would be economically attractive, which could be ready for commercialization before 1985, which appear decidedly uneconomic, and which require further study to determine their economic potential.

Specific closed-cycle projects are in the process of being started. Open-cycle systems will be developed to a technology readiness stage before proceeding to full-scale prototype development. Technical problems center around the effects of high temperatures and pressures. Materials, corrosion, erosion, and gas cleanup will be subjected to extensive study in the experimental facilities to be built.

Direct combustion. The objective is to develop and demonstrate on a commercial scale the direct combustion of high-sulfur coals without exceeding pollution standards. Fluidized-bed combustors containing sulfur oxide sorbents will be used in the burning of coal.

The program focuses on both atmospheric and pressurized, fluidized-bed boiler technology for powerplant and industrial/institutional heat uses. Initial efforts are concentrated on atmospheric systems having broad applications in both electric power generation and industrial heat and steam. Also, pressurized combustion is being examined to evaluate its full potential.

Direct combustion technology has near-term (1985) potential as an economic and efficient coal-fired alternative to existing boiler systems that use scrubbers for emissions control. Equipment based on this technology is expected to find a market both in new equipment installations and in the replacement of process heaters and boilers fired by natural gas. The latter applications are desirable because of continuing shortages in natural gas.

Fossil energy R&D

As the central research and systems studies area for all elements of fossil energy, the Fossil Energy Research Division conducts projects in the ERDA Energy Research Centers, National Laboratories, other government agencies, private industry, and universities. See box material for projects undertaken by this Division.

The general objective is to demonstrate on a near-commercial scale the more promising processes developed and evaluated in industry and government R&D programs. The fossil demonstration plant program provides the final step toward coal and shale utilization as future clean energy sources; it is partially funded through engineering, construction, and operation of processes on the verge of commercialization.

The first phase of each demonstration plant project (conceptual commercial plant process and mechanical design) will be funded entirely by ERDA. The second phase (detailed engineering, procurement and construction) and the third phase (operation) will be funded on a cost sharing 50/50 basis by the government and contractor.

As noted earlier, several demonstration plant projects have been defined. This program includes coal conversion to synthetic liquid fuels that range from heavy fuel oil to distillates; conversion to pipeline quality and fuel gas; and direct combustion of coal and integration of power production.

Magnetohydrodynamics (MHD)

An MHD electrical generation system utilizes a topping cycle combined with a bottoming conventional steam plant to achieve greater plant efficiencies. Essentially, MHD systems, both open and closed cycles, convert coal to a conducting hot gas that moves at high velocity through a magnetic field to produce power.

Basic feasibility of the open-cycle system has been established; it has been demonstrated that an MHD generator can operate by utilizing the products of direct coal combustion. However, engineering data for design and construction of pilot-scale facilities are not yet available.

The open-cycle program is structured to progress through three overlapping phases, the first focusing on specific development requirements leading to the design and construction of an engineering test facility (ETF). The second phase carries the ETF into an advanced engineering stage; the third phase covers possible design, construction, and operation of a commercial-scale plant.

Exploratory work also has been performed on closed-cycle systems; work has concentrated on heat exchangers and generator problems. Closed-cycle systems development is not as advanced as open-cycle systems. Current work is addressed to basic physical issues.

MHD research is aimed at demonstration of performance and prolonged operation of channels; development of components and integration of key subsystems; and integrated operation of MHD systems with electric utilities.

Petroleum, natural gas, oil shale

Petroleum and natural gas will continue to be the Nation's main fossil energy resources in the years ahead. ERDA's efforts are directed toward increasing the production of oil and gas by advanced production and recovery techniques and toward improving the efficiency of petroleum use and re-use.

Oil extraction efforts will demonstrate existing and improved secondary and tertiary recovery techniques rather than new refinery technology. Industry already has a broad technological base in refining.

The natural gas program is designed to stimulate the commercial production of natural gas from tight geologic formations containing vast quantities of fuel. These domestic reservoirs have remained untapped because they cannot be developed economically with current technology. Now being tested are advanced methods that include massive hydraulic fracturing, combinations of hydraulic and chemical explosive fracturing, and fracturing wells deviating from normal to intersect with natural fractures.

Fluid injection and fracturing methods for gas and oil extraction will be used in field demonstration projects. Solvent recovery methods for heavy oil and in-situ combustion methods for tar sands will be developed and demonstrated.

The goals of the oil shale program are: reducing the water requirements of the oil shale industry through in-situ processing; increasing the recoverable reserve base through improved production technology; and ensuring that environmental safeguards are built into the in-situ oil shale process. Advancing in-situ production of shale oil to commercial feasibility is targeted for the early 1980's.

The oil shale program emphasizes in-situ retorting rather than surfacing retorting, which is considered a proven technology. In addition, laboratory and bench-scale studies on composition and conversion of clean fuels from oil shale have been initiated to provide a technology base for improvements and new process development.

Summation

As the foregoing summary suggests, careful planning and long-term commitment by both the government and industry are essential to the attainment of energy self-sufficiency. Continued cooperation and accelerated scientific achievement will hasten that end and thus preserve a high quality of living and reinforce the goals of national security.

Philip C. White *is assistant administrator for Fossil Energy in the Energy Research and Development Administration. Prior to his appointment with ERDA, Dr. White was general manager of research. Standard Oil Co. (Indiana).*

Coordinated by LRE

How to put waste heat to work

Reprinted from ENVIRON. SCI. TECHNOL., **10,**
868 (September 1976)

Charles C. Coutant

Environmental Sciences Division
Oak Ridge National Laboratory
Oak Ridge, Tenn. 37830

The mutual evolution of ecological assessment, engineering designs, and regulatory considerations is making "thermal pollution" manageable. However, certain impacts associated with the cooling process remain as important technical and social issues.

In 1972, Arthur Levin of Battelle Memorial Institute, and his associates (*ES&T*, March 1972, p 224) reviewed ecological effects of thermal discharges to aquatic systems. While many of the points they raised are still appropriate, a number of dévelopments have changed the perspective of concern for the impacts of power plant cooling. Among these recent developments are

• completion of several large-scale field investigations of thermal effluents in which intake problems became apparent

• advances by the engineering profession in both predicting thermal discharge patterns and in designing outfalls that minimize zone of extreme temperatures

• publication of a National Academy of Science report on water quality criteria, including temperature

• passage of the 1972 water pollution control amendments with their emphasis on closed-cycle cooling

• the energy crisis, which has stimulated uses for waste heat

• a general maturing of the field of environmental impact assessment to the point that it has begun to consider risks of significant population or ecosystem damage, rather than concentrating on effects to individual organisms.

Heat

Previously, it was believed that the principal impact of power plant cooling systems on the environment came from the discharge of large quantities of heat. Laymen and professionals alike were conditioned by the history of water pollution control efforts to consider principally the substance(s) that emanated from the discharge pipe. For power plants this "substance" was heat, felt as temperature changes both in the cooling water itself and in the water body that receives the effluent. Heat was proclaimed to be a pollutant, and demands arose from many quarters to control "thermal pollution".

It is now understood that heated discharge is only one source of potential ecological impact from the cooling system of a power station. It is also understood that the principal engineering alternative to the traditional "once-through" cooling system, namely the cooling tower, has its own potential for influencing the environment. The area of impact may be in the terrestrial as well as in the aquatic environment. What is therefore necessary is an integrated approach to assessing ecological impacts and alternative engineering solutions for the overall power plant cooling system.

One reason for the early emphasis on temperature changes in aquatic ecosystems was the circumstance that most early studies of thermal effluents were conducted on small freshwater rivers in which temperatures remained high after effluent release. There, thermal effects were often dramatic and usually detrimental to the preferred ecosystem. For example, in the late 1950's, hot effluents from the Martin's Creek Power Plant (Pa.), studied by F. J. Trembley and others at Lehigh University, made striking changes in the composition of bottom fauna and attached algae of shallow riffles in summer, and attracted and killed many fish in the discharge canal in cooler months. Ruth Patrick, of the Philadelphia Academy of Natural Sciences, and her co-workers studied several small river locations, including the U.S. Energy Research and Development Administration's Savannah River Plant where production reactors heated the small streams of the swampy terrain to very high levels.

Studies of power plants (*ES&T*, March 1972, p 225) on large river systems and lakes, however, often failed to show such dramatic changes. Also, engineering designs that dispersed thermal effluents rapidly came into practice, and expanding knowledge of thermal effects provided boundary conditions for power plant designers to meet. This is not to say, however, that additional thermal data are not needed. On the contrary, there is still an inadequate understanding of behavioral responses to temperature, synergistic effects with toxicants, and long-term ecosystem effects of changed temperatures. Species-specific data on thermal responses, recognized as important for predicting impacts or setting limits for plant designs, are also lacking.

Entrainment and impingement

More recently, researchers began to notice effects related to the passage of water through the power plant system, rather than ecological damages at points of thermal discharge. Large fish and invertebrates were often impinged and killed on intake screens that had been designed to keep debris out of the condenser tubes. Small organisms, particularly larval stages of fish, were mutilated or thermally killed during their transit (entrainment) through pumps, heat exchange condensers, and piping. Whether screened out or entrained with the cooling water, organisms at the intake often seemed to fare worse than organisms at the discharge which had never been through the system.

The change in attention from thermal discharge to cooling water intake accompanied expansion of the utility industry to estuarine siting of steam electric stations. Small freshwater rivers generally have only limited amounts of planktonic organisms that would be susceptible to entrainment. Estuaries, however, are important spawning grounds and nursery areas for large numbers of aquatic species. Here, recirculating hydraulic patterns of fresh and salt water have encouraged evolution of drifting larvae. These drifting larvae cannot discriminate between the patterns of water flow that recirculate and nourish them in the estuary and those that draw them into power station intakes. As power stations grow both in size of individual units and in numbers of units on a given estuary, the probability increases that a larval fish will be entrained in a power station cooling system before it leaves the estuary.

An unfortunate result of early fixation on the thermal effluent component of cooling water impacts has been engineering designs that reduce discharge temperatures by increasing the volume of water pumped. Additional pumping volume, either through the condensers or as dilution flows in the discharge area, increases the numbers of organisms susceptible to impingement or physical damage of entrainment. On balance, the thermal effects are often less damaging, particularly in estuaries.

Ecological changes associated with the structures of cooling systems became apparent as intakes began to be looked at in detail and as discharge schemes made increasing use of elaborate effluent diffusers. Artificial shelters were created that attracted new species assemblages. Often, the structures at in-

Power plant cooling system's pollution sources

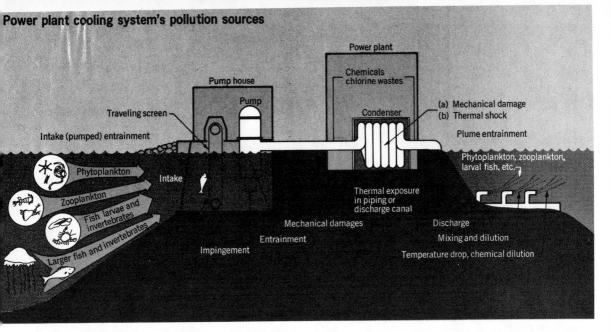

takes attracted fish and entrapped them in intake areas from which there could be no escape other than impingement on screens.

Plant structures also created some markedly different habitat conditions. For example, rock rip-rap and concrete changed areas of sandy beaches along lakes or the ocean into new habitats with unknown consequences for the ecosystem.

In addition to heat, chlorine is released in cooling waters either as slugs at intervals or as continuous additions at low levels. The chlorine is purposely added to reduce accumulations of biological slime that impairs heat transfer efficiency on condenser tubes. Larger fouling organisms, especially in marine waters, are prevented by chlorine from attaching to pumps and piping. The biocidal activity that is intended for specific target organisms necessarily acts on non-target organisms as well. Many of the ecological damages once ascribed to heat are now being more properly traced to chlorination.

Cooling towers

Cooling towers, as the principal non-aquatic alternative to "thermal pollution", have also come under scrutiny as sources of ecological damages. Despite their characterization as "closed cycle" systems, owing to the recycling of cooling water, these systems do discharge wastes to the environment.

Mechanical draft towers (in which a fan circulates air among water-covered wooden slats to speed evaporative cooling) blow a mist of cooling water droplets into the air as "drift". Also, all towers must release their cooling water to the environment as "blowdown" after several cycles through the system, in order to prevent excess buildup of salts as water is evaporated. Both of these releases involve chemicals that are added to the cooling water as corrosion inhibitors (such as chromates and organophosphate complexes) and biocides (principally chlorine). Terrestrial vegetation may suffer from airborne drift chemicals; aquatic organisms may received toxic quantities of blowdown chemicals. Organophosphorus compounds in blowdown can amount to significant inputs of phosphorus to waters already showing symptoms of advanced eutrophication caused by phosphorus enrichment.

Makeup water to compensate for evaporation and blowdown in cooling towers must be drawn from some water source at which problems of entrainment and impingement could arise; however, volumes are usually less than 10% of those required for open-cycle systems. Fogging by cooling tower plumes may cause safety hazards when towers are inappropriately located near airports or roads. Thus, cooling towers are not simple an-

swers to pollution problems, but are engineering schemes with potential impacts of their own that should be compared with impacts from open-cycle systems.

Probability of risk *vs* proof of effect

Current perspectives on power plant impacts are increasingly aimed at assessing probabilities of risk for aquatic populations, communities, and ecosystems. The field of pollution control has matured beyond the phase of simply demonstrating that there can be effects of potential pollutants on the organisms of the biosphere. It has entered the more difficult and demanding phase

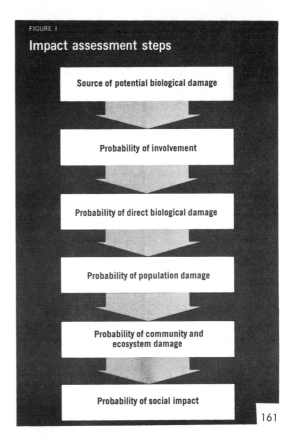

FIGURE I

Impact assessment steps

Source of potential biological damage

↓

Probability of involvement

↓

Probability of direct biological damage

↓

Probability of population damage

↓

Probability of community and ecosystem damage

↓

Probability of social impact

161

Productive conservation. *Tomatoes and beans thrive in this greenhouse warmed by waste heat*

of assessing the probabilities of risk associated with having particular human developments on particular types of water bodies.

This maturation of the assessment process intensifies the need for scientific and social evaluations and minimizes applicability of across-the-board, numerical limits imposed by national regulatory agencies. The requirement of the National Environmental Policy Act of 1969 that environmental impact assessment be conducted for major federal actions was evidence of this maturation. This is one reason why the National Academy of Sciences/National Academy of Engineering report, "Water Quality Criteria 1972", emphasized that it was presenting a method for analysis of thermal effects on aquatic life, not a set of rigid numerical guidelines.

A conceptual framework for impact analyses of power stations on aquatic life, which encompasses several stages and requires expertise in many disciplines, has evolved (Figure 1). Determining the probability of direct biological damage is merely one step leading toward assessment of risk to the ecosystem. The analysis must begin with careful attention to all sources of potential ecological damage (or change) from the facility, both in operating and construction phases. Then, information on the distribution and abundance of organisms in the ecosystem must be coupled with hydrologic and hydraulic assessments to determine the likelihood that key organisms or life stages will come in contact with the source of potential damage.

Once damages to the organisms are estimated, they must be translated into changes in population dynamics of the key species, particularly as yields to fisheries or possible declines to lower equilibrium levels or to extinction are concerned. At community and ecosystem levels, risks to normal structure and function, for example, change in diversity and types of species, and functional aspects such as energy flow and nutrient cycling are sought. Often, computer modeling is the best approach to examining population, community, and ecosystem effects, although such techniques are in their infancy. The assessment of the probability for social impact from any ecological changes is the final step.

Local conditions of the receiving systems are now seen to be as important to assessing probability of risk as are the general data on the direct biological effects of human activity. This fact is at variance with the long-held philosophy that standards for pollutants can be established for the nation as a whole, or even for entire states, based on representative scientific literature. To consider only the potential effect of an effluent or structure without looking at its realistic interactions with the receiving system is to see only half the story. Regulatory restrictions developed through analyses of probabilities of risk will necessarily

be *localized* restrictions, tailored to the particular receiving system (water body, regional air mass, and the like).

Assessment of risk, however, implies that one has some clear notion of what would be put at risk. One must clearly state the object or property believed to be at risk, and make social decisions regarding objectives for that object or property. Thus, chronic effects of power stations on an estuary can hardly be evaluated against vague notions of the "proper functioning" of the estuary ecosystem. On the other hand, if one defines "proper functioning" to mean, for example, continued yield of high numbers of striped bass to the commercial and sports fisheries, then there are clear objectives against which to measure the significance of any direct effects at the power station. Certainly, the objectives chosen would need the backing of either scientific justification or social desire.

Such objectives vary from location to location across the country. Enhancement of largemouth bass populations in rivers and reservoirs, for example, is seen as a prime goal in the "bass country" of the American southeast; yet this species is considered an undesirable competitor with salmon in the Pacific Northwest. As the objectives differ, so do the assessments of risk and the restrictions that are selected to minimize risk in a particular location.

While there now seem to be clear objectives for environmental impact assessments of power plant cooling systems, one must concede that present environmental knowledge is inadequate for a comprehensive assessment. Knowledge of direct biological effects exceeds present understanding of what these combined effects mean to populations. The most comprehensive models of population dynamics of important species falter when interactions among species are questioned.

"Ecosystem Analysis" is a new framework for determining ecosystem-level impacts (which received deserved attention in the intensive International Biological Program activities of the past 5 years), but ecosystem understanding has yet to advance to where useful predictions can be made for installations such as power stations. Impact assessment can proceed only as far as the understanding of environmental systems is advanced through research and analysis. To enhance this understanding, much-needed studies are presently being funded by the Energy Research and Development Administration, the Nuclear Regulatory Commission, the Electric Power Research Institute, the Environmental Protection Agency, and many electric utilities.

Waste heat management

There is a growing desire to put waste heat to use rather than to disperse it to the environment in ever more costly cooling devices. Some of these uses are ecological. It seems clear to most people that society must learn to conserve energy; the great national debate now is over how to conserve. Rather than regarding conservation according to the dictionary definition (to keep from being damaged, lost, or wasted), most people think of conservation as reduction in energy use. Lights are turned off, car pools are formed, and so on. Such forms of *reductive energy conservation* may momentarily slow energy use, but the long-term trend will still be upward. However, a better answer may lie in *productive energy conservation*, in which degraded energy is put to further productive uses. For instance, heat that is reused does not have to be generated with additional fuel.

A principle that is emerging from ecosystem research is that complex mature ecosystems have efficient devices for keeping energy within the realm of productivity with only a gradual loss as heat. While the initial conversion of solar energy to organic molecules is not very efficient, elaborate biochemical systems in organisms and diverse species assemblages in ecosystems effectively utilize a "low grade", dispersed radiant energy source and keep converted energy at work as long as possible. Human society could emulate the efficiency. It is encouraging to see analyses of net energy use being developed; there could be a start toward developing energy-efficient social systems.

Productive uses of waste heat range from moderate management of thermal discharges to complete engineering control of the flow of heat for industrial uses or building heating. In certain areas, experimental or production greenhouses are making use of waste heat. The most often discussed ecological uses are in aquaculture where temperature control at near-optimum growth temperatures for the cultured species can markedly increase protein output, as compared to uncontrolled culture systems. Species that require years to grow to market size in nature can be harvested within one year under thermal aquaculture. For example, catfish have been successfully grown in thermal culture systems at the Tennessee Valley Authority's Gallatin Steam plant in Tennessee, and by a commercial grower at a power station in Nebraska. Oysters are cultivated in warm discharge waters of the Long Island Lighting Company's Northport plant. Obtaining sufficient food for rapid growth rates, the near-impossibility of using significant quantities of the nation's waste heat for such purposes, and the institutional problems of linking food producers with electric utility companies seem to be the main discouraging notes for such enterprises.

A promising "use" for waste heat is in development of multipurpose cooling reservoirs. Wherever sufficient land is available, cooling reservoirs offer distinct advantages in costs and social benefits over cooling towers. Power plant cooling waters appear to enhance recreational fishing on small reservoirs by providing longer growing seasons for game fish, by concentrating fish and fishermen in discharge areas in cooler months, when catch rates are improved, and by improving the ratio of game fish to non-game fish. Whether the effect is caused by added heat *per se*, or by induced current patterns that prevent summer deoxygenation near the bottom has not been settled; the benefits appear real, however.

The key to a successful multipurpose reservoir appears to be in designing and managing a nearly closed reservoir system that incorporates the needs of the key species (for example, largemouth bass) as well as those of the power station. Indiscriminate discharge of heated waters to natural systems is replaced by an integrated, managed system designed to provide maximum recreational opportunity as well as optimum heat dissipation for the power plant. Many such man-made systems are already operating; the researcher's task is to determine why they function well, so that design criteria can be established for new ones.

Verification of predictions

There is an increasing pragmatism concerning verification of dire predictions of thermal and other power plant effects. For about 20 years, researchers, regulators, and lay environmentalists have hypothesized a long list of possible ecological consequences of cooling water use. It seems appropriate now to determine which of these hypothetical consequences occur and which present any significant risk to populations, ecosystems or social values.

Field research is at a stage at which reasonably detailed comparisons between hypotheses and observed results at many power stations across North America and in Europe should be able to be made. Environmental impact statements for nuclear power plants, prepared since 1971 by the U.S. Atomic Energy Commission's Office of Regulation (now the Nuclear Regulatory Commission), are massive sets of hypotheses that are now being evaluated, based on monitoring data that are becoming available. Many suspected long-term effects can be sought at smaller fossil-fueled power stations that have operated for many years. In the tradition of the scientific method, ecologists should be able to view objectively both the hypotheses and the results, and to arrive at conclusions that can genuinely aid the next generation of power plant designs.

An appropriate guideline for the future is that there is some environmental impact from anything done. An engineering system as complex as power station cooling is bound to have many sources of potential environmental change associated with it. For years, only the thermal component was a matter of concern. The perspective has now broadened with regard to the open-cycle system, but there is a trend toward similar short-sightedness concerning potential impacts from cooling towers. Instead of searches for a panacea, a careful analysis of the relative impacts of alternative cooling schemes at a chosen location is needed, as well as more careful selection of sites that will present minimal environmental complications. Environmental impact itself should also be measured against other goals and needs of society.

There are uses of waste heat from power stations that can turn part of the environmental problem into a social benefit. It is ironic that increasingly complex devices are designed for dumping waste heat into the atmosphere while the world suffers an energy shortage. Multipurpose cooling lakes, thermal aquaculture, and greenhouse heating offer possibilities of enhancing ecological productivity in ways desirable to man. Disposal of waste heat from electricity generation should be integrated with energy requirements of the region. In many cases, overall energy efficiency might be improved if there were to be less efficiency in electricity production, leaving a higher temperature discharge more suitable for other uses.

Additional reading

Coutant, C. C., Biological Aspects of Thermal Pollution. II-Scientific basis for water temperature standards at power plants. *CRC Crit. Rev. Environ. Contr.* **3** (1), 1–24 (1972).

Goodyear, C. P., Coutant, C. C., and Trabalka, J. R., Sources of Potential Biological Damage from Once-Through Cooling Systems of Nuclear Power Plants. ORNL/TM-4180, Oak Ridge National Laboratory, Oak Ridge, Tenn., 1974.

Stratton, C. L., and Lee, G. F., Cooling Towers and Water Quality. *J. Water Pollut. Contr. Fed.* **47** (7), 1901–1912 (1975).

Hana, S. R., and Pell J., (Eds.). *Cooling Tower Environment— 1974.* CONF-740302, U.S. Energy Research and Development Administration, Technical Information Center, Oak Ridge, Tenn., 1975.

Yarosh, M. M., et al., Agricultural and Aquacultural Uses of Waste Heat. ORNL-4797, Oak Ridge National Laboratory, Oak Ridge, Tenn., 1972.

Charles C. Coutant *is leader of the Power Plant Effects Project at Oak Ridge National Laboratory (Oak Ridge, Tenn.). A general aquatic ecologist by training, Coutant has specialized in analyses of man's impacts on aquatic life.*

Coordinated by JJ

Reprinted from ENVIRON. SCI. TECHNOL., **10**, 735 (August 1976)

Energy recovery from pulping wastes

Ecodyne evaporators eliminate water discharge, recover spent liquor, and increase process efficiency

The Port Angeles division of ITT Rayonier (Wash.) produces various grades of pulps, including those for ordinary paper, acetate, and viscose rayon. About 500 tons/day are produced and distributed as raw materials for manufacturers around the world.

In the past, spent sulfite liquor from the pulp mill that included other wood solubles, degraded cellulose, unreacted ammonium bisulfite and SO_2, were discharged into the Strait of Juan de Fuca, where it was quickly dispersed by tidal action. Now, after a $35 million mill expansion program, Rayonier is recovering materials and heat energy to improve process efficiency.

When the state of Washington established a new pollution and environmental protection code, the company first considered switching the mill to a soda-base process that was considered to be more compatible with spent liquor recovery. Rayonier completed an engineering study for such conversion and even obtained proposals for necessary new equipment, but costs were much higher than anticipated. Therefore, Rayonier decided to stay with its ammonia-base sulfite process.

The mill uses a fully bleached ammonia-base sulfite process from batch digestors. In the digestor, cooking acid contains 7–8% total SO_2 as either free SO_2 or ammonium bisulfite. Pressures and temperatures vary with the pulp being produced. The cooked pulp is discharged into blow-pits and transferred to red stock washers where about 90% of the residual red stock liquor is removed from the pulp.

This red liquor contains about 10% of the dissolved solids from the process, and was dispersed in the Strait.

Operational change

Key to the recovery operation is seven spent liquor evaporator bodies that produce the most efficient and rapid concentration of spent pulping liquor solids to a combustible state.

Ecodyne Corporation's Unitech Division (Union, N.J.) designed and built an evaporator system that would concentrate spent red liquor for burning and SO_2 recovery. Vapor recompression pumps were recommended that could economically reduce steam at boiler pressure for other uses in the mill, while at the same time transfer the energy difference to the liquor vapors being pressurized.

Rayonier recovery utilities superintendent Randall Starr explains that up to this point, the turbogenerators and reducing stations satisfied low pressure steam requirements. "With the new system, the compressor turbines act also as reducing stations," he said.

The design was chosen to provide only as much vapor recompression as could be compatible with steam reduction needs. The remainder of the evaporation could then be accomplished in a four-body steam-energized evaporator system.

How it works

The system's design called for an overall evaporation rate of approximately 560 000 lbs/h, with an extra 10% for catch-up or pull-away capacity. The

evaporation capacity of the 3-stage vapor recompression system was 450 000 lbs/h. This portion of the system was not overdesigned because the flexibility in the vapor recompression is built into the system with variable loadings on the steam turbines. The multiple-effect evaporations operate between approximately 110 000 and 160 000 lbs/h evaporation rate, over a range of 25–52% solids, thus providing the extra 10% catch-up capacity.

To accommodate the mill's variable low-pressure steam requirements, the vapor compression pre-evaporator section was designed as three completely separate falling-film units. If low-pressure steam requirements are low or if one body is taken off line for wash, the entire system associated with the body, including compressor and turbine drive, is also taken off line.

Compressed vapor is energy source

In the 3-stage Vapor Compression Falling-Film (VRFF) system, the heating medium is the vapor removed from the red liquor and increased in both temperature and pressure by a turbine-driven vapor compressor. When the spent liquor enters the VRFF units from the SO_2 stripper, it contains about 10.7% solids. Effluent from the third stage of the VRFF system is 25–30% solids and is ready for the four-body Preheat Falling-Rising-Film (PFR) evaporator system. Vapors from evaporation at 15 psia are compressed to approximately 21.5 psia. The saturation temperature is increased from 213 °F to about 232 °F.

Because the boiling temperatures and liquor viscosities are slightly different in each of the three stages, the actual volume of vapor produced in each stage is correspondingly different. The advantage of turbine compressors is apparent here. As compressor vapor inlet volume decreases to maintain the desired compressor outlet pressure, the turbines slow. Operating the units at reduced capacity is not a problem, and reduced evaporation rates caused by scaling or fouling of the heat transfer interfaces usually do not pose a problem in maintaining nominal pressures and temperatures. If inlet vapor volumes drop too low, the system will shut down automatically.

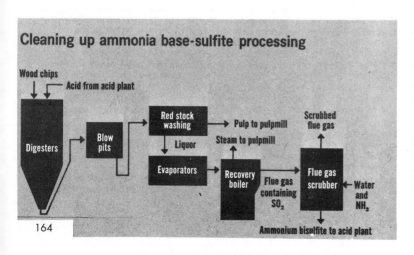

Cleaning up ammonia base-sulfite processing

Wood chips →
Acid from acid plant →
Digesters → Blow pits → Red stock washing → Pulp to pulpmill
Red stock washing → Liquor → Evaporators
Steam to pulpmill
Evaporators → Recovery boiler → Flue gas containing SO_2 → Flue gas scrubber ← Water and NH_3
Scrubbed flue gas
Ammonium bisulfite to acid plant

How trash is being turned into useful heat

The John Deere Co. uses pyrolysis and heat recovery to meet 75% of its needs for paint drying ovens

Reprinted from ENVIRON. SCI. TECHNOL., **10**, 860 (September 1976)

The John Deere Horicon Works is a division of Deere and Company, one of the largest 100 corporations in the U.S., with annual sales of more than $2.96 billion. The Horicon Works produces Deere's line of lawn and garden equipment and snowmobiles.

Its Horicon Works generates some 3300 tons of trash each year. The mix is 50% lumber, 10% plastic, and 40% miscellaneous. In the past, the corrugated waste was baled and sold. On a contract basis, today it would cost the company $38 400 a year to dispose of its waste, but when the Horicon landfill was closed by the local authorities in 1974, the company ventured into a field of new technology.

To combat the natural gas shortage and rising fuel bills, the company started burning its own trash and garbage—generating enough heat energy to save more than 14.6 million ft^3 of natural gas annually.

Dwayne D. Trautman, manager of process and tool engineering for the Deere plant says, "In the process, the company will save $52 000 a year in both fuel bills and hauling costs. Considering that the entire installation for recovering this otherwise wasted energy will cost $110 000, the company will realize a substantial return (45%) on its initial investment—a rather unusual payoff for a new technology."

Trautman continues, "By combining pyrolysis with heat recovery, we are bringing about the 'marriage' of two known technologies and thus have dis-

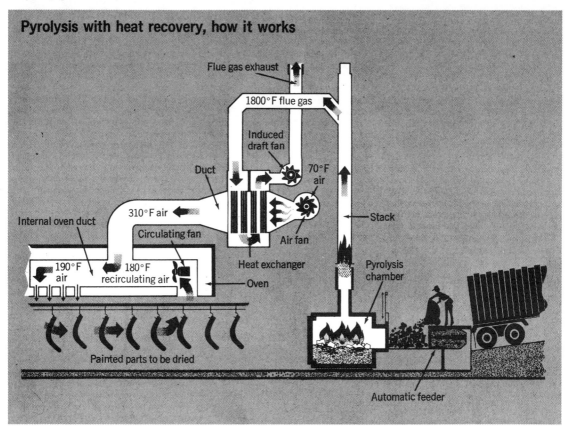

Pyrolysis with heat recovery, how it works

Flue gas exhaust

1800°F flue gas

Induced draft fan

Duct

70°F air

310°F air

Internal oven duct

Circulating fan

Stack

Air fan

190°F air

180°F recirculating air

Heat exchanger

Pyrolysis chamber

Oven

Painted parts to be dried

Automatic feeder

covered a new source of energy right on our doorstep—one that we had previously been paying people to take away and burn in the ground. At today's price for oil, and projected prices for gas, the garbage and trash we used to haul away is now worth between $7–10/ton in recoverable energy."

How it works

The heat is created in a pyrolysis waste disposal system designed by Kelley Co. (Milwaukee, Wis.) in which a combustible gas is generated in the primary burning chamber. As a result, the process literally "burns up its own smoke", and emissions are far below air pollution control standards set by environmental protection agencies.

Through its representative, Paul Reilly Co., the Kelley Co. also supplied the heat recovery unit that generates 3.9 million Btu/h, providing heat for 30 000 ft^2 of paint drying oven area. This heat cut Deere's consumption of natural gas by more than 14.6 ft^3/year, or 75% of its total yearly gas consumption in the oven.

Kelley Co. vice president Robert Pfleger estimated that 720 trillion Btu recovered annually from the nation's 198 million tons of combustible solid waste would heat 2.4 million homes for a year in a northern climate. He says, "Every Btu saved by industry through the use of on-site heat recovery from the pyrolysis of solid waste saves energy for use in our homes. It also helps industry hold the line on costs and thus keeps inflation under control."

Savings

The pyrolytic system has been burning 2000 lbs of trash each day since September 1974. The unit, at that time, eliminated an expenditure of $33 000 a year for hauling wastes to the nearest landfill.

According to the estimate of Wisconsin Power and Light Co. (Madison), suppliers of gas to the Horicon Works, this recovered energy would be enough to heat more than 90 homes in Wisconsin's climate for a year.

Supervisor of Engineering Services Karl Eberle projected that the dollar savings in fuel cost alone for the first full year of operation (1976) would be more than $19 000. For 1977 and 1978, the savings will be $2000 and $27 000, for a 3-yr total savings in fuel costs alone of $70 000. This is approximately the entire cost of the heat recovery unit. Even greater annual savings will be realized as gas prices continue to increase.

The only gas used now by the Horicon Works is for a pilot light in the incinerator's afterburner, and the remaining 25% of the energy needed directly to heat their drying oven, according to George Gibert, director of mechanical engineering, Wisconsin Power & Light Co.

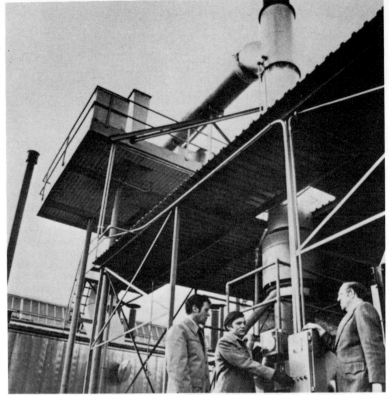

Heat source. *Its fuel will be trash and garbage*

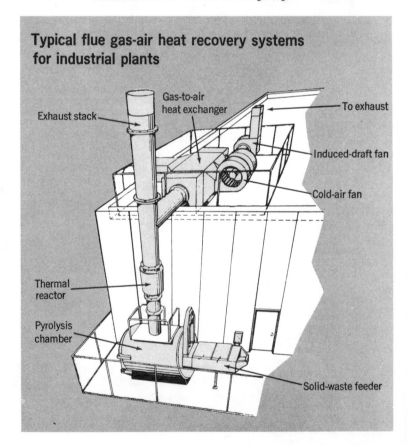

Typical flue gas-air heat recovery systems for industrial plants

Gas-to-air heat exchanger

Exhaust stack

To exhaust

Induced-draft fan

Cold-air fan

Thermal reactor

Pyrolysis chamber

Solid-waste feeder

Books

Equipment

Electrostatic Precipitation of Fly Ash. Harry J. White. vii + 63 pages. APCA Reprint Series. Air Pollution Control Association, P.O. Box 2861, Pittsburgh, PA 15230. 1977. $6 ($4.50 for APCA members), paper.

This collection contains articles on fly ash and furnace gas characteristics; precipitator design; equipment; case histories and problems; and hot precipitators. Resistivity problems are also covered.

Particulates and Fine Dust Removal: Processes and Equipment. Marshall Sittig. xv + 605 pages. Noyes Data Corp., Mill Rd. at Grand Ave., Park Ridge, N.J. 07656. 1977. $48, hard cover.

What is the latest technology in this field? This book gives a complete answer to that question, and emphasizes reduction of such emissions at the source. Complete, up-to-date information concerning the nature of the problem, and practical technology and patents aiming at its solution are given.

Air Pollution. Vol. 4, 3rd ed. Arthur C. Stern, Ed. xxii + 946 pages. Academic Press, Inc., 111 Fifth Ave., New York, N.Y. 10003. 1977. $49.50, hard cover.

This book is the fourth of five volumes, and is part of a standard reference work. It deals principally with engineering practicalities. Among subjects discussed are mist elimination; electrostatic precipitation; industries; combustion; source control; selection of devices; and related matters in abating air pollution.

Air Pollution Control and Design Handbook, Part I. Paul N. Cheremisinoff and Richard A. Young, Eds. x + 704 pages. Marcel Dekker, Inc., 270 Madison Ave., New York, N.Y. 10016. 1977. $39.50, hard cover.

Air pollution was brought on by our technological wizardry and exacerbated by the twentieth-century lifestyle. But the problem is now being attacked with legal and economic dedication. This book was especially written for those managers faced with pollution problems from stationary sources; it offers analyses and recommendations of experts in several problem areas. The stress is always on the pratical aspects of pollution abatement.

Handbook of Ventilation for Contaminant Control. Henry J. McDermott. viii + 368 pages. Ann Arbor Science Publishers Inc., P.O. Box 1425, Ann Arbor, MI 48106. 1976. $29.50, hard cover.

Requirements for safety in the workplace are becoming ever stiffer. This includes the air the working people must breathe. Written by an industrial hygienist, this book explains principles of ventilation; how to design ventilation to meet OSHA requirements; instructions for handling toxic materials, and many other pertinent topics.

Control of Air Pollution Sources. Joseph M. Marchello. vii + 630 pages. Marcel Dekker, Inc., 270 Madison Ave., New York, NY 10016. 1976. $65.60, hard cover.

Many are the sources of air pollution, and many are the techniques needed to control it. This book covers many aspects of air-quality management, fate of pollutants in the air, control systems, energy conversion, costs, and monitoring, as well as other pertinent topics. It is Volume 7 of the publisher's Chemical Processing and Engineering series.

The Environmental Control Industry. Kenneth Ch'uan-k'ai Leung and Jefferey A. Klein. 138 pages. Ann Arbor Science Publishers Inc., P.O. Box 1425, Ann Arbor, MI 48106. 1976. $14.95, paper.

This report analyzes many aspects and factors of this industry. It discusses the types of firms one finds in the field, and takes an exhaustive look at markets for equipment in air and water pollution, and solid waste. The authors, financial analysts, prepared the report for the Council on Environmental Quality.

The Claus Process. 50 pages. R. H. Chandler Ltd., P.O. Box 55, Braintree, Essex CM7 6HD, England. 1976. $15, paper.

Many government regulations now require, or will call for reduction of sulfur emissions to the atmosphere to very low levels. With new technology, more efficient stripping of sulfur from tail gases becomes possible. This bibliography reviews 66 literature references and provides abstracts of 95 patents in this technology from 9 countries.

Air Pollution Control: Processes, Equipment, Instrumentation. Lexington Data, Inc., Box 311, Lexington, Mass. 02173. 1972. $250.

Report indexes and abstracts of 1400 U.S. patents issued from 1967 through 1971 on this subject. Of the 1400 patents issued, only 180 dealt with internal combustion engine emissions. For another example, only 52 of the patents were aimed at NO_x control.

Management and Engineering Guide to Economic Pollution Control: A General Approach to Industrial Waste Problems With Case Histories. David Krofchak. 124 pages. Clinton Industries, Inc., Publishing Div., P.O. Box 1208, 32880 Dequindre Ave., Warren, Mich. 48092. 1972. $10, paper.

Attempts to describe the fundamentals of an economic approach to the solution of difficult industrial waste treatment problems. Illustrates how management can turn seemingly money-losing pollution abatement programs into cost-saving improvements to production processes. Also reference source for engineers.

Scrubber Handbook. xxvi + 791 pages. National Technical Information Service, U.S. Dept. of Commerce, Springfield, Va. 22151. 1972. $13.50, paper.

Represents an effort to present the best available engineering information on scrubbers. Covers fundamental principles and then covers generally applicable design methods. Handbook was prepared by Ambient Purification Technology, Inc. (Riverside, Calif.) under contract to EPA.

Emissions from Continuous Combustion Systems. Walter Cornelius, William G. Agnew, Eds. x + 479 pages. Plenum Publishing Corp., 227 W. 17th St., New York, N.Y. 10011. 1972. $25, hard cover.

Describes current basic and applied research directed toward solving the problem of emissions from continuous combustion systems. Proceedings of a symposium sponsored by General Motors Research Laboratories.

Recent Advances in Air Pollution Control. 530 pages. American Institute of Chemical Engineers, 345 E. 47th St., New York, N.Y. 10017. 1974. $15 ($7.50 for AIChE members).

Describes technological advances toward purifying the environment and achieving national standards on air quality. Two sections cover national emission standards and their impact on industry; three discuss emission species; and the remaining covers miscellaneous topics, including pollution control as a process design variable. Ask for S-137.

Air Pollution Technology. Dean E. Painter. xii + 283 pages. Reston Publishing Co., Inc., P.O. Box 547, Reston, Va. 22090. 1974. $13.95, hard cover.

Intended both as a college text for persons who plan to become involved in air pollution control technology and as an interdisciplinary program to inform all people of their responsibilities in conquering and controlling air pollution. Contains material condensed from numerous pamphlets, federal registers, and other documents of current origin. Each chapter lists references.

Control of Air Pollution in the USSR.

N. F. Izmerov. 157 pages. Office of Publications and Translation, World Health Organization, 1211 Geneva 27, Switzerland. 1973. $3.15, paper.

Traces the historical development and presents the status of air pollution control in the USSR. The book examines the various sources of air pollution and the devices used to control them. It gives a detailed description of the ways in which air pollution control is organized and implemented in the USSR. Special attention is given to surveillance.

Handbook of Environmental Control.

Richard G. Bond, Conrad P. Straub, and Richard Prober, Editors. Vol 1: Air Pollution. xii + 576 pages. Chemical Rubber Co., 18901 Cranwood Parkway, Cleveland, Ohio 44128. 1972. $36, hard cover.

First of a four-volume series that brings together information in tabular form useful in evaluating the environment. It considers the interrelationship of man and his environment, as well as aquatic and terrestrial aspects of the ecosystem. Concerned entirely with air pollution, this volume contains sections on air pollutants, their effects, emission sources, and control measures.

Handbook of Environmental Control.

Richard G. Bond, Conrad P. Straub, and Richard Prober, Editors. Vol 2: Solid Waste. ix + 580 pages. Chemical Rubber Co., 18901 Cranwood Parkway, Cleveland, Ohio 44128. 1973. $31, hard cover.

Second of a four-volume series that brings together information in tabular form useful in evaluating the environment. Focusing entirely on solid wastes, this volume contains sections on sources and composition, effects, and controls and management. It considers the interrelationship of man and his environment, as well as pollution effects on aquatic and terrestrial aspects of the ecosystem.

Automotive Emission Control.

92 pages. R. H. Chandler Ltd., P.O. Box 55, Braintree, Essex CM7 6HD, England. 1975. $125, paper.

This book is Volume IV of the series, and it covers patents, with abstracts over the second half of 1973. Coverage includes mufflers for treating exhaust gases, catalysts, afterburners, and engine modifications aimed at reducing automotive pollution. Patents covered are mostly from Britain, France, Germany, and the U.S.

Fine Particulate Control Technology.

H. M. Englund and W. T. Beery, Eds. vii + 214 pages. Air Pollution Control Association, 4400 Fifth Ave., Pittsburgh, Pa. 15213. 1975. $14, hard cover; $12, paper ($3 discount for APCA members).

Emphasis is on fabric filtration, wet scrubbing, and electrostatic precipitation, particularly for fine particles. These subjects were discussed in papers presented during various APCA symposia held at Boston (fabric filtration), San Diego, Calif. (fine particle scrubbing) and Pensacola Beach, Fla. (electrostatic precipitation) during 1974.

Sulfur Dioxide Removal From Waste Gases.

A. V. Slack and G. A. Hollinden. xii + 294 pages. Noyes Data Corp., Mill Road at Grand Ave., Park Ridge, NJ 07656. 1975. $36, hard cover.

What can be done about SO_2 and its sources? This book updates a previous edition published in 1971, and gives the latest answers to this question. New technology is fully described, and technology of 1971 that has been superseded has been omitted. Methods utilizing throwaway or recovery processes for SO_2 control are exhaustively described, as well as absorbent chemistry, catalytic oxidation/reduction, and control economics.

Cracking Down, Oil Refining and Pollution Control.

Gregg Kerlin and Daniel Rabovsky. 478 pages. Council on Economic Priorities, 84 Fifth Ave., New York, NY 10011. 1975. $140 (non-profit organizations should contact CEP for special price), paper.

Oil refining is considered one of the prime sources of environmental pollution. This book tells who has the best and worst air and water pollution control records, what compliance technology has been developed, and how pollution control can result in cost, material, and energy savings. It also tells how control varies as to region, and why government enforcement is the single greatest pollution control determinant.

Odor Control and Olfaction.

James P. Cox. 500 pages. Pollution Sciences Publishing Co., P.O. Box 175, Lynden, WA 98264. 1975. $50, hard cover.

Odors are a part of the environment, and they can sometimes be pleasing, sometimes offensive. However, the field of odor control has been neglected.

This book fills that gap, and gives full data concerning over 2000 osmogenes, as well as a patent study from 1890 to the present. Means of odor control are covered, as well as properties of odorous materials. The book can be a legal guide to odor pollution offense. A comprehensive glossary is given.

Industrial Control Equipment for Gaseous Pollutants.

Louis Theodore and Anthony Buonicore. Volume I, 225 pages; Volume II, 168 pages. CRC Press, Inc., 18901 Cranwood Parkway, Cleveland, Ohio 44128. 1975. Vol. I, $39.95; Vol. II, $29.95, hard cover.

These volumes provide fundamentals and design principles of industrial control equipment for gaseous pollutants. Practical applications are given. Many different systems for gaseous pollutant control, along with their design principles, are thoroughly discussed. Environmental effects of gaseous pollutants are also included.

Industries

Introduction to Pulping Technology.

12 cassette tapes and 20-chapter study guide. Publications Order Department, Technical Association of the Pulp and Paper Industry, One Dunwoody Park, Atlanta, GA 30341. 1976. $195 ($165 to TAPPI members).

This home study course covers every facet of pulping technology from wood structure and anatomy to end products. Among topics of environmental interest are acid recovery and by-products, paper and board recycling, and pollution problems and abatement.

Air Pollution Control in the Steel Industry.

H. M. Englund and W. T. Berry, Eds. 58 pages. Publications Department, Air Pollution Control Association, 4400 Fifth Ave., Pittsburgh, PA 15213. 1976. $6 ($4.50 to APCA members), paper.

This book is one of the APCA reprint series that features what APCA considers to be outstanding papers from its Journal. Coke oven emission controls are discussed, as are acid gases, Claus plant operations, coke oven gas sulfur removal, occupational hazards of exposure to coal tar pitch volatiles, energy consumption, and other related topics.

Pollution Control in Metal Finishing.

Michael R. Watson. xi + 295 pages. Noyes Data Corp., Noyes Bldg., Park Ridge, N.J. 07656. 1973. $36, hard cover.

Book is part of Pollution Technology Review series and is based on authoritative government reports and U.S. patents. It attempts to clarify the ways and means open to metal finishers who must keep their pollution down to a minimum. Important processes are interpreted and explained by examples.

Pollution in the Electric Power Industry: Its Control and Costs. David L. Scott. xvi + 104 pages. Lexington Books, Heath & Co., Lexington, Mass. 02173. 1973. $10, hard cover.

Seeks to determine accurate pollution abatement costs for the electric power industry, and demonstrates it is the consumer who must pay these costs. Book says it is wrong to assume that the increasing expense of producing electric power is due entirely to environmental concerns since fuel and labor costs, capital expenditures, and interest rates also are increasing.

Air Pollution Control Patent Information. Periodical. The McIlvaine Co., 2970 Maria Ave., Northbrook, Ill. 60062. $96/yr.

This periodical utilizes a system of patent summaries on 3 × 5 cards issued bi-weekly, and which can be filed for easy retrieval in a file box McIlvaine provides. Two sets of cards are given—one to be filed by control method, and the other by name of assignee. Periodically, cards are issued that index patents by emission source. The cards contain comprehensive data from the *Official Gazette,* including drawings.

Transportation

Emissions from Internal Combustion Engines and Their Control. D. J. Patterson, N. A. Henein. vii + 355 pages. Ann Arbor Science Publishers, Inc., P.O. Box 1425, Ann Arbor, Mich. 48106. 1972. $18.75, hard cover.

Covers the chemical and chemical engineering aspects of the subject as well as mechanical engineering. Delineates the fundamentals of combustion and emission formation in both homogeneous and heterogeneous combustion systems. Intended as an introductory text in combustion engine emissions and their control. Designed for the engineer, research worker, or student who is concerned with the theory and practice of engine and vehicle emission control.

Coal and Energy

Coal Desulfurization. Robert A. Meyers. xii + 254 pages. Marcel Dekker, Inc., 270 Madison Ave., New York, N.Y. 10016. 1977. $29.75, hard cover.

Sulfur in coal can be pyritic or organic, or in some combination of these. Maybe it can be removed before the coal is burned; indeed, programs aimed at such removal are going forward at full speed. But sulfur removal must be not only technologically feasible, it must also be economical. This book takes a comprehensive look at sulfur removal technology.

Fluidization Technology. Dale L. Keairns, Ed. Vol. 1, xiii + 466 pages; Vol. 2, xii + 608 pages. Hemisphere Publishing Corp., 1025 Vermont Ave., N.W., Washington, DC 20005. 1976. Vols. 1 and 2, $75, both hard cover.

Atmospheric and pressurized fluidized-bed combustion might offer hope for environmentally acceptable use of fuels that would normally be "dirty." These volumes cover such combustion of coal, oil, waste, and other materials; limestones; calcination; management; and other pertinent topics. They comprise papers from the Asilomar International Fluidization Conference sponsored by the Engineering Foundation and held in 1975.

Deep Coal Mining: Waste Disposal Technology. William S. Doyle. x + 392 pages. Noyes Data Corp., Mill Road. at Grand Ave., Park Ridge, NJ 07656. 1976. $36, hard cover.

Those who are familiar with coal mining areas know the waste piles and their hazards of fires and acid drainage very well. This book, based on 19 government reports and 7 patents, describes methods of controlling and preventing this pollution. Treatment and neutralization of acid drainage, control of fires and resultant air pollution, and recovery of materials, as well as reclamation of refuse bands themselves, are discussed in detail.

Coal Conversion Technology. I. Howard Smith and G. J. Werner. xx + 133 pages. Noyes Data Corp., Mill Rd., at Grand Ave., Park Ridge, NJ 07656. 1976. $24, hard cover.

The conversion of coal to clean fuels is the subject of much discussion at this time, but what is the actual state-of-the-art? This book, originally prepared as an "in-house" report for Millmerran Coal Pty. Ltd. (Brisbane, Australia), discusses more than 100 processes in various stages of development. Among topics emphasized are desulfurization, demineralization, and depolymerization of coal to produce clean, easily-ignitable gaseous or liquid products.

Coal Technology. 126 pages. C-E Tyler Industrial Products, 7887 Hub Parkway, Cleveland, Ohio 44125. 1975. $45, spiral-bound.

This book contains proceedings of a 1975 series of conferences on the status of coal in meeting U.S. demands for clean energy. C-E Tyler sponsored these conferences which covered the SYNTHOIL process, technology for clean fuel applications, SO_2 regulations affecting coal users, chemical desulfurization of coal, the SYNTHANE process, water quality management, and other related, relevant, timely topics.

Coal Utilization Symposium—Focus on SO_2 Emission Control. 220 pages. National Coal Association, The Coal Building, 1130 17th St., N.W., Washington, D.C. 20036. 1974. $8.00, paper.

Book contains 22 papers selected by the Association's program committee, by government and industry experts involved in air pollution research and control. Methods of removing sulfur oxides from flue gases and other desulfurization techniques are carefully examined. Multiple approaches reflect complexity of the problem and diversity of emission sources.

Thermal Processing of Municipal Solid Waste for Resource and Energy Recovery. Norman J. Weinstein and Richard F. Toro. Ann Arbor Science Publishers, Inc., P.O. Box 1425, Ann Arbor, MI 48106. 1976. $20.

This work is a critical analysis of thermal processing of solid waste for energy recovery and resource recycling. Pyrolysis processes, thermal processing, costs, and resource and energy recovery are discussed in depth. Also included are chapters on air pollution control and liquid and solid residue disposal.

SPECTRUM: An Alternate Technology Equipment Directory. 64 pages. Alternate Sources of Energy, Inc., Route 2 Box 90A, Milaca, MN 56353. 1976. $2, paper.

SPECTRUM is a catalog of over 400 special items of equipment from almost all fields of alternate energy technology. It covers conservation, solar, energy-saving architecture, organic fuels, wast disposal, wind, hydro, and others. Product descriptions, specifications, uses, and suppliers are given.

Energy Technology Handbook. Douglas M. Considine, Ed. xxvii + 1857 pages. McGraw-Hill Book Co., 1221 Ave. of the Americas, New York, N.Y. 10020. 1977. $49.50, hard cover.

"You name it, this book has it." Petroleum, gas, coal bioconversion, chemical fuels, nuclear, solar, geothermal—they are all covered. Environmental problems arising from energy production/use are also extensively discussed.

Solar Energy: Technology and Applications. J. Richard Williams. 176 pages. Ann Arbor Science Publishers, Inc., P.O. Box 1425, Ann Arbor, Mich. 48106. 1977. $9.95, hard cover.

This book is a revised edition that discusses many solar energy aspects. It tells how to build a heating/hot water system, and how to select the proper collectors. It also discusses industrial and agricultural applications, air conditioning, ocean-thermal power, wind, geosynchronous power plants, and other timely topics.

Wind Technology Journal. Periodical. Herman M. Drees, Ed. Wind Technology Journal, P.O. Box 7, Marstons Mills, Mass. 02648. 1977. $15/year for American Wind Energy Association members; $20/year for non-members.

This journal emphasizes technologies that are logical extensions of those within the domain of wind energy conversion. Among subjects presented are engineering of wind energy systems; blade design and fabrication; and other technological aspects of wind energy conversion and utilization.

Energy Technology IV: Confronting Reality. 496 pages. Government Institutes, Inc., 4733 Bethesda Ave., Bethesda, Md. 20014. 1977. $28.

This work comprises presentations by well-known experts on energy at the 4th Energy Technology Conference, held at Washington, D.C., in March. For instance, there is a forecast that coal use will triple, and nuclear fusion, wind, biomass, and the like, will provide only 10% of U.S. needs by the year 2000. Experts were from a cross-section of government, industry, academia, and other groups.

Energy from the Wind/Supplement I. Barbara L. Burke and Robert N. Meroney. Colorado State University, Engineering Research Center, Foothills Campus, Fort Collins, Colo. 80523. 1977. $10, loose-leaf.

Prepared in the same format as the basic volume, the supplement contains over 1100 new references to books, conference proceedings, articles, and technical reports on wind power; most of these new references were published between 1973–77. A cumulative author and subject index is included.

Capturing the Sun Through Bioconversion. Conference Proceedings. Conference Coordination Office, The Washington Center, 1717 Massachusetts Ave., N.W., Washington, DC 20036. 1976. $18.

The conference was held last spring, and its proceedings contain material contributed by 146 representatives of environmental groups, government, and industry. The proceedings list prime sources of bioconversion materials, and tells how they can be used. An estimate of ability to meet 10–50% of the world's energy requirements early next century, through bioconversion, is advanced.

Economic and Technical Feasibility Study for Energy Storage Flywheels. Stock No. 052-010-00486-7. Superintendent of Documents, U.S. Government Printing Office, Washington, DC 20402. 1976. $10.

Flywheels store energy and do not pollute. In this study, the state-of-the-art of flywheels is set forth; prospects for technological improvement are examined; and needs for performance enhancement are outlined. ERDA, who prepared this report, estimates that flywheels could save 1.4 million bbl of oil, cumulatively, by 1995, and $7 billion in the transportation and utilities sectors.

Introduction to Energy Technology. Marion L. Shepard et al. ix + 300 pages. Ann Arbor Science Publishers, Inc., P.O. Box 1425, Ann Arbor, MI 48106. 1976. $12.50, hard cover.

Like it or not, the era of cheap, abundant energy is fast drawing to a close. This book discusses the technology of energy and its development, as well as solutions to the energy "crisis" with minimal stress on the biosphere. Problems and potentials of fossil, nuclear, and renewable energy sources are covered.

Wind Power. Daniel M. Simmons. x + 300 pages. Noyes Data Corp., Mill Road at Grand Ave., Park Ridge, N.J. 07656. 1975. $24, hard cover.

Wind is a "clean", renewable energy source. However, information concerning this energy source is scattered and difficult to bring together. This book, nevertheless, has gathered these data together, and apprises the reader of what wind energy systems are available, the state-of-the-art in the field, and methods of storing wind energy. Site selection is discussed, as is the international scene. Detailed machine design data are given.

Control of Environmental Impacts from Advanced Energy Sources. 326 pages. Superintendent of Documents, U.S. Government Printing Office, Washington, D.C. 20402. 1974. $4.

New and advanced sources of energy will have considerable technological and environmental effects. The EPA points out major problems associated with geothermal energy, oil shale, underground coal gasification, process water, exotic pollutants, and land. The study was done by Stanford Research Institute (Menlo Park, Calif.). Ask for EPA-600/2-74-002.

ACS titles which have not appeared in ES&T.

Catalysts for the Control of Automotive Pollutants. *James E. McEvoy, Ed.* Advances in Chemistry Series No. **143**. 199 pages. 1975. $19.95.

Current research by auto makers, catalyst companies, universities, and chemical and petroleum companies on all aspects of catalytic conversion to reduce automotive emissions. Emphasis is on analytical methods, mechanisms of catalytic removal, and catalysts themselves. Specific topics examined in fourteen papers include variation of selectivity, catalyst poisoning, the nature of the catalyst support, and others.

Removal of Trace Contaminants from the Air. *Victor R. Deitz, Ed.* ACS Symposium Series No. **17**. 207 pages. 1975. $17.25.

Sixteen chapters provide critical and in-depth coverage of air pollution characterization and removal. The collection stresses interactions among particulates and gas phase contaminants; pesticides; occupational contaminants; cigarette smoke and aerosol filtration; sulfur dioxide; trace gas adsorption; nitrogen oxides; and high ozone concentrations.

Approaches to Automotive Emissions Control. *Richard W. Hurn, Ed.* ACS Symposium Series No. **1**. 211 pages. 1974. $14.00.

Nine chapters spotlight current developments toward the goals of lower emissions and greater fuel economy; the impact of automotive trends and emissions regulations on gasoline demand; gaseous motor fuels; fuel volatility, the pre-engine converter; and low emissions combustion engines.

Coal Desulfurization: Chemical and Physical Methods. *Thomas D. Wheelock, Ed.* ACS Symposium Series No. **64**. 332 pages. 1977. $25.00.

This volume covers research on industrial processes and methods for removing sulfur from coal. Both chemical and physical extraction methods that are used currently or that are being adopted currently by industry are reviewed. Twenty-three chapters are divided into four groups covering sulfur in coal and its determination; physical methods for removing sulfur from coal; extraction of sulfur from coal by reaction and leaching; and removal of sulfur by pyrolysis, hydrodesulfurization, and other gas-solid reactions.

Analysis of Petroleum for Trace Metals. *Robert A. Hofstader, Oscar I. Milner, and John H. Runnels, Eds.* Advances in Chemistry Series No. **156.** 189 pages. 1976. $21.50.

Fifteen chapters deal with sampling procedures, standards, storage and stability, contamination, sample preparation, and directions for detecting particular metals. Major oil companies address themselves to the different methods available — including increasingly popular electroanalytical methods — and emphasize variations of atomic absorption spectroscopy, the principle technique used in the book.

Trace Elements in Fuel. *Suresh P. Babu, Ed.* Advances in Chemistry Series No. **141.** 216 pages. 1975. $17.25.

The latest research results on these often-toxic emissions cover their origin, the quantities in which they escape into the atmosphere, determination methods, and physiological effects. Specifically, fifteen chapters detail mineral matter and trace elements in coal; coal pretreatment and combustion; mercury and trace element mass balance, and environmental toxicology.

Trace Elements in the Environment. *Evaldo L. Kothny, Ed.* Advances in Chemistry Series No. **123.** 149 pages. 1973. $13.75. Paper, $8.00.

Nine chapters examine the geochemical cycle of trace elements in the environment. Boron, zinc, and selenium are discussed, as well as atmospheric pollutants, marine aerosol salt and dust, particulates, inorganic aerosols, S, V, Zn, Cd, Pb, Se, Sb, Hg. Includes methods of identification, separation, and measurement.

Pollution Control and Energy Needs. *Robert M. Jimeson and Roderick S. Spindt, Eds.* Advances in Chemistry Series No. **127.** 249 pages. 1973. $19.00.

Nineteen papers focus on energy demands vs. primary fuel supplies and the effectiveness of technologies that have been developed to meet environmental regulations. Topics include natural energy reserves, control of SO_x and NO_x, H-Oil desulfurization, sulfur oxides removal from stack gases, RC/Bahco system, effect of desulfurization methods on ambient air quality, and several available processes.

Index

Z